BIOCHEMISTRY OF CARBOHYDRATES

BIOCHEMISTRY OF CARBOHYDRATES

By

Dr. S.K. Prasad

School of Studies of Zoology & Biotechnology

Vikram University

Ujjain

DISCOVERY PUBLISHING HOUSE PVT. LTD.

NEW DELHI-110 002

Published by:

DISCOVERY PUBLISHING HOUSE PVT. LTD.
4383/4B, Ansari Road, Darya Ganj
New Delhi-110 002 (India)
Phone : +91-11-23279245, 43596065, 23253475
E-mail : discoverybooksindia@gmail.com
discoverypublishinghouse@gmail.com
namitwasan9@gmail.com
web : www.discoverypublishinggroup.com

***First Published:* 2010**

***Reprinted:* 2022**

ISBN: 978-81-8356-610-0

Biochemistry of Carbohydrates

Printed at:
Infinity Imaging Systems
Delhi

Preface

Biochemistry today has made spectacular progress in unraveling the mysteries of animate nature. This progress has allowed us to gain deeper insight into the principles of vital activity and has to a very significant extent stimulated the development of applied disciplines, especially medicine. The present title *"Biochemistry of Carbohydrates"* is intended for those who wish to understand living organisms, especially man. Biochemistry is essential for this purpose, but it would be almost impossible for a student to survey on his own the massive body of existing knowledge, constantly augmented by a remarkable torrent of brilliat discoveries. The purpose of the book, then is to organize our knowledge into something that can be comprehended in a relatively short time and still convey a reasonable complete picture of the chemical structure and function of man. Readability without sacrifice of coverage has been a prime goal. An important device in gaining that goal is to keep attention constantly focused on function, with repeated use of rationalization to show that the chemical facts are not isolated, but part of a whole.

Biochemistry has two major goals as a fundamental science, it treats the vital functions from the standpoint of physical chemistry, and as an applied discipline, it points out practical applications for the wealth of scientific knowledge it has acquired. The dual purpose has been, as far as possible, take into account in the writing of this book. We have also tried to summarize our pedagogical experience in teaching biochemistry to students specializing in medicine and pharmacy. The material of this book has been organized according to the principle of functionally in order to trace the close relationship between the functions of the living organism and the structures of its constituents molecules as well as the chemical and physico-chemical process in which they are involved.

The aim of this book is to present a core of biochemical knowledge that is desirable for undergraduate and postgraduate students and also those involved in the field of medical, microbiology, biotechnology and pharmaceutical. Every attempt has been made to keep abrest of the advances in the subject and at the same time to include the fundamentals.

To make the work more comprehensive and informative, the author has consulted many authoritative books, research journals, abstracts, monographs etc. He is grateful to all those great scholars whose work are cited or substantially reproduced.

There can be no claim to originality except in the manner of treatment and much of the information has been obtained from the books and scientific journals available in the different libraries.

The author expresses his thanks to his friends and colleagues whose continue inspirations have initiated him to bring out this book.

The author expresses his gratitude to Mr. Wasan and Staff of M/s Discovery Publishing House Pvt. Ltd., for their whole hearted co-operation in the publication of this book.

In the mean time, the author will remain sincerely responsible for any shortcomings of the book and be grateful to the readers for their suggestions and constructive criticism for the continuous betterment of the book. He takes this opportunity to appeal to the readers to send their suggestions straightway to his publisher.

Author

CONTENTS

CHAPTER

1 Carbohydrates

Carbohydrates are aldehydes or ketones of higher polyhydric alcohols or components that yield these derivatives on hydrolysis. They occur naturally in plants (where they are produced photosynthetically), animals and microorganisms and fulfil various structural and metabolic roles. Monosaccharides are the simplest carbohydrates and they often occur naturally as one of their chemical derivatives, usually as components of disaccharides or polysaccharides. Difficulties are encountered in the qualtitative and quantitative analysis of carbohydrate mixtures because of the structural and chemical similarity of many of these compounds, particularly with respect to the stereoisomers of a particular carbohydrate. As a consequence, many chemical methods of analysis are unable to differentiate between different carbohydrates.

Analytical specificity may be improved by the preliminary separation of the components of the mixture using a chromatographic technique prior to quantitation and techniques such as gas-liquid and liquid chromatography are particularly useful. However, the availability of purified preparations of many enzymes primarily involved in carbohydrate metabolism has resulted in the development of many relatively simple methods of analysis which have the required specificity and high sensitivity and use less toxic reagents. The preparation of the sample prior to its analysis will depend upon the nature of both the sample and the analytical method chosen and may involve the disruption of cells, homogenization and extraction procedures as well as the removal of protein or other interfering substances. It may be necessary to prevent the decomposition and degradation of the carbohydrate content during such treatments or during storage by the addition of antibacterial agents such as thymol or merthiolate, or substances such as fluoride ions, which will inhibit the enzymic transformation of the carbohydrates.

GENERAL STRUCTURE AND FUNCTION

Monosaccharides are composed of carbon, hydrogen and oxygen and have the general formula of $C_nH_{2n}O_n$. They are usually classified according to the number of carbon atoms which they contain and the nature of the reactive carbonyl group, *i.e.* an aldehyde or ketone. The number of carbon atoms in a monosaccharide can range from three to nine although those most frequently occurring are composed of six or less, of which the hexoses are particularly significant. The monosaccharides are often represented as a linear structure but owing to the reactive nature of the carbonyl group, those composed of five or more carbon atoms usually exist in cyclic forms and are present as such in the more complex carbohydrate molecules.

Table 1.1. Examples of Commonly Occurring Monosaccharides

General class	*Number of carbon atoms*	*Formula*	*Trivial name*	
			Aldose	*Ketose*
Trioses	3	$C_3H_6O_3$	Glycerose (glyceraldehyde)	Dihydroxyacetone
Tetroses	4	$C_4H_8O_4$	Erythrose	Erythrulose
Pentoses	5	$C_5H_{10}O_5$	Ribose	Ribulose
Hexoses	6	$C_6H_{12}O_6$	Glucose	Fructose

Stereoisomerism

Although almost all compounds of biological origin contain some asymmetric carbon atoms, the phenomenon of stereoisomerism is particularly important in the case of carbohydrates. The simplest aldose, glyceraldehyde, has only one asymmetric carbon atom or chiral centre (carbon 2), and therefore only two possible arrangements of the OH and H of the CH_2OH unit are possible. If the OH group is shown as projecting to the left on the penultimate carbon atom in the straight chain representation of the molecule, then by convention the molecule is called the L isomer, while if the OH group is represented as projecting to the right it is known as the D isomer. These two stereoisomers of the same carbohydrates are enantiomers or mirror images of one another. All carbohydrates can exist in either of these two forms and the prefix of D or L only refers to the configuration around the highest numbered asymmetric carbon atom.

Enantiomers have the same name (e.g. D-glucose and L-glucose) and are chemically similar compounds but have different optical properties. The majority of naturally occurring monosaccharides, whether they be aldoses or ketoses, are of the D configuration.

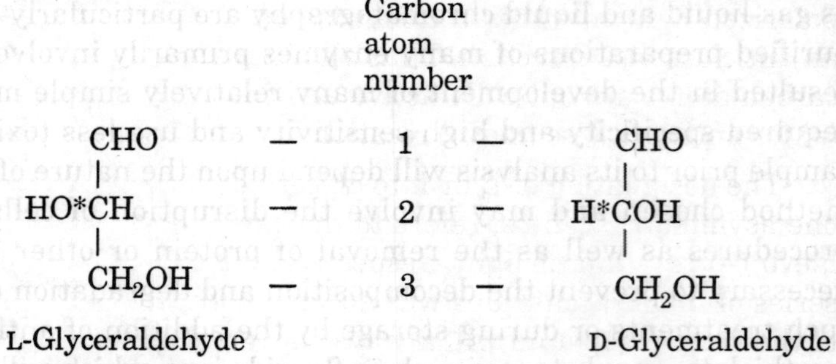

Fig. 1.1. Enantiomers of glyceraldehyde. Enantiomers are mirror images of each other and are chemically similar but will cause a beam of plane polarized light to rotate in opposite directions. Glyceraldehyde has only one asymmetric centre (*) and the designation of D or L is determined by the orientation of the O and OH groups about this carbon atom. For carbohydrates with more than one asymmetric carbon atom, the prefix D or L refers only to the configuration about the highest numbered asymmetric carbon atom, although in such enantiomers the configuration about all the asymmetric centres will also be reversed.

The two isomers of glyceraldehyde are the parent compound of what are known as the D and L series of the aldoses. A similar series of carbohydrates can be derived from the ketoses but because the parent compound dihydroxyacetone does not contain an asymmetric carbon atom, the series is built up from the ketotetrose, erythrulose, which is formed when a CHOH group is added to dihydroxyacetone. Each time the number of carbon atoms is increased by one (with the addition of a CHOH group) a new carbohydrate is produced which has a chemically distinct form and a completely different name. Because this group may be added as either HO—C—H or H—C—OH, an extra asymmetric (chiral) centre is created within the molecule, allowing it to exist in two isomeric forms.

These two new isomers which differ at only one asymmetric centre are called epimers. and each epimer has its own name. Each series is built up by adding additional CHOH units between carbon 1 and carbon 2 to produce increasing numbers of isomers, the actual number being 2^n, where *n is* the number of additional chiral centres. Thus there are eight stereoisomers of the D-aldohexoses (Fig. 1.3), eight stereoisomers of the L-aldohexoses, and four stereoisomers of both the D-ketohexoses and the L-ketohexoses. The presence of an asymmetric carbon atom confers the property of optical activity on the molecule, enabling it to cause the rotation of a beam of plane polarized light in either a clockwise or an anticlockwise direction. Thus all naturally occurring carbohydrates containing asymmetric carbon atoms are optically active and can be designated (+) for clockwise (dextro) rotation or (–) for anticlockwise (laevo) rotation.

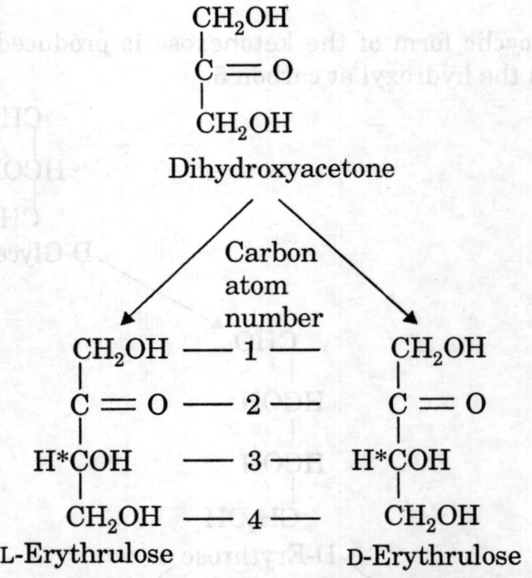

Fig. 1.2. Dihydroxyacetone and enantiomers of erythrulose. The triose, dihydroxyacetone, is the parent compound of the ketoses because it has the lowest number of carbon atoms but it does not contain an asymmetric centre. Therefore the D and L series of the ketoses are built up from the two enantiomers of the tetrose, erythrulose, which has one asymmetric carbon atom (*).

The designation of D or L to glyceraldehyde, the simplest monosaccharide, which has only one asymmetric centre, refers to the capability of these two optical isomers for dextro (+) or laevo (–) rotation of light, respectively. It does not follow, however, that a member of a D series of monosaccharides will automatically be dextrorotatory, although the D and L forms of the same compound will always show opposite directions of rotation. When equal amounts of dextro- and laevorotatory isomers are present, as they are in a synthetically prepared compound, the resulting mixture has no optical activity, is designated DL and is called a racemic mixture.

Structure

An aldehyde and alcohol group can react together to form a hemiacetal, and when this occurs within an aldohexose, a five- or six-membered ring structure is formed. Both pentoses and ketohexoses form similar cyclic structures and it is as such that hexoses and pentoses exist predominantly in nature, with only traces of the aldehyde or ketone forms being present at equilibrium. The cyclic hemiacetal of an aldohexose is formed by reaction of the aldehyde at carbon 1 with the hydroxyl group at carbon 4 or 5 producing an oxygen bridge. Similarly the cyclic form of the ketohexose is produced by reaction of the ketone group at carbon 2 with the hydroxyl at carbon 5.

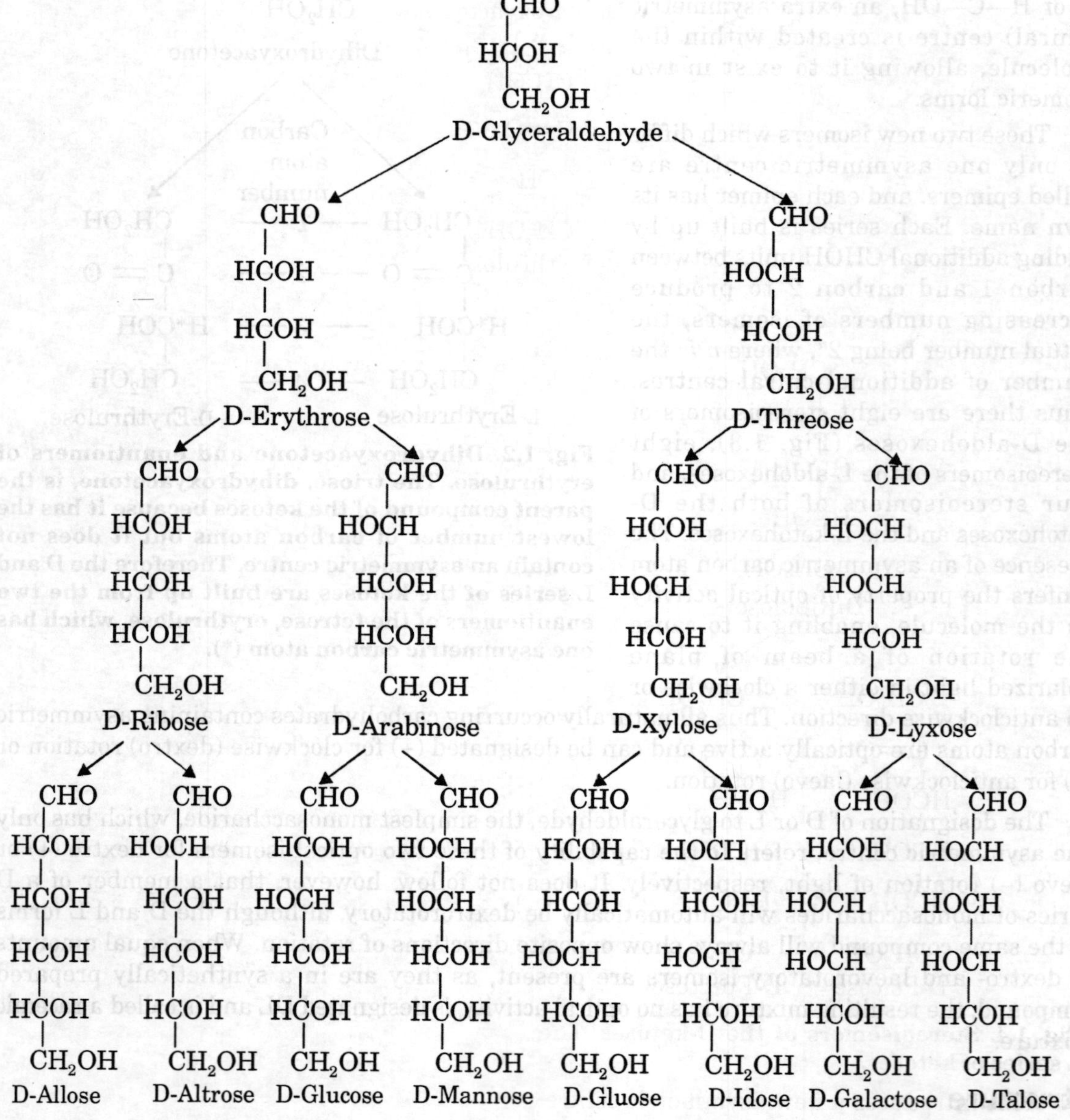

Fig. 1.3. Stereoisomers of the D-aldoses. D-Ribose and D-arabinose differ only in their configuration about a single carbon atom (carbon 2) and are examples of epimers. Diastereoisomers are stereoisomers which are not enantiomers of each other but are chemically distinct forms, the eight D-hexoses being examples. Some, however, are also epimers of each other, for example D-allose and D-altrose. The number of aldoses in the L series is equal to that of the D series and each compound is an enantiomer of one in the other series.

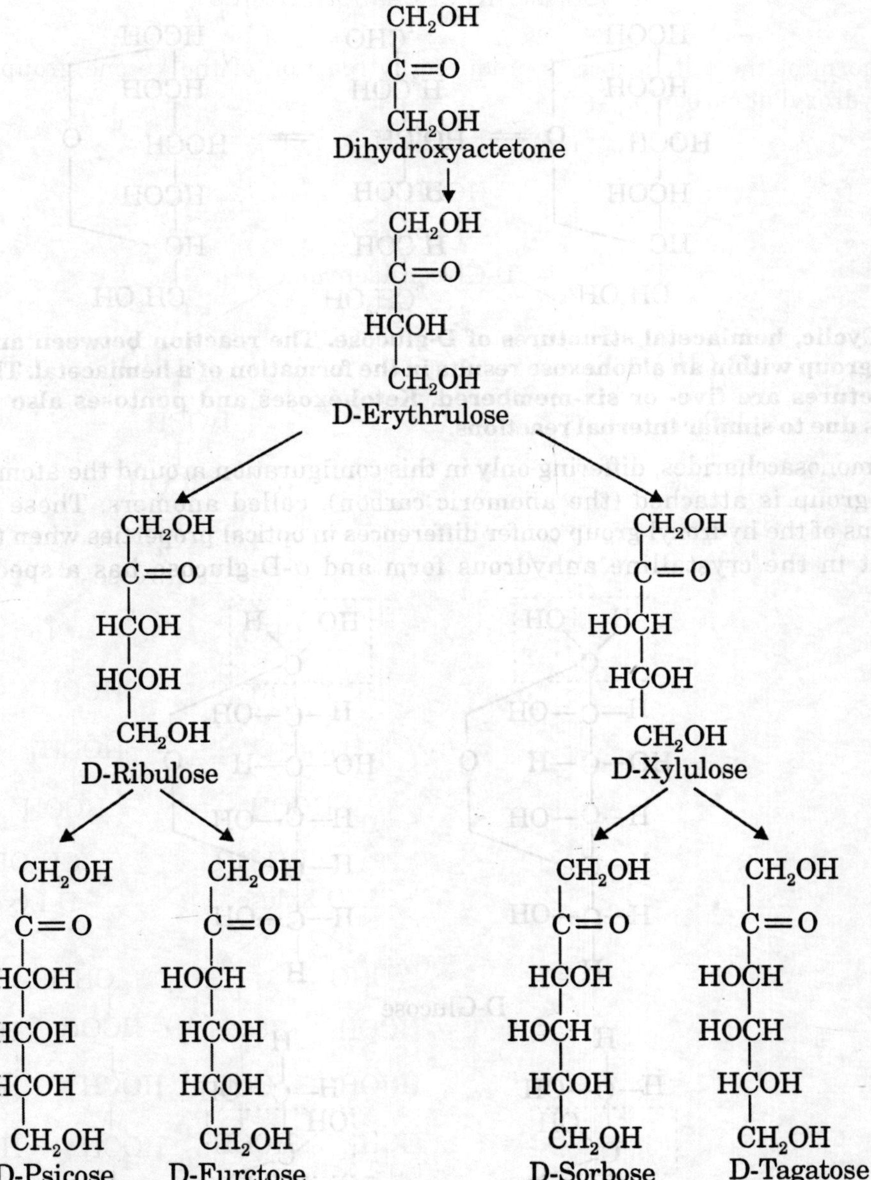

Fig. 1.4. Stereoisomers of the D-ketoses. There will be an equal number of isomers in the L series of ketoses.

This introduces a new asymmetric centre at carbon 1 and increases the possible number of isomers by a factor of two. The configuration of this new hydroxyl group (the glycosidic hydroxyl group at carbon 1) relative to the oxygen bridge results in two additional isomers of the original carbohydrate which are designated as either the *a* form (OH group represented on the same side of the carbon atom as the oxygen bridge) or the β form (OH group represented on the opposite side of the carbon atom to the oxygen bridge).

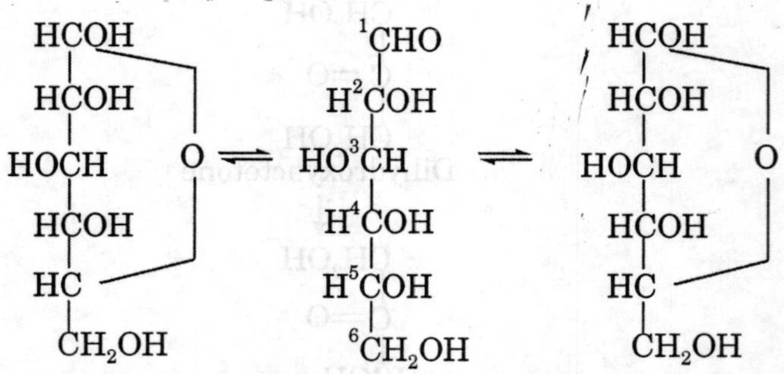

HCOH | ^{1}CHO | HCOH
HCOH | H^{2}COH | HCOH
HOCH O ⇌ | HO^{3}CH | ⇌ HOCH O
HCOH | H^{4}COH | HCOH
HC | H^{5}COH | HC
CH_2OH | $^6CH_2OH$ | CH_2OH

Fig. 1.5. Cyclic, hemiacetal structures of D-glucose. The reaction between an alcohol and aldehyde group within an aldohexose results in the formation of a hemiacetal. The only stable ring structures are five- or six-membered. Ketohexoses and pentoses also exist as ring structures due to similar internal reactions.

Such monosaccharides, differing only in this configuration around the atom to which the carbonyl group is attached (the anomeric carbon), called anomers. These two possible orientations of the hydroxyl group confer differences in optical properties when the substance is present in the crystalline anhydrous form and α-D-glucose has a specific rotation

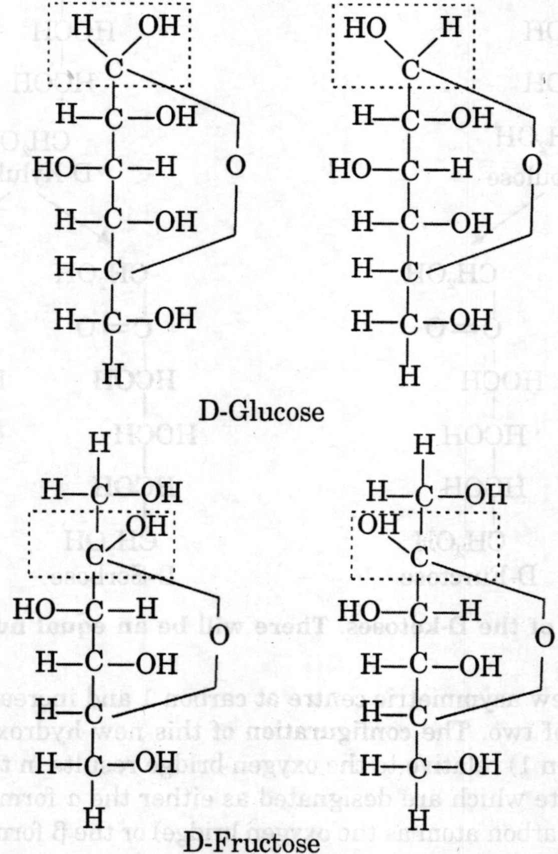

Fig. 1.6. Anomeric forms of D-glucose and D-fructose. The alpha and beta anomers are named with reference to the configuration of the glycosidic hydroxyl group associated with the oxygen bridge.

of + 113° whereas that of β-D-glucose is +19.7°. However, either form in aqueous solution gives rise to an equilibrium mixture which has a specific rotation of + 52.5°, with approximately 36% being in the α form and 64% in the β form, with only a trace present as the free aldehyde.

Because it takes several hours for this equilibrium to be established at room temperature, any standard glucose solution for use with a specific enzyme assay (e.g. glucose by the glucose oxidase which is specific for β-D-glucose) should be allowed to achieve equilibrium before use, so that the proportions of each isomer will be the same in the standard and test solutions. Enzymes that accelerate the attainment of this equilibrium are called mutarotases and can be incorporated in assay reagents in order to speed up the equilibrium formation. The cyclic forms adopted by the hexoses and pentoses can be depicted as symmetrical ring structures called Haworth projection formulae, which give a better representation of the spatial arrangement of the functional groups with respect to one another.

The nomenclature is based on the simplest organic compounds exhibiting a similar five- or six-membered ring structure, namely furan and pyran. The cyclic five-membered structures of sugars are called furanoses and the six-membered rings are called pyranoses. Thus the name given to the five-membered ring formation adopted by D-glucose is D-glucofuranose and the six-membered ring structure is D-glucopyranose. Only traces of glucose occur in the furanose form in solution, the pyranose form being more stable.

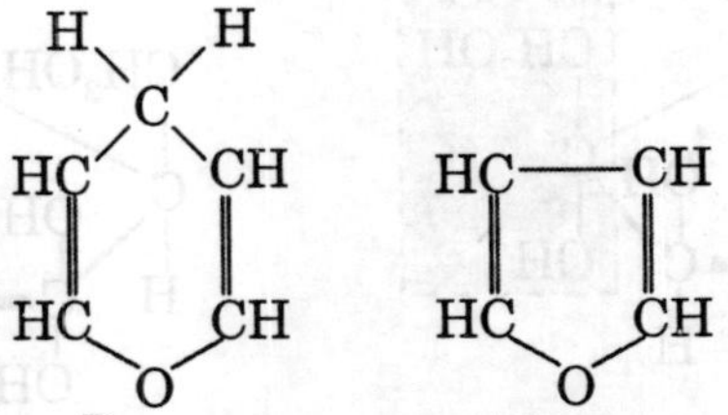

Fig. 1.7. Six- and five-membered cyclic ethers. The stable ring structures which are adopted by hexoses and pentoses are five- or six-membered and contain an oxygen atom. They are named as derivatives of furan or pyran, which are the simplest organic compounds with similar ring structures, e.g. glucofuranose or glucopyranose for five or six-membered ring structures of glucose respectively.

Galactose, however, as well as some other aldohexoses, does exist in appreciable amounts in the furanose form in solution and as a constituent of polysaccharides. Ketoses may also show ring configuration and the ketohexose, fructose, exists as a mixture of furanose and pyranose forms with the latter form predominating in the equilibrium mixture. However, the configuration of fructose when it occurs as a constituent of a disaccharide or a polysaccharide is usually the furanose form. In the symmetrical ring structures depicted by the Haworth formulae, the ring is considered to be planar and the configuration of each of the substituent groups represented as being either above or below the plane. Those groups that were shown to the right of the carbon backbone in the linear representation are drawn below the plane of the ring and those that were originally on the left are drawn above the plane.

An exception is the hydrogen attached to the carbon 5, the hydroxyl group of which is involved in the formation of the oxygen bridge, which is drawn below the plane of the ring, despite the fact that in the linear representation of the D form it is shown to the left. The orientation of the hydroxyl group at carbon 1 determines the α or β designation, and is shown below the plane for the α form and above the plane for the β form.

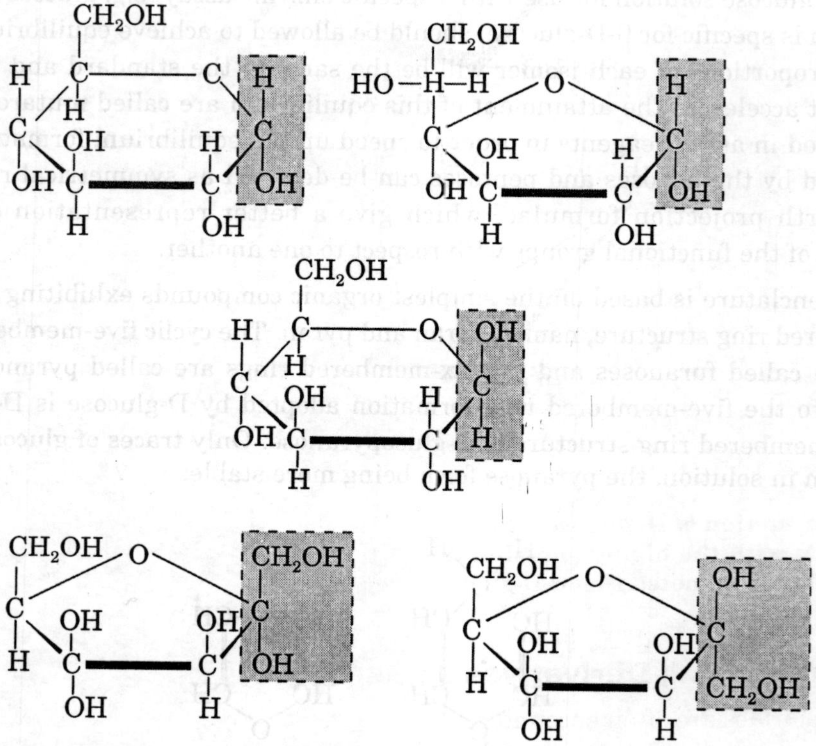

Fig. 1.8. Haworth projection formulae. The ring is considered to be planar with the substituent groups projecting above or below the plane. The thickened lines represent the portion of the ring that is directed out of the paper towards the reader. The alpha and beta anomeric forms are shown with the hydroxyl group at carbon 1 below or above the plane of the ring respectively.

The planar Haworth projection formulae bear little resemblance to the shape of the six-membered pyranoses that actually adopt a non-planar ring conformation comparable to that of cyclohexane. The chair form is the most accurate representation of the molecule and the configuration of the ring substituents may be described as either equatorial (in the plane of the ring) or axial (perpendicular to the ring). Although two conformations are possible for each simple sugar, in general the form with the least axial constituents is the more stable. It should be noted that with the conversion of β-D-glucopyranose to α-D-glucopyranose, an axial hydroxyl group becomes evident and the number of equatorial and axial ring substituents of the stereoisomers of the aldohexoses will vary. The adopted conformation is of importance in carbohydrate chemistry because the axial or equatorial configuration influences the reactivity of the molecule.

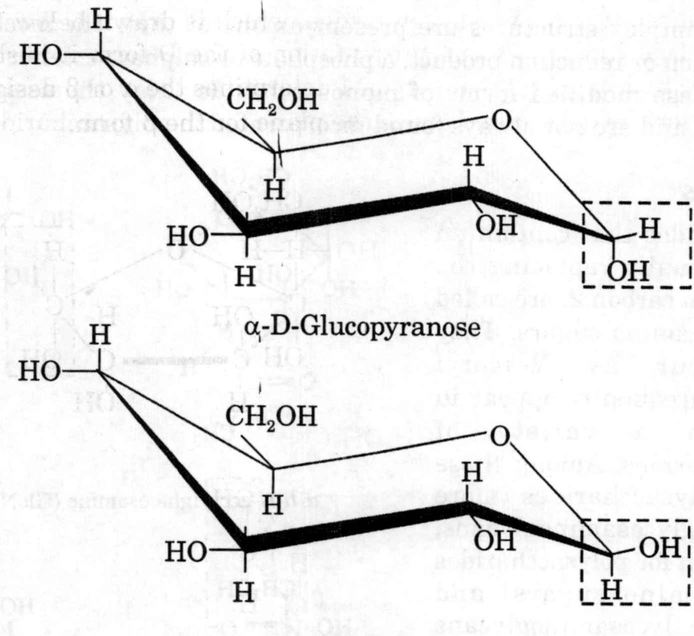

Fig. 1.9. Conformation of D-glucose. The pyranoses adopt a non-planar ring conformation and the chair form with the highest number of equatorial rather than axial hydroxyl groups is favoured. It should be noted that α-D-glucopyranose, in contrast to β-D-glucopyranose, has an axial hydroxyl group.

Naturally Occurring Derivatives of Carbohydrates

Monosaccharides are often encountered biochemically as components of complex molecules, in which they are linked to other residues that may or may not be monosaccharides. Such naturally occurring carbohydrates vary in size and complexity ranging from disaccharides, which as the name implies are composed or two monosaccharide units, to the large complex polysaccnarides. Homopolysaccharides are polymers of one monosaccharide whereas heteropolysaccharides are composed of more than one type of monosaccharide and may also involve non-carbohydrate residues.

Glycoproteins are mainly protein but contain between 5 and 10% carbohydrate by weight. Related molecules which are predominantly polysaccharide are normally referred to as proteoglycans. A whole range of different structures are found that vary in size, proportion of carbohydrate and the nature of the carbohydrate residue. They are prevalent in mammalian tissues, especially mucous secretions, and many important proteins, including some enzymes, serum proteins, pituitary hormones and blood group substances, are glycoproteins. Among the important glycolipids, which are combinations of carbohydrate and lipid, are the cerebrosides and gangliosides. These are constituents of brain and nervous tissue and are usually considered with lipids because they are water-insoluble.

Water-soluble polymers of high relative molecular mass containing lipid and carbohydrate, known as lipopolysaccharides, are found in bacterial cell. In many cases the monosaccharides found in these complex structures are present as one of their chemical derivatives, which may be an oxidation or reduction product, a phosphate or sulphate ester or an amino derivative, etc. However, these modified forms of monosaccharides may themselves have important biochemical roles and are not always found incorporated in polysaccharides.

Glycosamines

Monosaccharides that contain an amino group, usually replacing the hydroxyl group on carbon 2, are called glycosamines or amino sugars. They commonly occur as *N*-acetyl derivatives and frequently appear in this form in a variety of heteropolysaccharides. Among these are the mucopolysaccharides (more correctly called glycosaminoglycans, the accepted term for polysaccharides that contain amino sugars) and peptidoglycans. Glycosaminoglycans are found in animal connective tissues and body fluids and are complex structures that also contain uronic acids. Peptidoglycans are components of bacterial cell walls and structures vary between species. They are composed of *N*-acetyl derivatives linked to short chains of amino acids. Several antibiotics contain glycosamines, including erythromycin, which is an *N*-methyl derivative and carbomycin, in which the uncommon 3-amino derivative of ribose is found.

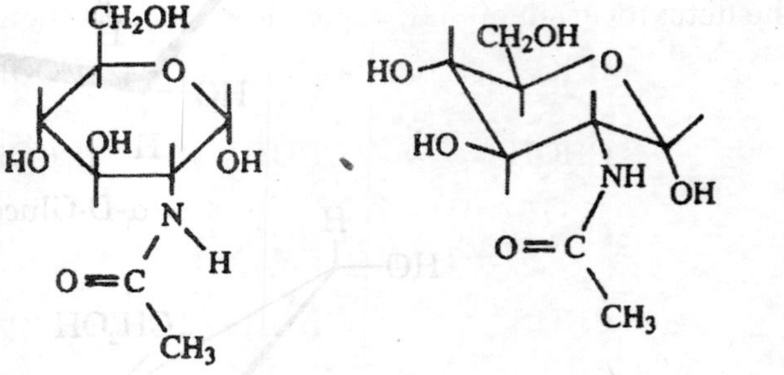

CH_2OH, O, HO, OH, OH, N, H, O=C, CH_3

α-D-*N*-acetylglucosamine (GlcNAc)

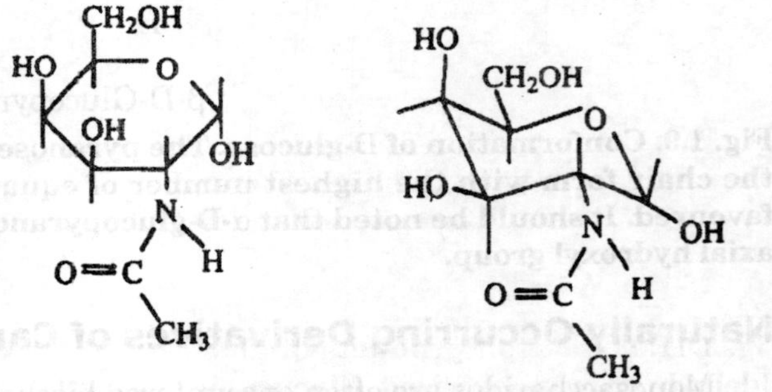

CH_2OH, HO, O, OH, OH, N, H, O=C, CH_3

α-D-*N*-acetylgalactosamine (GalNAc)

Fig. 1.10. *N*-Acetyl derivatives of monosaccharides.

Esters

The phosphate esters and, to lesser extent, the sulphate esters of monosaccharides are very important naturally occurring derivatives. Metabolism of carbohydrates involves the formation and interconversion of a succession of mono-saccharides and their phosphate esters of which glucose-1-phosphate and fructose-6-phosphate are important examples. The sulphate esters of monosaccharides or their derivatives (usually esterified at carbon 6) are found in several polysaccharides, notably chondroitin sulphate, which is a constituent of connective tissues.

Oxidation Products

There are three possible classes of sugar acids which may be produced by the oxidation of monosaccharides. The aldonic acids are produced from aldoses when the aldehyde group at carbon 1 is oxidised to a carboxylic acid. If, however, the aldehyde group remains intact and only a primary alcohol group (usually at carbon 6 in the case of hexoses) is oxidised then a uronic acid is formed. Both aldonic and uronic acids occur in nature as intermediates in carbohydrate metabolism and the uronic acids are important constituents of the glycosaminoglycans. Glucuronic acid, formed from glucose, plays a particularly useful role in the detoxification of many compounds before their excretion by the kidneys.

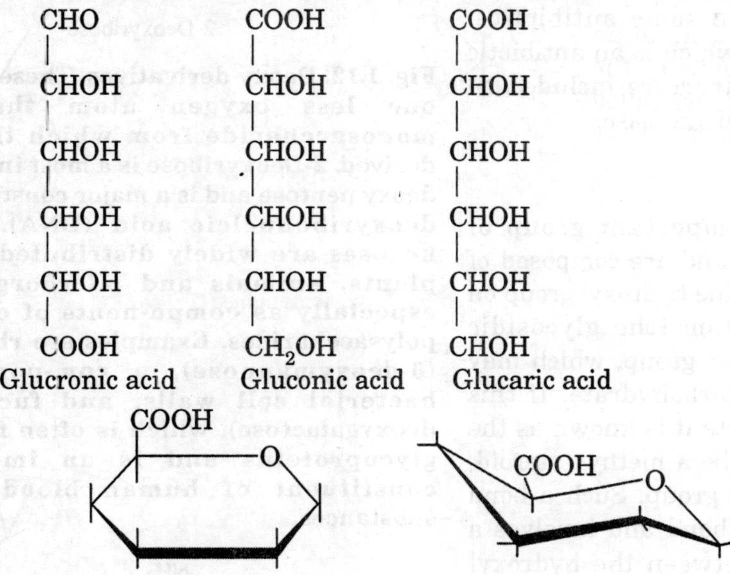

Glucuranoic acid

Fig. 1.11. Oxidation products of glucose. Gluconic acid is an aldonic acid formed when the aldehyde group is oxidized. Glucuronic acid, a uronic acid, is a result of oxidation of the primary alcohol group. When both the aldehyde and the primary alcohol groups are oxidized, glucaric acid is formed, which is an aldaric acid.

It is capable of assisting their removal by forming esters with an alcohol or phenolic group and many drugs and acids are excreted in the urine conjugated with glucuronic acid in this way. Aldaric acids are formed when stronger oxidizing conditions are employed, such as nitric acid, when both the aldehyde and primary alcohol groups are oxidized.

Glycosylamines

Members of this group of compounds, which includes the extremely important nucleosides, are formed when the carbon atom of a monosaccharide, or often its deoxy derivative (Fig. 1.12), is linked directly to the nitrogen of a nitrogenous base, including the amino group of an amino acid in a peptide chain, with the loss of the hydroxyl group. Such *N*-glycosidic linkages must not be confused with the bonding in a glycoside, which is through an oxygen atom. Those nucleosides found in the nucleic acids DNA and RNA involve the joining of ribose of deoxyribose to a purine or a pyrimidine base.

One such nucleoside is adenosine, in which a nitrogen of adenine is linked to carbon 1 of the pentose, ribose. In this form it is a component of RNA but as a phosphorylated derivative of adenosine (e.g. ATP), which is a high energy compound, it fulfils an important role in metabolism. The dinucleotides NAD and NADP are two cofactors necessary for many enzymic transformations and these also contain *N*-glycosides of ribose phosphate. Other important nucleosides are found in some antibiotics, puromycin, for example, which is an antibiotic of fungal origin whose structure includes *N*-acetyl ribose linked to a purine base.

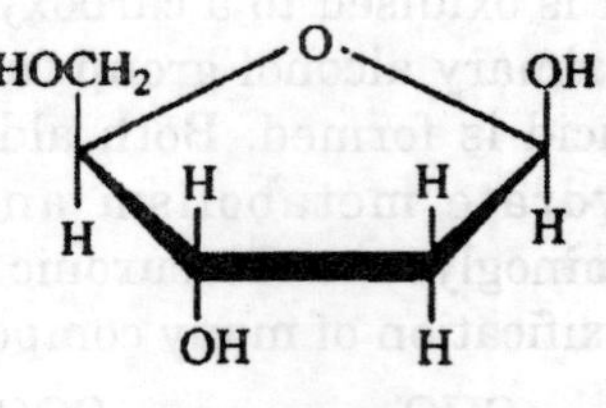

Fig. 1.12. Deoxy derivatives. These contain one less oxygen atom than the monosaccharide from which they are derived. 2-Deoxyribose is a most important deoxy pentose and is a major constituent of deoxyribonucleic acid (DNA). Deoxy hexoses are widely distributed among plants, animals and microorganisms especially as compo-nents of complex polysaccharides. Examples are rhamnose (6-deoxymannose), a com-ponent of bacterial cell walls, and fucose (6-deoxygalactose), which is often found in glycoproteins and is an important constituent of human blood group substances.

Glycosides

Glycosides are an important group of carbohydrate derivatives and are composed of a carbohydrate linked via the hydroxyl group on the anomeric carbon atom (the glycosidic hydroxyl group) to another group, which may or may not be another carbohydrate. If this group is not a carbohydrate it is known as the aglycone group and may be a methyl, steroid, glycerol or more complex group. Such a bond is known as a glycosidic bond and involves a condensation reaction between the hydroxyl group at the anomeric carbon of the cyclic form of the carbohydrate and an alcohol group of the other molecule. It is in this way that the monosaccharide units are linked in disaccharides and polysaccharides and also how carbohydrates may be linked to proteins via the hydroxy group of an amino acid, *e.g.* serine.

Glycosides can vary considerably in their composition and may be found as constituents of plant and animal tissues, where they fulfil a variety of roles. Various extracts containing specific glycosides are used as drugs, dyes and flavourings. The glycosidic bond may be classified as either an alpha or beta linkage depending upon the orientation of the glycosidic hydroxyl group, but a complete description of the linkage should also state the carbon atom through which the oxygen bridge is formed and the ring type of the monosaccharide, *i.e.* pyranose or furanose.

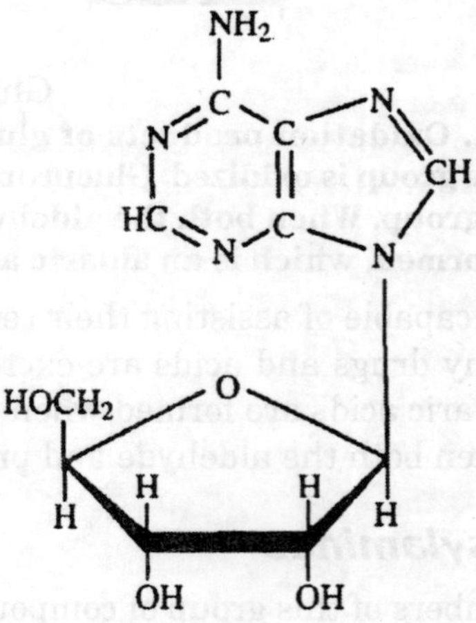

Fig. 1.13. A glycosylamine. Adenosine is a nucleoside and is an example of a glycosylamine in which the nitrogen atom of the purine, adenine, is linked directly to carbon 1 of β-D-ribofuranose.

Thus the complete specification of maltose (a disaccharide composed of two glucose units with an alpha linkage) is α-D-glucopyranosyl-(1→4)-D-glucopyranose. The structure of cellobiose, which is also composed of glucopyranose units but involves a beta linkage, is also shown for comparison. The nature of the glycosidic linkage between two monosaccharides, or the modified forms in which they naturally occur, is an important feature of the structure of di-, oligo and polysaccharides, especially with respect to their capacity to act as substrates for enzymes.

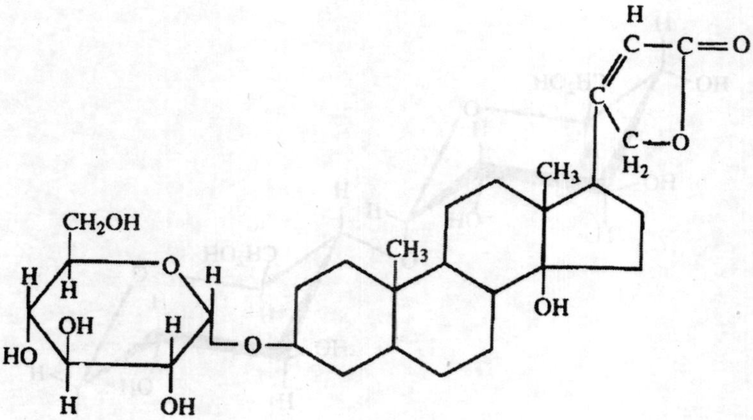

Digitoxenin

Fig. 1.14. A steroid glycoside. Digitoxenin is a cardiac-stimulating drug. It is a steroid glucoside in which a compound with a steroid structure is linked to α-D-glu-copyranose by condensation of the anomeric hydroxyl group on the carbohydrate and the alcohol group at position 5 on the steroid nucleus.

The action of those hydrolases which break glycosidic bonds depends on the nature of the bond as well as the two substances that it links. An example of such stereospecificity is shown by the capacity of human salivary pancreatic amylase to split the α glycosidic linkages between the glucose units in starch but not the β linkages of cellulose. However, the glycosidic hydrolases, although being generally dependent upon the composition of the glycoside, may act on several similar glycosides.

Thus testicular hyaluronate glycanohydrolase (EC 3.2.1.35) will cleave β(1→4) linkages between *N*-acetyl-D-glucosamine and D-glucuronic acid residues which make up the glycosaminoglycan, hyaluronic acid, but will also degrade chondroitin sulphates, which are composed of D-glucuronic acid linked to sulphated *N*-acetyl-D-galactosamine. Disaccharides and polysaccharides form a high proportion of the diet, and are present in natural and processed foods; starch, the polysaccharide storage form of glucose found in potatoes and grain, forms a major proportion of the total. The digestion of these dietary carbohydrates relies on the capability of certain enzymes, present in the saliva and intestinal secretions, to hydrolyse the α glycosidic linkage between the monosaccharide units. Thus the polysaccharide, is enzymically degraded with the eventual release of the monosaccharides, which are then

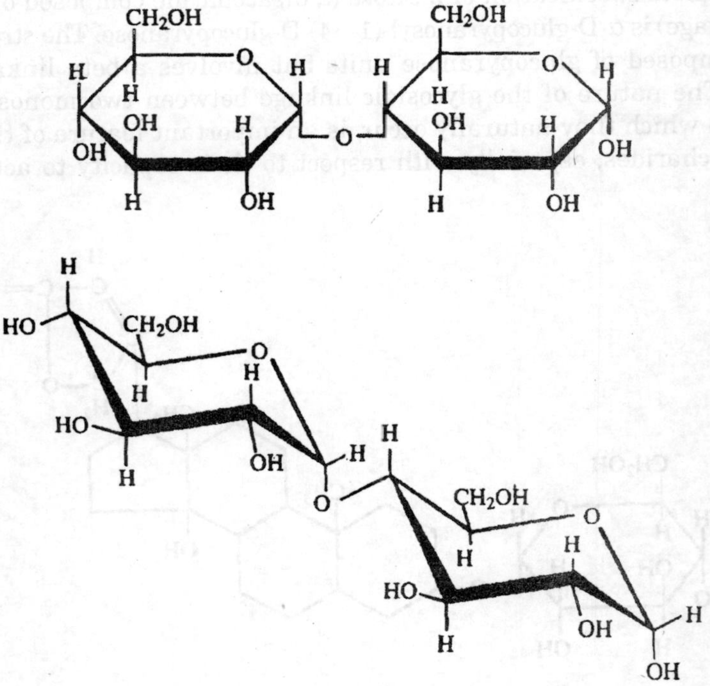

Maltose α (1→ 4) glycosidic linkage

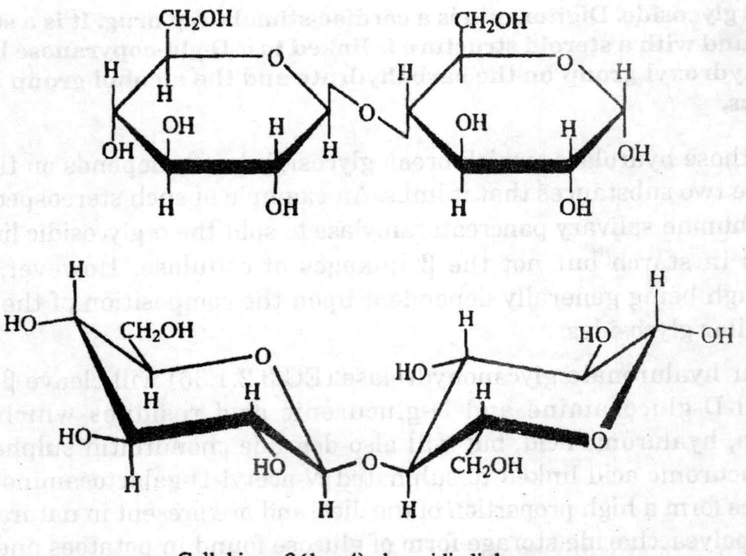

Cellobiose β (1→ 4) glycosidic linkage

Fig. 1.15. Structure of disaccharides. Maltose and cellobiose are both disaccharides which contain only glucose units, the difference in their structures lying in the way the glucose units are joined. The glycosidic linkage in maltose is α(1→4), whereas that in cellobiose is β(1→4).

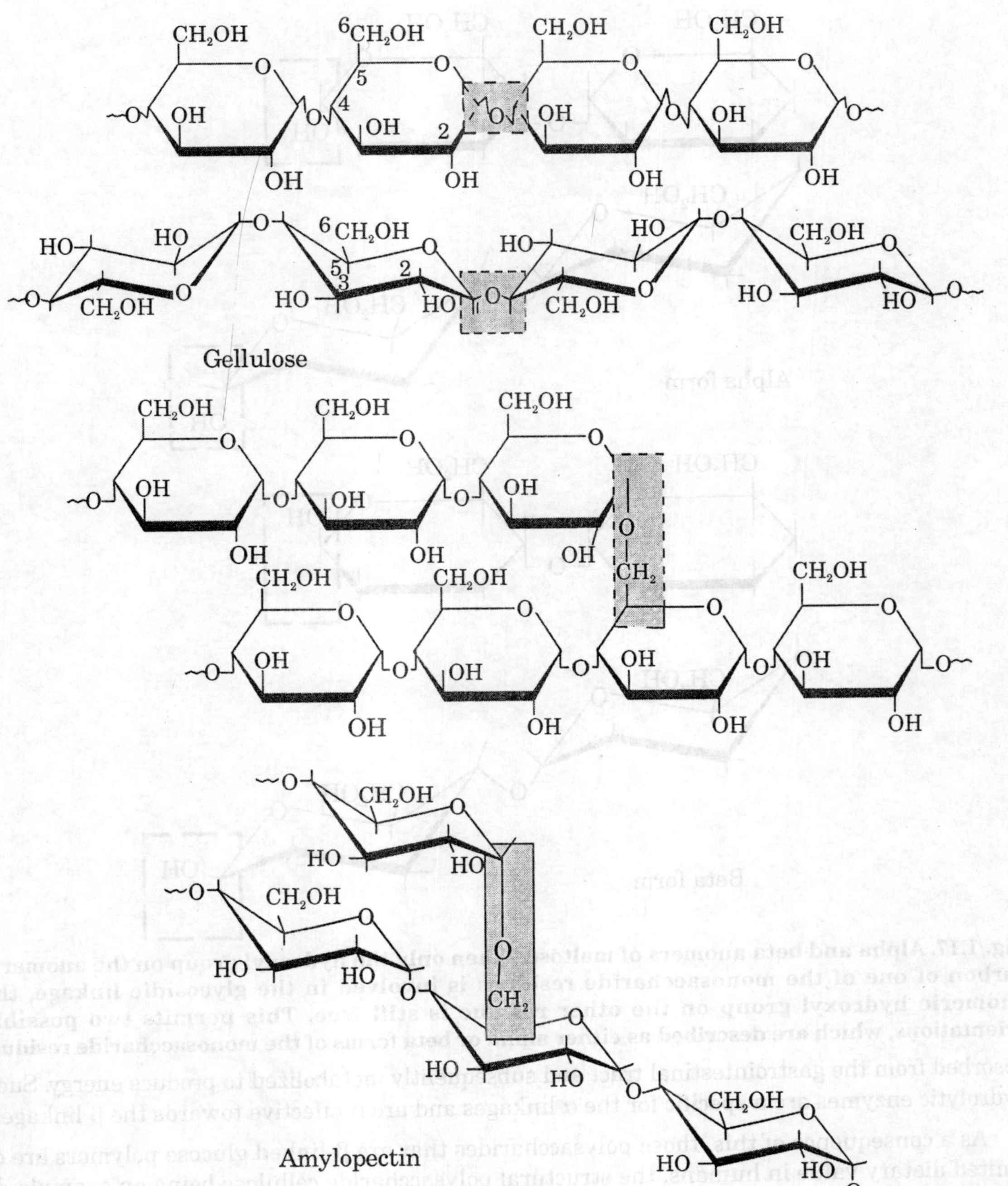

Fig. 1.16. Structures of glucopyranose homopolysaccharides. Cellulose is a linear structure of the glucopyranose units linked β(1→4). Starch consists of amylose, which has a linear α(1→4) structure, and amylopectin, which has α(1→6) branch points on the linear α(1→4) chains. Glycogen has a similar structure to amylopectin, but with a greater degree of α(1→6) branching.

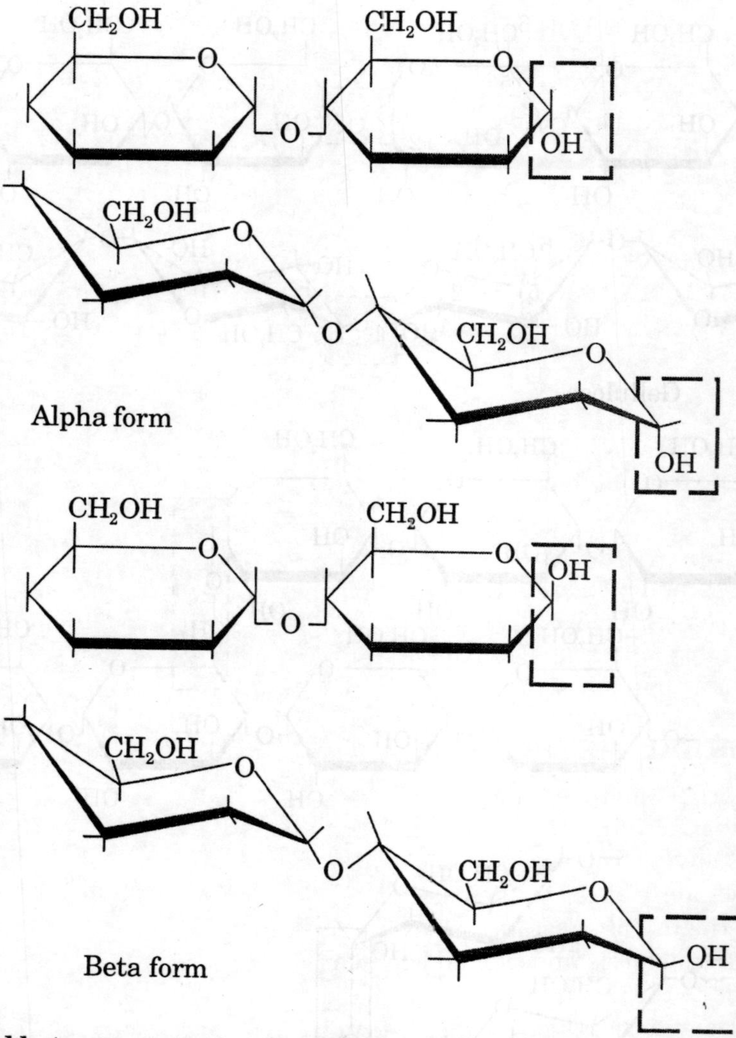

Alpha form

Beta form

Fig. 1.17. Alpha and beta anomers of maltose. When only the hydroxyl group on the anomeric carbon of one of the monosaccharide residues is involved in the glycosidic linkage, the anomeric hydroxyl group on the other residue is still free. This permits two possible orientations, which are described as either alpha or beta forms of the monosaccharide residue.

absorbed from the gastrointestinal tract and subsequently metabolized to produce energy. Such hydrolytic enzymes are a specific for the α linkages and are ineffective towards the β linkages.

As a consequence of this, those polysaccharides that are β-linked glucose polymers are of limited dietary value in humans, the structural polysaccharide cellulose being an example. It is, however, a substantial and effective nutritional component of the diet of ruminants, which have a high bacterial population in the rumen capable of metabolizing cellulose with the release of glucose. Polysaccharides also serve as energy reservoirs and the catabolism of glycogen, which is an α-linked polymer of glucose similar to amylopectin and is found in animal muscle and liver tissue, fulfils this function.

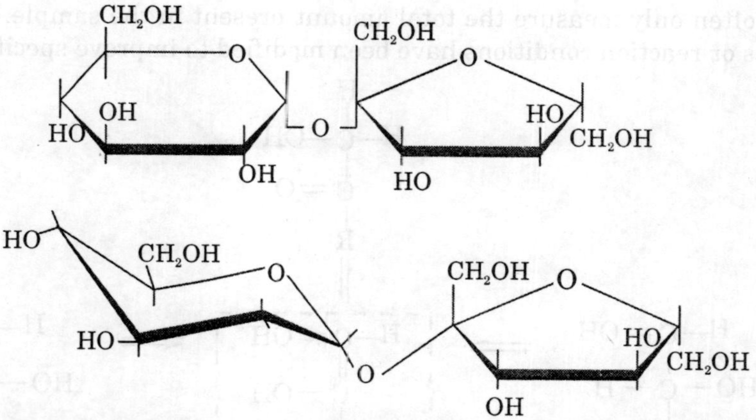

Fig. 1.18. Structure of sucrose. Sucrose is a disaccharide which has only one possible structural form, there being no free anomeric hydroxyl group.

When the glycosidic linkage involves the anomeric carbon of only one of the monosaccharides, the free aldehyde group on the other monosaccharide can have two possible orientations. Thus, α and β forms of such a compound are possible. If the two monosaccharides are linked by an oxygen bridge between the anomeric carbon of each, as is the case with sucrose, the possibility of a free aldehyde or ketone group remaining is eliminated and only one structural form is possible which does not have reducing properties.

CHEMICAL METHODS OF CARBOHYDRATE ANALYSIS

Many of the earliest methods available for the measurement of carbohydrates were based on their chemical reactivity and involved the addition of a particular reagent with the subsequent formation of a coloured product. Although the cheapness and technical simplicity of these procedures contributed to their previous popularity, they are inherently non-specific and often involve the use of substances that are now recognized as hazardous. It is essential that the hazard assessment is undertaken before using such techniques. The lack of specificity is largely attributed to the fact that many mono-saccharides have similar chemical composition and properties, although their biological significance varies widely.

The numerous isomeric forms in which the hexoses exist, all having identical chemical composition, with some even showing interconversion to an entirely common structure in alkaline solution, demonstrate the difficulties encountered in attempting to measure an individual member of this group in a sample containing others. Chemical methods are not even capable of differentiating effectively between classes of monosaccharides because the reactions of the carbonyl group are used. This function is common to all monosaccharides and thus they may give similar or sometimes identical reaction products. Nevertheless, providing that their limitations are appreciated, the use of chemical methods is justified in a variety of situations.

Reactions of the Carbonyl Group

The reactions of the carbonyl group form the basis of many qualitative methods for the detection of carbohydrates and several have been used quantitatively. These are general methods that often only measure the total amount present in the sample. However, in some cases, reagents or reaction conditions have been modified to improve specificity.

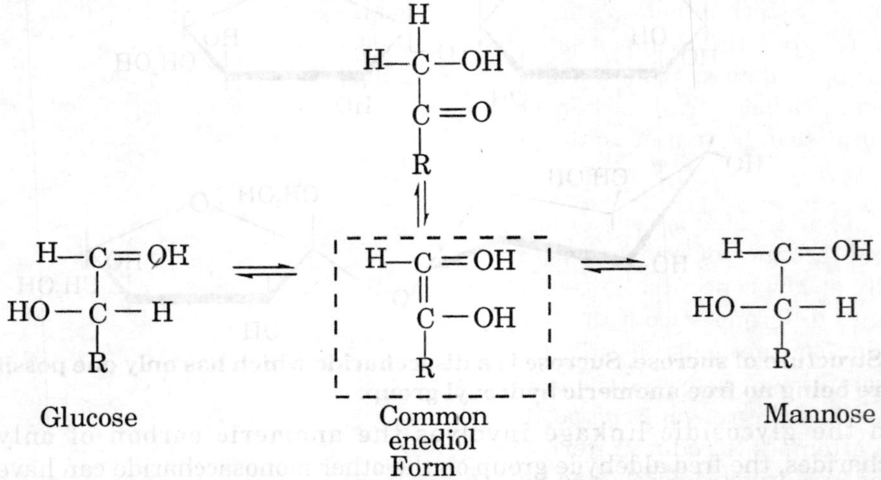

Fig. 1.19. Interconversion of glucose, mannose and fructose in weakly alkaline solution.

Reduction Methods

Carbohydrates that have a potentially free aldehyde or ketone group exist in solution at equilibrium with the enediol form. At a slightly alkaline pH this conversion is favoured and the resulting enediol is an active reducing agent. Reduction methods can be used for disaccharides provided that the aldehyde or ketone group of at least one of the monosaccharides has not been eliminated in the glycosidic bond. Sucrose is an example of a disaccharide in which the anomeric carbon atoms of both monosaccharides are involved in the glycosidic bond and the reducing power is lost.

However, this distinction between reducing and non-reducing disaccharides can sometimes be used to advantage in qualitative tests. One of the commonest methods involves the reduction of cupric ions (Cu^{2+}) to cuprous ions (Cu^{+}), which in alkaline solution form yellow cuprous hydroxide, which is in turn converted by the heat of the reaction to insoluble red cuprous oxide (Cu_2O). In the qualitative tests based on this reaction, the production of a yellow or orange-red precipitate indicates the presence of a reducing carbohydrate.

It is necessary to keep the cupric salts in solution and to this end Benedict's reagent incorporates sodium citrate while Fehling's reagent uses sodium potassium tartrate. Under carefully controlled reaction conditions, the amount of cuprous oxide formed may be used as a quantitative indication of the amount of reducing carbohydrate present, although different carbohydrates will result in the formation of different amounts of cuprous oxide. The methods of measuring the amount of cuprous oxide formed are numerous but the one most frequently used involves the reduction of either phosphomolybdic acid or arsenomolybdic acid by the cuprous oxide to lower oxides of molybdenum.

The intensity of the coloured complex produced is related to the concentration of the reducing substances in the original sample. The colour produced with arsenomolybdic acid is more stable and the method is more sensitive than with phosphomolybdic acid. The

neocuproine method for the measurement of the cuprous oxide is more sensitive than the phosphomolybdic acid reagent and uses 2,9-dimethyl-1, 10-phenanthroline hydrochloride (neocuproine), which produces a stable colour and is specific for cuprous ions. Although a variety of oxidants other than copper salts have been used **ferricyanide** is the only other one of note. Ferricyanide ions (yellow solution) are reduced to ferrocyanide ions (colourless solution) by reducing carbohydrates when heated in an alkaline solution. The concentration of the carbohydrate can be related to the decrease in absorbance at 420 nm.

$$Fe(CN_6)^{3-} \rightarrow Fe(CN_6)^{4-}$$

The precision of this type of method in which quantitation involves inverse colorimetry (*i.e.* the absorbance decreases with increasing concentration of the analyte) is questionable, especially at low concentrations of the analyte, because of the difficulty of measuring slight absorbance differences from the high blank reading. In addition to the lack of specificity of such reduction methods already mentioned, non-carbohydrate reducing substances present in the sample will also react similarly resulting in positive error. Over the years, workers have modified the reagent composition, sample preparation and even the shape of the test-tubes in attempts to reduce interference. Such names as Fehling, Benedict, Nelson, Somogy, Folin and Wu are still associated with these reduction methods. Such considerations are of little consequence nowadays, when reduction methods are less frequently used and the more specific and precise enzymic or chromatographic methods are preferred.

Reactions with Aromatic Amines

Various aromatic amines will condense with aldoses and ketoses in glacial acetic acid to form coloured products whose absorbance maxima are often characteristic of an individual carbohydrate or group. The use of different aromatic amines or absorbance measurements at alternative wavelengths gives a degree of specificity for individual sugars. The carcinogenic nature of some aromatic amines has ruled out their use as laboratory reagents. Aromatic amines that have been used include *o*-toluidine, *p*-aminosalicylic acid, *p*-arninobenzoic acid, diphenylamine and *p*-aminophenol.

Their ability to react preferentially with a particular carbohydrate or class of carbohydrate is often useful, e.g. *p*-aminophenol, which shows some specificity for ketoses compared with aldoses and is useful for measuring fructose. These reagents have proved particularly useful for the visualization and identification of carbohydrates after separation of mixtures by paper or thin-layer chromatography, when colour variations and the presence or absence of a reaction aid the interpretation of the chromatogram.

Reactions with Strong Acids and a Phenol

When heated with a strong acid, pentoses and hexoses are dehydrated to form furfural and hydroxymethylfurfural derivatives respectively (Fig. 1.20), the aldehyde groups of which will then condense with a phenolic compound to form a coloured product. This reaction forms the basis of some of the oldest qualitative tests for the detection of carbohydrates, *e.g.* the Molisch test using concentrated sulphuric acid and α-naphthol. By careful choice of both the reaction conditions and the phenolic compound used, it may be possible to produce a colour that is characteristic of a particular carbohydrate or related group, so giving some degree of specificity to the method. Thus, Seliwanoff's test uses hydrochloric acid and either resorcinol or 3-indolylacetic acid to measure fructose with minimal interference from glucose.

The colour produced by pentoses with orcinol (Bial's reagent) or *p*-bromoaniline is sufficiently different from that produced by hexoses to permit their quantitation in the presence of hexoses. However, none of the methods based on the formation of furfural or its derivatives can be considered to be entirely specific. There may be non-carbohydrate substances present in a biological sample that will decompose on heating under the acidic conditions and will react in a similar manner to a carbohydrate. Glucuronic acid is an example and is often present in abundance in urine and may give a false positive reaction for carbohydrates.

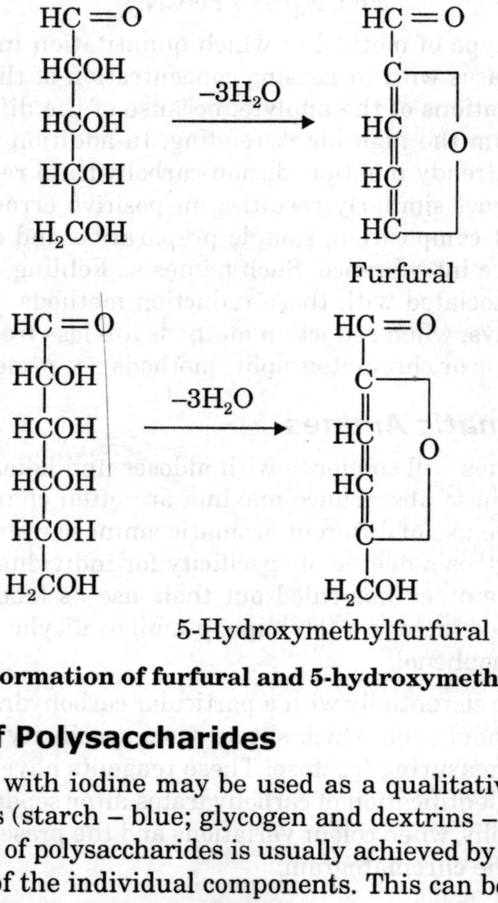

HC=O, HCOH, HCOH, HCOH, H_2COH —($-3H_2O$)→ HC=O, C, HC, HC, HC, O (Furfural)

Furfural

HC=O, HCOH, HCOH, HCOH, HCOH, H_2COH —($-3H_2O$)→ HC=O, C, HC, HC, C, O, H_2COH (5-Hydroxymethylfurfural)

5-Hydroxymethylfurfural

Fig. 1.20. Formation of furfural and 5-hydroxymethylfurfural.

Structural Studies of Polysaccharides

Although the reaction with iodine may be used as a qualitative test for the presence of some homopolysaccharides (starch – blue; glycogen and dextrins – red; cellulose and inulin - no colour), the quantitation of polysaccharides is usually achieved by hydrolysis of the glycosidic linkages with the release of the individual components. This can be accomplished by heating at 60 °C with concentrated hydrochloric acid for 30 minutes followed by the quantitation of the monosaccharide using a suitable method. However, this acid hydrolysis procedure may result in the destruction of some carbohydrates and milder conditions using acetic acid or lower concentration of hydrochloric acid (0.1–2.0 mol l^{-1}) are often preferable.

Methanolysis under mild conditions will yield the methyl glycosides, which are especially suitable for analysis by gas-liquid chromatography. Enzymic hydrolysis of specific linkages is also of great value in the investigation of the structure and composition of hetero- and homopolysaccharides, and in such studies it may also be important to measure any non-carbohydrate moieties or monosaccharide derivatives that have been released by chemical or enzymic hydrolysis.

Thus the uronic acid residues may be determined in the analysis of certain glycosaminoglycans as may the lipid component of a glycolipid and the sialic acid of a glycoprotein. Structural investigations into the degree of branching and into the position and nature of glycosidic bonds and of non-carbohydrate residues in polysaccharides may include periodate oxidation and other procedures such as exhaustive methylation. X-ray diffraction and spectroscopic techniques such as nuclear magnetic resonance and optical rotatory dispersion also give valuable information especially relating to the three-dimensional structures of these polymers.

Periodate Oxidation

Periodate oxidation involves the simultaneous oxidation and cleavage of carbon to carbon bonds that have adjacent free hydroxyl groups or an aldehyde or ketone group adjacent to a hydroxyl group. The action of periodic acid is represented by the following equation :

$$\begin{array}{c} R \\ | \\ H-C-OH \\ | \\ H-C-OH \\ | \\ R \end{array} + IO_4^- \longrightarrow 2R-CHO + H_2O + IO_3^-$$

The reaction with aldoses and ketoses is different and the procedure may be used to distinguish between them. Free monosaccharides and glycosides do not react in an identical manner with periodic acid and this reveals information about the structure, sequence and linkage of monosaccharides or their derivatives in a polysaccharide. The oxidation may be performed at pH 5.0 by addition of an aqueous solution of sodium metaperiodate. The reaction is allowed to proceed in the dark at 4 °C and after 24 hours the excess periodate is destroyed by the addition of ethylene glycol. The aims in periodate oxidation are to elucidate the number of neighbouring hydroxyl groups by estimating the number of moles of periodate consumed and to determine the structure of the moiety remaining after the reaction.

The amount of periodate used in the reaction may be determined in several ways, including titrimetric and spectrophotometric methods. If the products or periodate oxidation are reduced with sodium borohydride, polyalcohols are produced which may be readily hydrolysed under mildly acidic conditions and the reaction products can be determined. The identification of these products gives considerable information about the linkages between the polysaccharide components and also their sequence in the overall structure. The laboratory procedures and the interpretation of the results of periodate oxidation studies are complicated exercises and this brief account is intended only as a general introduction to the topic and further details, if required, should be sought from a more specific text.

ENZYMIC METHODS OF CARBOHYDRATE ANALYSIS

There are a large number of enzymes that are capable of modifying carbohydrates or carbohydrate derivatives, and that may be used in various analytical methods. The hydrolytic enzymes, which break glycosidic linkages, are useful in the study of disaccharide or polysaccharide structure and in methods for quantitation. Such enzymes will hydrolyse the glycosidic linkages between the monosaccharide residues and release the individual components for further analysis. The enzyme is chosen bearing in mind the nature of the glycosidic linkage involved, which may not be unique to one particular disaccharide or polysaccharide.

Thus α-glucosidase will hydrolyse both the α(1→4) linkage of maltose and α(1→2) linkage of sucrose, resulting in the release of glucose in both cases. Enzymic methods for the

Table 1.2. Examples of glycosidases.

Hydrolysing enzyme	*Glycosidic linkage*	*Trivial name of substrate*	*Hydrolysis products*
β-Galactosidase (EC 3.2.1.23)	β(1→4)	Lactose	D-Galactose. D-glucose
α-Glucosidase (EC 3.2.1.20)	α(1→4)	Maltose	D-Glucose. D-glucose
β-Fructofuranosidase (EC 3.2.1.26)	β(1→2)	Sucrose	D-Glucose. D-fructose
α-Galactosidase (EC 3.2.1.22)	α(1→6)	Melibiose	D-Galactose. D-glucose
Amyloglucosidase (EC 3.2.1.3)	α(1→4) α(1→6)	Glycogen	D-Glucose
Cellulase (EC 3.2.1.4)	β(1→4)	Cellulose	D-Glucose

quantitation of monosaccharides are employed when a higher degree of specificity is required than can be achieved by the majority of the chemical methods. They often enable the quantitation of one stereoisomer in the presence of others and can often differentiate between the α and β anomeric forms. Absolute specificity of an enzyme for only one substrate is rare and there may be several monosaccharides present in a sample that can be acted on to varying degrees by the same enzyme.

Hexokinase (EC 2.7.1.1), although used in a method for the measurement of glucose, is not specific for that substrate and will catalyse the phosphorylation of other hexoses. The specificity of a particular enzyme may also vary according to the source from which it was prepared (*e.g.* microbial, fungal, animal or plant origin) and the method of commercial preparation can affect the purity and thus may influence the results obtained when the enzyme is used. Although it is now possible to purchase many enzymes in a highly purified form, the possibility of the presence of small amounts of an unwanted enzyme, whose substrate is also present in the sample being analysed, should not be discounted.

Assay of Glucose using Glucose Oxidase

The flavoprotein enzyme, glucose oxidase (EC 1.1.3.4), which may be obtained from *Aspergillus niger,* catalyses the oxidation of β-D-glucose by atmospheric oxygen to produce D-gluconolactone, which is converted to gluconic acid with the production of hydrogen peroxide :

$$\beta\text{-D-glucose} + O_2 + H_2O \rightarrow \text{D-gluconic acid} + H_2O_2$$

The oxidation of α-D-glucose occurs at less than 1% of the rate of oxidation of the β anomer. Because these two forms exist in solution in equilibrium in the proportion of 36% (α) and 64% (β), mutarotation of the α to the β form must be allowed to reach equilibrium in the sample and standards for consistent results.

The inclusion of aldose-1-epimerase (glucomutarotase) (EC 5.1.3.3) in the glucose oxidase reagent will permit rapid restoration of the α-β equilibrium, effectively enabling the reaction to go to completion. The rate of oxidation of other monosaccharides (e.g. galactose, mannose, xylose, arabinose and fructose) by glucose oxidase has been shown to be negligible or zero but some derivatives of glucose do react slightly, e.g. 2-deoxy-D-glucose shows a reaction rate of less than 5% of that with β-D-glucose. Quantitation of glucose using glucose oxidase is achieved by measurement of either the hydrogen peroxide formed or the oxygen consumed during the reaction, both of which are proportional to the β-D-glucose content of the sample.

Measurement of the Hydrogen Peroxide Formed

Spectrophotometric methods : The methods that were originally developed for routine use were colorimetric procedures in which the hydrogen peroxide formed was measured by monitoring the change in colour of a chromogenic oxygen acceptor in the presence of the enzyme peroxidase (EC 1.11.1.7). Such chromogens are colourless in their reduced form but exhibit characteristic colours when oxidized, and *o*-tolidine and *o*-dianisidine were among the substances originally used.

However, owing to their potential carcinogenic nature, they have been superseded by other, less toxic, chemicals displaying similar chromogenic properties and offering the methodological advantages of faster reaction time and the production of a more stable coloured product. While glucose oxidase is highly specific for β-D-glucose, the colorimetric determination of hydrogen peroxide is far less specific and significant errors may be introduced in this second stage of the assay reaction. Difficulties will arise if the glucose oxidase preparation is contaminated with catalase, which destroys the hydrogen peroxide by converting it to oxygen and water. In addition some substances, such as ascorbic acid, glutathione and haemoglobin, interfere with the reaction by competing with the chromogen as hydrogen donors.

However, some of the more recently introduced chromogens are said to minimize these effects and the reaction involving the peroxidase-catalysed oxidative coupling of 4-amino-phenazone and phenol to produce a coloured complex is widely used. Another commonly used chromogen is 2,2′-azino-di-(3-ethyl-benzthiazolone sulphonic acid), which provides a simple and sensitive assay method.

Procedure 9.1 : Quantitation of glucose using glucose oxidase

Reagents

Glucose oxidase reagent

Glucose oxidase (EC 1.1.3.4) 3000 units (50 μkatal)

Peroxidase (EC 1.11.1.7) 5000 units (85 μkatal)

2,2′-Azino-di-(3-ethyl-benzthiazolone) sulphonic acid (ABTS) 1.0 g

Phosphate buffer (0.1 mol l^{-1}) pH 7.0, 1 litre

Method

To 4.0 ml glucose oxidase reagent add 0.1 ml sample.

Mix and allow to react at 30°C for exactly 30 min.

Measure the absorbance at 560 nm.

Standard

A series of standard solutions of glucose (0–20 mmol l^{-1}) should be treated in exactly the same manner as the sample and a calibration graph drawn using the results.

Calculation

The concentration of glucose in the sample is read off the calibration graph using the absorbance value obtained for the sample.

There are numerous commercially produced kits and dry reagent test devices which are available for the determination of glucose using glucose oxidase. They contain all the required reagents although their composition may vary between manufacturers, especially with respect to the chromogenic oxygen acceptor that is used.

Electrochemical methods : The electrochemical measurement of the hydrogen peroxide produced forms the basis of instruments often referred to as glucose analysers. Several are commercially available and although the design varies from one manufacturer to another, a common feature of those that amperometrically measure the hydrogen peroxide produced is the use of glucose oxidase in an immobilized form. This is often incorporated in an enzyme electrode which is surrounded by a small chamber of buffered reagents into which the sample is introduced.

Other similar biosensor devices have more recently been developed which demonstrate improved specificity and linear range. Alternatively, the immobilized enzyme may be packed in a bed permitting continuous sample analysis, for example of a process stream. The sample is passed through the bed of immobilized enzyme and the hydrogen peroxide produced is monitored amperometrically. Dual channels permit simultaneous analysis of glucose and lactose or glucose and sucrose, the disaccharides being hydrolysed enzymically and the glucose content measured.

Thus in the analysis of sucrose, an immobilized sucrase breaks the glycosidic linkage to yield glucose and fructose and an immobilized mutarotase ensures the conversion of α-D-glucose to β-D-glucose for its measurement by the glucose oxidase method. The measurement of lactose is achieved similarly using β-galactosidase as the hydrolytic enzyme.

Measurement of oxygen consumed : Alternatively, the initial oxidation of glucose can be monitored and this is most easily achieved by measuring the amount of oxygen consumed during the reaction. An electrochemical method using a polarographic Clark oxygen electrode has been used and the first oxygen electrode to be described for the measurement of glucose in 1962 contained soluble glucose oxidase held between cuprophane membranes. Recent modifications have resulted in the incorporation of a variety of forms of immobilized glucose oxidase into the electrode.

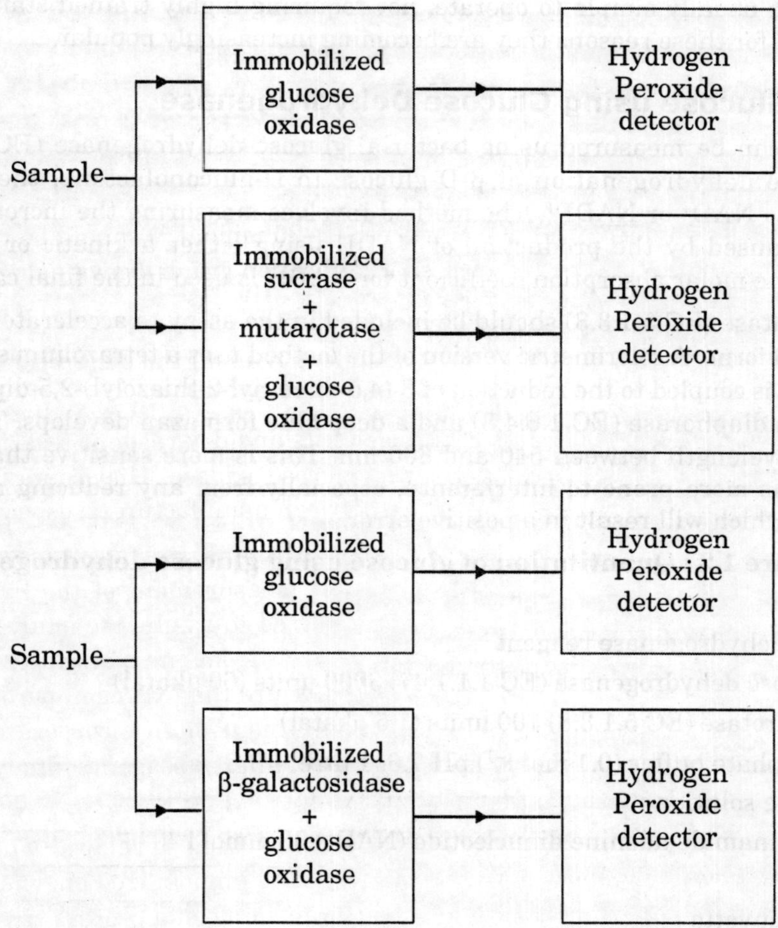

Fig. 1.21. Schematic diagram of the simultaneous continuous automated analysis of glucose and sucrose or glucose and lactose mixtures.

Catalase is a common contaminant of glucose oxidase preparations and will result in erroneous measurement of oxygen consumption owing to the regeneration of oxygen from the hydrogen peroxide produced in the reaction. This error can be prevented by destroying the hydrogen peroxide as it is produced, by adding ethanol, which results in the formation of acetaldehyde :

$$H_2O_2 + \text{ethanol} \longrightarrow \text{acetaldehyde} + H_2O$$

Iodide and ammonium molybdate are also used and rapidly reduce the hydrogen peroxide to water :

$$H_2O_2 + 2I^- + 2H^+ \xrightarrow[\text{molybdate}]{\text{ammonium}} I_2 + 2H_2O$$

Several types of glucose analysers are commercially available and although the purchase of such an instrument involves considerable initial financial outlay, it will have advantages of speed of analysis (seconds), low reagent volumes and very small sample size (e.g. 10 µl). Glucose analysers are usually simple to operate, not requiring highly trained staff to carry out the analysis, and for these reasons they are becoming increasingly popular.

Assay of Glucose using Glucose Dehydrogenase

Glucose can be measured using bacterial glucose dehydrogenase (EC 1.1.1.47), which catalyses the dehydrogenation of β-D-glucose to D-gluconolactone, the hydrogen being transferred to NAD^+ or $NADP^+$. The method involves measuring the increase in absorbance at 340 nm caused by the production of NADH using either a kinetic or fixed time assay technique. The molar absorption coefficient for NADH is used in the final calculation.

A mutarotase (EC 5.1.3.3) should be included in the assay to accelerate the conversion of the α to the β form. A colorimetric version of the method uses a tetrazolium salt. The oxidation of the NADH is coupled to the reduction of 3-(4,5-dimethyl-2-thiazolyl)-2,5-diphenyltetrazolium bromide by a diaphorase (EC 1.6.4.3) and a deep blue formazan develops. The absorbance is read at a wavelength between 540 and 600 nm. This is more sensitive than the ultraviolet method but is more prone to interference, especially from any reducing agents present in the sample, which will result in a positive error.

Procedure 1.2 : Quantitation of glucose using glucose dehydrogenase

Reagents

Glucose dehydrogenase reagent

Glucose dehydrogenase (EC 1.1.1.47) 3000 units (60 μkatal)

Mutarotase (EC 5.1.3.8) 100 units (1.5 μkatal)

Phosphate buffer (0.1 mol l^{-1}) pH 7.6, 1 litre

Coenzyme solution :

Nicotinamide adenine dinucleotide (NAD^+) 30 mmol l^{-1}

Method

Mix in a cuvette

2.6 ml glucose dehydrogenase reagent

0.2 ml sample

Record the 'baseline' absorbance at 340 nm

Initiate the reaction by adding :

0.2 ml NAD^+ solution

Allow to react until the absorbance at 340 nm is maximal and stable (15-10 min).

Calculation

(*a*) Calculate the concentration of NADH formed using the molar absorption coefficient for NADH at 340 nm (6.22×10^3 mol^{-1} cm^{-1})

$$\text{Concentration of NADH (mol } l^{-1}) = \frac{\text{Final absorbance} - \text{Baseline absorbance}}{6.22 \times 10^3}$$

(*b*) Calculate the concentration of glucose in the original sample

$$\text{Glucose concentration (mol l}^{-1}) = \frac{\text{Concentration of NADH formed} \times 3.0}{0.2}$$

Assay of Glucose using Hexokinase

Another enzyme used for the measurement of glucose is hexokinase (EC 2.7.1.1) which catalyses the phosphorylation of glucose to produce glucose-6-phosphate with adenosine triphosphate as the phosphate donor and magnesium ions as an activator. The rate of formation of glucose-6-phosphate can be linked to the reduction of NADP by the enzyme glucose-6-phosphate dehydrogenase (EC 1.1.1.49). This indicator reaction can be monitored spectrophotometrically at 340 nm or fluorimetrically:

$$\text{glucose} + \text{ATP} \xrightarrow{\text{hexokinase}} \text{glucose-6-phosphate} + \text{ADP}$$

$$\text{glucose-6-phosphate} + \text{NADP}^+ \xrightarrow{\text{glucose-6-phosphate dehydrogenase}} \text{6-phosphogluconate} + \text{NADPH} + \text{H}^+$$

The enzyme hexokinase is, however, not specific for glucose and is capable of converting some other hexoses to their corresponding 6-phosphate derivatives. Additionally, the specificity of the enzyme may vary slightly depending on its source. Yeast hexokinase will catalyse the phosphorylation of a number of other hexoses as well as glucose including D-mannose, D-mannose, D-fructose, 2-deoxy-D-glucose and D-glucosamine.

This provides an assay system for mannose or fructose if the initial reaction is linked to a suitable second reaction. In the determination of glucose, although there is a lack of specificity of the hexokinase, the overall assay is very specific for glucose because the linking enzyme is specific for glucose-6-phosphate and will react with neither fructose-6-phosphate nor mannose-6-phosphate without the incorporation of phosphoglucose isomerase (EC 5.3.1.9) to convert it to glucose-6-phosphate. Thus it is a requirement for the specific glucose assay that the hexokinase preparation is not contaminated with phosphoglucose isomerase.

Miscellaneous Methods for the Measurement of Hexoses

Two other commonly occurring hexoses which are usually found as components of polysaccharides or combined with other molecules in complex structures are galactose and fructose and, in a similar manner to other monosaccharides, enzymic methods are available for their measurement. An enzymic method for the measurement of fructose using hexokinase was described earlier, together with the method for mannose and glucose. Galactose oxidase (EC 1.1.3.9) catalyses the oxidation of β-D-galactose in a similar manner to the oxidation of β-D-glucose by glucose oxidase and forms the basis of an identical quantitative method.

The specificity is less than that of glucose oxidase and although glucose is not a substrate, other monosaccharides and glycosides, including L-altrose, D-talose, D-galactosamine, melibiose and raffinose, may also be oxidized depending upon the source of the enzyme. An assay using galactose dehydrogenase (EC 1.1.1.48), which catalyses the conversion of D-galactose to D-galactonolactone in the presence of the coenzyme NAD^+, is more specific and enzyme preparations are available for which the only other substrates are α-L-arabinose and β-D-fucose. The generation of NADH is conveniently monitored at 340 nm and permits quantitation of the galactose:

$$\text{D-galactose} + \text{NAD}^+ \xrightarrow{\text{galactose dehydrogenase}} \text{D-galactonolactone} + \text{NADH} + \text{H}^+$$

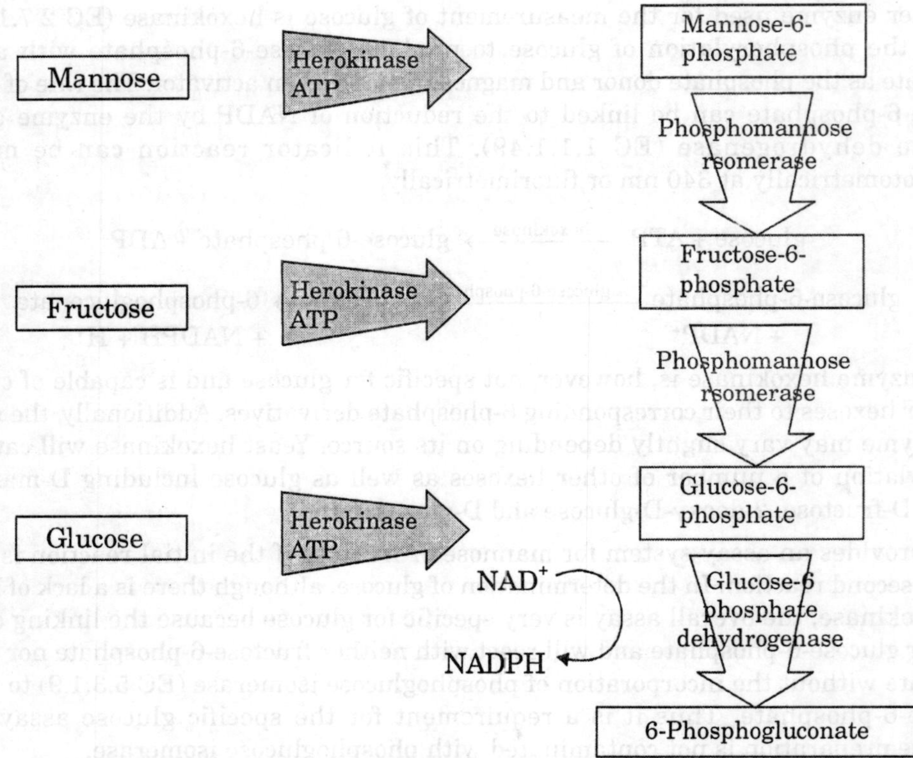

Fig. 1.22. Assay systems for mannose, fructose and glucose using hexokinase. Mannose, fructose and glucose may all be assayed independently using hexokinase and the appropriate additional enzymes. All the reactions can be monitored by the increase in absorbance at 340 nm as NADP* is reduced to NADPH.

SEPARATION AND IDENTIFICATION OF CARBOHYDRATE MIXTURES

Historically, techniques such as the formation of osazones and the demonstration of fermentation have contributed significantly to the separation and identification of carbohydrates. Observation of the characteristic crystalline structure and melting point of the osazone derivative, prepared by reaction of the monosaccharide with phenylhydrazine, was used in identification. This method is not completely specific, however, because the reaction involves both carbon atoms 1 and 2 with the result that the three hexoses, glucose, fructose and mannose, will yield identical osazones owing to their common enediol form.

Fermentation tests are based on the ability of yeast to oxidize the sugar to yield ethanol and carbon dioxide, although only the D-isomers are fermentable and only relatively few of these. Modern chromatographic techniques are, however, much more acceptable and paper and thin-layer techniques are useful for routine separation and semi-quantitation of carbohydrate mixtures, although GLC or HPLC techniques may be necessary for the more complex samples or for quantitative analysis.

Paper and Thin-layer Chromatography

Both ascending and descending paper chromatographic techniques have been used and, when thin-layer supports are employed, the use of either silica gel or cellulose is applicable. As the number of carbohydrates present in the sample is often small, the careful choice of solvent will generally make it unnecessary to perform the two-dimensional separations that are often needed when large numbers of substances, such as amino acids, are present. Reference solutions of each carbohydrate can be made up in concentrations of approximately 2 g l^{-1} dissolved in an isopropanol solvent (10% v/v in water) and samples of about 10μl should give discernible spots after separation.

Solvent Systems

Several monophasic solvent systems are useful for the separation of carbohydrate mixtures, and in all those listed in Table 1.3 the smallest solute molecules have the fastest mobility. Thus pentoses have higher R_F values than hexoses, followed by disaccharides and oligosaccharides. The distance moved by different oligosaccharides is a reflection of the number of monosaccharide units of which they are composed, with the smallest molecules again moving the furthest.

However, in general, the distances moved by all classes of carbohydrates are small and although modification of solvent composition may result in greater overall mobility, the relative differences between the components is still low and it may be necessary to run the solvent off the end of the support to achieve a satisfactory separation. In these circumstances the distance moved by glucose is used as a reference and is given the value R_F = 100 in any solvent system. The migration of another carbohydrate is reported as its R_g value:

$$R_g = \frac{\text{distance moved by substance}}{\text{distance moved by glucose}} \times 100$$

Locating Reagents

A variety o reagents can be used for visualization of the separated components and it may be useful to run duplicate chromatograms and use a different stain on each one to assist in identification of unknown spots. The most commonly used reagents make use of the chemical reactions of carbohydrates already described in the section on quantitative methods and appropriate safety precautions must be taken when using the various locating reagents. The actual reagent composition may be modified either in terms of concentration of components or by substitution of one chemical for another very similar one, although the principle of the reaction with a carbohydrate remains unaltered. Such variations in reagent composition may be advantageous in promoting the production of characteristic colours for different carbohydrates either within a class (e.g. hexoses with aniline diphenylamine phosphate reagent) or by differentiating on a broader basis (e.g. pentoses from hexoses, or aldoses from ketoses). In certain situations the use of a reagent that incorporates a specific enzyme may be advocated and it will be necessary to be aware of any apparent lack of enzyme specificity and to have a knowledge of all the substrates on which the enzyme will act.

Table 1.3 Some monophasic solvents for thin-layer chromatography of carbohydrates

Solvent Composition	*Proportions**	*Comments*
Ethyl acetate Pyridine Water	60 30 20	Commonly used. Gives good separation of pentose and hexoses. Will resolve glucose and galactose (C)
n-Butanol Pyridine Water	60 40 30	Many variations in composition may be used to increase or decrease overall mobilities (C)
Formic acid Methyl ethyl ketone Tertiary butanol Water	15 30 40 15	Gives good separation of monosaccharides and disaccharides (C)
Ethyl acetate Ethanol Pyridine Acetic acid Water	70 10 10 10 10	Useful for separation of pentoses and hexoses (SG)
n-Butanol Acetone Acetic acid Water	35 35 10 20	May be useful if amino acid separations are also performed. R_F values tend to be low, and glucose and galactose are not resolved. Use second after solvents containing pyridine to remove it (C)
n-Butanol Acetic acid Water	2 1 1	Separates monosaccharides and disaccharides. Expecially useful for sugar acids. Triple development useful for oligosaccharides (SG)
n-Butanol Acetic acid Diethyl ether Water	9 6 3 1	Separates monosaccharides and disacc-harides. Especially useful for methyl glycosides and sugar alcohols. Gives good separation with mono-, di-, tri- and oligosaccharides. Used as second solvent after *n*-butanol/acetic acid/water (SG)

*Proportions of constituents can be varied.
Recommended supported media: (C), cellulose; (SG), silica gel.

Table 1.5 gives examples of some useful locating reagents, the composition of which may be modified for dipping or spraying techniques, the colours usually appearing after heating at 100°C for 5-10 min. The reaction of acid locating reagents may be impaired if any pyridine present in the chromatographic solvent is not completely removed. This is particularly important when paper and cellulose thin layers are used, although with silica it is not so critical. The pyridine is absorbed by cellulose and cannot be removed completely by oven drying, although the problem can be overcome by dipping the dried chromatogram in *n*-butanol and re-drying before applying the locating reagent or by employing a second solvent system which includes an alcohol, usually *n*-butanol.

Table 1.4. Colour reactions of common mono- and disaccharides

Carbohydrate	*Locating reagents*		
	Naphthoresorcinol	*4-Aminobenzoic acid*	*Aniline diphenylamine phosphate*
Ribose	—	Red/brown	Blue/green
Xylose	—	Violet	Green/blue
Arabinose	—	Red/brown	Blue/green
Xylulose	Green/brown	Red/brown	Blue/green
Glucose	—	Brown	Grey/blue
Galactose	—	Brown	Grey/blue
Fructose	Red	Pink*	Orange/brown
Sucrose	Red	Brown*	Brown
Lactose	—	Brown	Blue

*Indicates that the reagent shows poor sensitivity for that carbohydrate.

Gas-liquid Chromatography

Gas-liquid chromatography may be the method of choice when it is necessary to identify or to quantitate one or more carbohydrates especially when they are present in small amounts. Although this technique is often used because it is possible to resolve carbohydrates with very similar structures, the fact that α and β anomers and pyranose and furanose forms of the same carbohydrate all give separate peaks is sometimes a disadvantage.

Gas chromatography must be carried out using volatile derivatives of the carbohydrates and although many have been studied (e.g. *O*-methyl ethers, *O*-acetyl ethers, trimethylsilyl-*O*-methyl oximes and *O*-trimethylsilyl ethers) using a variety of stationary phases, the *O*-trimethylsilyl (TMS) derivatives are probably used most frequently, although circumstances may dictate the use of an alternative.

Such TMS derivatives are simple to prepare and have been suc-cessfully applied to a wide variety of compounds. It should be noted that, in general, carbohydrates require only weak silylating conditions, otherwise random isomerization will occur with the production of spurious peaks making interpretation of the chromatographic trace very difficult. The use of a mixture of HMDS (hexamethyldisilazane), TMCS (trimethylchlorosilane) and pyridine (2:1:10) is recommended although carbohydrates combined with or containing amino, phosphate, carboxylic groups or nucleic acids will require a more powerful silylating agent and BSA (*N, O*-bis(trimethylsilyl) acetamide) or BSTFA (*N, O*-bis(trimethylsilyl) trifluoroacetamide) in conjunction with TMCS (trimethylchlorosilane) and pyridine are widely used.

The choice of a stationary phase will depend upon the nature of the carbohydrates to be separated and whereas an OV-17 column (phenylmethyl polysiloxy gum) may give satisfactory isomeric separations, a non-polar phase such as OV-1 (methylpolysiloxy gum) may be more useful for a wider range of carbohydrates.

L-Fucose and D-galactose each gave three separate peaks corresponding to the furanoside and the alpha and beta methylpyranosides. D-Mannose, D-glucose and W-acetyl-glucosamine (GlcNAc) each gave two peaks due to their alpha and beta pyranosides. D-Mannitol, which was used as an internal standard, and *N*-acetylneuraminic acid (AcNeu) gave single peaks. Under these chromatographic conditions, complete resolution of all the components was not

Table 1.5 Examples of locating reagents suitable for the TLC of carbohydrates

Reagent	*Composition and use*	*Reaction principle*	*Comments*
Concentrated sulphuric acid	Concentracted acid **Spray**	Dehydration of carbohydrates	General locating reagent for all classes of carbohydrate. Some slight colour differences for different classes (yellow/brown/black). Not suitable for use on cellulose plates.
Orcinol—sulphuric acid	200 mg orcinol in 100 ml 10% sulphuric acid **Spray**	Formation of furfural and derivatives with heat and acid which condense with a phenol	Detects mono-, di-, tri- and oligosaccharides. Colour variations of green/purple/brown
Naphthoresorcinol	200 mg naphthoresorcinol in 3.2 ml 90% orthophosphoric acid in 100 ml methanol **Dip**	Formation of and derivatives with head and acid which condense with a phenol	Reacts with ketoses. Colours (red/brown) fade at room temperature but not at –20°C. Residual pyridine from solvent will interfere.
4-Aminobenzoic acid	1.4 g 4-aminobenzoic acid + 3.2 ml 90% orthophosphoric acid in 100 ml methanol **Dip**	Reaction with aromatic amine in hot acid	Very useful for many different carbohydrates. Very sensitive for pentoses. Colours (red/brown) can be preserved by coating with vinyl from aerosol
Anilinediphenylamine phosphate	1 ml aniline + 1g diphenylamine in 100 ml acetone. Add 10 ml 85% orthophosphoric acid **Dip**	Reaction with aromatic amine in hot acid	Not as sensitive as some reagents but colour variations are useful in interpretation (grey/blue/green/brown)

achieved and some had identical retention times, e.g. β-D-methylgalactopyranoside and α-D-methylglucopyranoside.

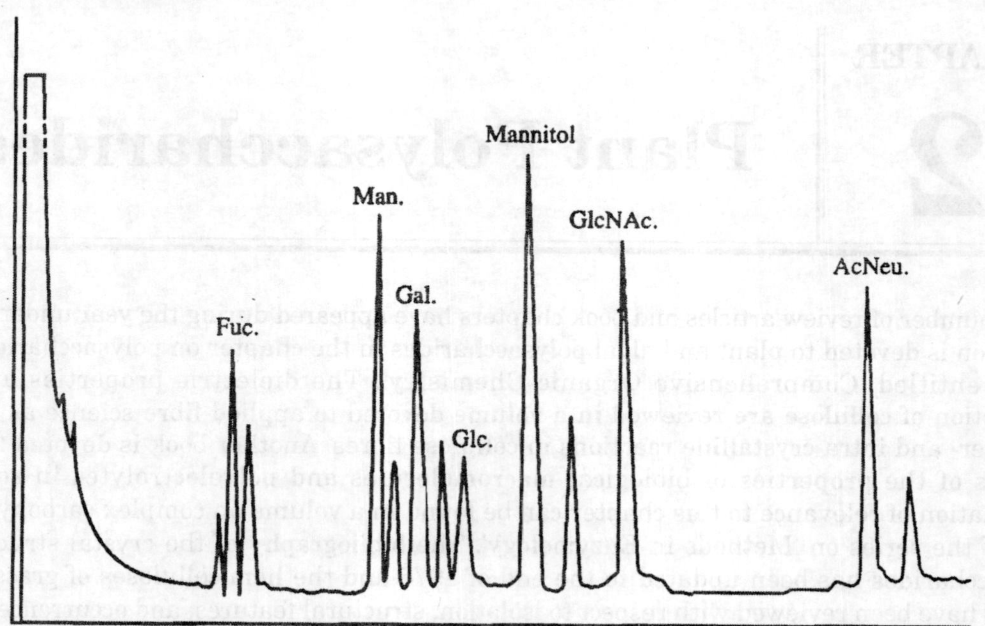

Fig. 1.23. Separation of equimolar concentrations of methylglycosides by gas-liquid chromatography. The analysis was performed on an OV-1 stationary phase using a temperature gradient from 120 to 220°C.

High Performance Liquid Chromatography

Separation and quantitation of carbohydrate mixtures may be achieved using HPLC, a method that does not necessitate the formation of a volatile derivative as in GLC. Both partition and ion-exchange techniques have been used with either ultraviolet or refractive index detectors. Partition chromatography is usually performed in the reverse phase mode using a chemically bonded stationary phase and acetonitrile (80:20) in 0.1 mol l^{-1} acetic acid as the mobile phase. Anion- and cation-exchange resins have both been used. Carbohydrates form anionic complexes in alkaline borate buffers and quaternary ammonium anion-exchange resins in the hydroxyl form can be used for their separation. Problems caused by rearrangement reactions in alkaline solution have been minimized by the use of boric acid/ glycerol buffers at pH 6.7. It is possible to separate monosaccharides at this pH by elution with such a buffer to which sodium chloride has been added. This is omitted for di- and trisaccharides and a weaker boric acid/glycerol buffer is used which extends their elution times and permits good resolution.

Sulphonated cationic exchange resins with metal counter-ions are also useful for carbohydrate analysis. Water or mildly acidic eluents are normally used under temperature-controlled conditions, e.g. 85°C. Resins are available in a variety of forms. The calcium or lead form is generally recommended for monosaccharides and disaccharides whereas, for oligosaccharides, the silver or sodium form is preferred.

CHAPTER

2 Plant Polysaccharides

A number of review articles and book chapters have appeared during the year under report. A section is devoted to plant and algal polysaccharides in the chapter on polysaccharides in a series entitled 'Comprehensive Organic Chemistry'. The dielectric properties and d.c. conduction of cellulose are reviewed in a volume devoted to applied fibre science as well as the inter- and intra-crystalline reactions in cellulose fibres. Another book is devoted to some aspects of the properties of biological macromolecules and polyelectrolytes in solution. Information of relevance to this chapter can be found in a volume on complex carbohydrates, part of the series on 'Methods in Enzymology'. The bibliography of the crystal structure of polysaccharides has been updated to the end of 1975 and the hemicelluloses of grasses and cereals have been reviewed with respect to isolation, structural features, and occurrence within the plantstuff.

STARCHES

The C n.m.r. spectra of the previously reported comb-like derivatives prepared by orthoester- and Helferich-type condensations of D-glucose on to amylose have been recorded. It was known from monomeric studies that the orthoester method favoured the β-D-glycosidic bond, whereas the Helferich method favoured the α-D-glycosidic bond. The spectra confirmed that 4,6-disubstitution occurred on the D-glucose residues of amylose and that a degree of polymer linearity existed but the spectra also indicated that extensive, if not exclusive, β-D-glucosylation occurred with both condensation methods. Differential thermal analysis has been applied to an examination of the thermal behaviour of starch and foodstuffs.

Potato starch showed an endotherm at *ca.* 65°C which was not influenced by either the rate of heating, or the concentration and the initial amount of starch. Similar analyses, applied to intact potato, Indian lotus, taro, and sweet potato, were distinct from each other but were similar to the starch from each plant except that the isotherm was shifted by 3–7 °C to the higher temperature region. The absorption and circular dichroism spectra of the blue colour produced by tri-iodide ions in amylose have been characterized. Four fundamental Raman lines were observed for I_3^- bound to amyloses which were not dependent on the DP in the range 20-100 or the excitation wavelength. Concanavalin A does not react with amylose but does react with the amylose-iodine complex although the complex can be dispersed by shaking.

The complex is non-specific as it cannot be displaced by concanavalin A-specific sugars. A simplified determination of amylose in milled rice has involved an initial dispersion in dilute alkali. The amylose-iodine complex at pH 7.5 was estimated from the absorption at 630 nm. Ultrafiltration was considered to be a suitable method for a hydrodynamic study of the amylose-iodine complex formation. No partially complexed amylose molecules were present in the permeate and so it was considered that complex formation took place instantaneously in a whole molecule and not successively in different zones of the molecule.

The conformation of amylose in aqueous solution is entirely that of loose and extended helices. The formation of the iodine complex is visualized as a trapping of iodine atoms by the contraction of the loose helical macro-molecule into tight and stiff helices. The specific volumes of the crystalline and amorphous portions of starch have been determined. All the values were corrected by extrapolation to zero moisture content.

The amorphous portion gave values very similar to those for cellulose, being *ca.* 0.673 for potato, waxy corn, and waxy rice starch and *ca.* 0.684 for sweet potato, tapioca, corn, and wheat starch. Digital oscillator densimeter techniques have been used to determine the density of dispersions of starch, glycogen, dextrin, amylopectin, and amylose in water at 40 °C but at different concentrations. The conformation of amylose in aqueous solution was determined by light-scattering sedimentation-equilibrium measurements and was found to be dependent on the molecular weight. When the molecular weight was outside the 'dissolving gap' ($6.5 \times 10^3 < M_r < 1.6 \times 10^5$) the molecule behaved as a random coil while inside the range the molecules aggregated to a rigid coil.

The conformation of the rigid coil was that of a double helix and resembled β-type (retrograded) amylose. When potato starch granules were heated with a limited amount of water (water : starch < 1.5) two endothermic transitions were observed on differential scanning calorimetry. The lower endothermic transition always occurred at 66 °C and was the only transition observed when excess water was present. The temperature of the higher transition was inversely proportional to the water content and decreased in intensity with increasing water content.

The appearance of the higher endothermic transition allowed the stoicheiometry for full hydration of 14 water molecules per hexose unit to be determined. Synthetic amyloses have been prepared from maltopentaose and D-glucose 1-phosphate by the action of potato phosphorylase. The products were characterized by the analysis of the products of β-amylolysis and were used as standards in the determination of the molecular weight of starches by gel chromatography. To elucidate the most probable types of chelation that exist in the chains of amylose and cellulose, studies have been undertaken on molecular models which represent, in part, the maltose and cellobiose entities in the polymers.

These models were some intra- and extra-cyclic acetals of the 2-cyclohexyloxytetrahydropyran type, which contained primary or secondary hydroxy-groups whose covalent disposition, relative to the acetal function, were analogus to those of OH-2, OH-3′, OH-6, and OH-6′ of amylose and celluse. Most of the diastereoisomers, containing at least two asymmetrical carbon atoms were separated by gel permeation chromatography and their structures were confirmed by ^{1}H and ^{13}C n.m.r. spectroscopy. It was possible, by i.r. spectroscopy, to observe absorption bands corresponding to chelation in rings containing 5, 6, 7 or 8 members.

In maltose, methyl b-maltoside, and amylose chelation was observed between OH(2) and OH(3′) and possibly between OH(2) and O(1). A Raman spectroscopic study of the amylosemonoglyceride complex, using monostearin as the model compound, indicated that the hydrocarbon conformation inside the amylose helix was ordered as if in a crystalline state. Three helical turns existed for each hydrocarbon chain. The average chain length of corn starch has been shown to be reduced by γ-irradiation as well as by acidic hydrolysis, the extent being dependent on the dose rate.

These was no evidence for the formation of dialdehydes by ring cleavage of the D-glucose residues. The degrees of branching of commerical D-glucose syrups and D-glucose syrups prepared by reverse osmosis have been investigated. The results of methylation and enzymatic degradation analyses showed no significant differences. The acidic and thermal stabilities of unmodified pea starch are destroyed on preparing the acetate or monophosphate derivatives. These properties are similar to those of corn starch.

The thermal stability was not affected on acetylation of a cross-linked pea starch. In this paper, a unique purification step was outlined that removed damaged starch granules along with pea hull fragments. A higher concentration of oval or irregularly shaped granules were found in the starch from *Vicia faba* than corn starch. The granules also had a higher amylose content, gelatinization temperature range, and water-binding capacity, but a lower swelling power and solubility than either wheat or corn starch. Starches have been prepared from finger millet (*Eleusine coracana*) and Foxtail millet (*Setaria italica*) and were characterized by their low solubility. The mature starches were susceptible to α-anylase action. The amylose contents varied between 15.5 and 17.5%, but successive leaching with water and alkali gave amylose components with different molecular weights and β-amylolysis limits.

The results suggested that the amylose fraction was heterogeneous. Recent studies on oat starch showed that the iodine affinity, solubility, and λmax of the iodine complex were all lower than for other cereal starches. The concentration of fractions between the size of amylose and amylopectin was higher than in wheat and the β-amylolysis limit was lower, indicating that the starch was more highly branched. Flours, prepared from the five legumes, navy, pinto, faba, mung bean, and lentil all had similar starch and moisture contents but the starch content was lower than that of wheat flour.

The legume starch granules are oblong and do not have such a wide variation in size as wheat flour granules. The range of amylose contents was lower than that for wheat starches. The acetylacted wheat starch derivative had a higher molecular weight than the legume starches. A method for the separation of the amylose and amylopectin components of cotton leaves, but which should be applicable to other plants, has been reported. After extraction with aqueous dimethysulphoxide, the amylose component was separated from the amylopectin and cell-wall particles by chromatography on macroporous agarose while the amylopectin was purified by its interaction with concanavalin A. It is now generally accepted that the digestion of starch by many mammalian species involves, firstly, the action of pancreatic α-amylase to give oligosaccharide end-products which are hydrolysed further to D-glucose by the enzymes of the small intestine brush border.

The enzymology of these reactions has been reviewed with respect to both nutritional and clinical implications. The controlled hydrolysis of corn starch with resin-bound α-amylase has yielded a modified starch of high molecular weight in yields of *ca.* 93%. The amylose component was hydrolysed preferentially and due to the *exo* nature of the enzyme, oligosaccharides of DP 1-8 were isolated.

The modified starch had a lower tendency to retrograde. The hydrolysis of gelatinized modified and unmodified starch with porcine pancreatic α-amylase has been studied. Modification of the starch with hydroxypropyl and acetate groups reduced the extent of digestion, acetylation being better than hydroxypropylation. Cross-linking of the unmodified starch with phosphate groups also lowered the extent of digestion. The ability of native and dioxan-extracted starch from potato and maize to bind Ca^{2+} ions influenced their digestibility *in vitro* with pancreatic α-amylase. With maize starch, the digestibility was improved on extraction but lowered by Ca^{2+} ions, whereas the digestibility of potato starch was improved with both native and extracted starch. Scanning electron microscopy showed that numerous pin holes could be observed on the surface of starch granules when attacked by various amylases.

It was also noticed that these pores penetrated into the inner layers of the granules during enzymic action and in some granules, a terraced or step-shaped appearance in their inner portions was observed. These internal characteristics are probably indicative of a layered internal structure in the granules. The other observations from scanning electron microscopy were the presence of striated structures on the surface of starch granules from banana, lily, and lotus which had been attacked by pancreatin. Rapid heating to 200 °C and rapid cooling of starch with a 20% moisture content resulted in a product which could be gelatinized at room temperature.

The thermally modified starch was a better substrate than the native starch for saccharification with D-glucoamylase. The action of Q-enzyme on a model amylodextrin which contained a single-branch linkage has been studied. It was concluded that the enzymatic process for synthesizing the branch linkage in amylopectin is a random action of the Q-enzyme on a complex, possibly a double helix, formed between two (1 → 4)-linked α-D-glucan chains.

The action pattern predicts a novel arrangement of the unit chains in amylopectin. The intestinal surface membrane from rats contains both (1→4)-α- and (1 → 6)-α-D-glucosidase (-glucanase) activities and these enzymes have been given the trivial names of 'sucrase' and 'isomaltase', respectively. The concurrent action of the sucrase and isomaltase active sites of the hybrid sucrase-isomaltase enzyme complex has been studied on the hexasaccharide (1). Hydrolysis occurred by sequential removal of D-glucose residues from the non-reducing terminal end of the molecule through the pentasaccharide (2), tetrasaccharide (3), maltotriose, maltose, and finally D-glucose. The free isomaltase monomer could hydrolyse each step outlined, but displayed maximal specificity for the hydrolysis of (3) to maltotriose. The free sucrase was active against all 1,4-linked residues but could not hydrolyse (3) to maltotriose. Instead of acting synergystically, the two α-D-glucosidase sites appear to cleave α-limit dextrins by their complementary specificity.

α-D-Glc*p*-(1 → 4)-α-D-Glc*p*-(1 → 4)-α-D-Glc*p*
1
↓
6
α-D-Glc*p*-(1 → 4)-α-D-Glc*p*-(1 → 4)-D-Glc*p*
(1)

α-D-Glc*p*-(1 → 4)-α-D-Glc*p*
1
↓
6
α-D-Glc*p*-(1 → 4)-α-D-Glc*p*-(1 → 4)-D-Glc*p*
(2)

α-D-Glc*p*
1
↓
6
α-D-Glc*p*-(1 → 4)-α-D-Glc*p*-(1 → 4)-D-Glc*p* → Maltotriose → Maltose → D-Glucose
(3)

The retrogradation of amylose in solutions of enzymatically liquified starch depended on the degree of starch hydrolysis and on the manner of preparing the solution. Even at DE values of 20-30%, retrogradation occurred in solutions of <5% and >50% concentration.

The subcellular localization of the starch biosynthetic and degradative enzymes of spinach leaves was carried out by measuring the distribution of the enzymes in a crude chloroplast pellet and in separated components of a protoplast lysate. The enzymes which were involved in the syntheses of starch were detected in the chloroplasts, whereas some of the enzymes involved in the degradation of starch were mainly in the soluble protein fraction but were also found in the chloroplast. The digestion pattern of the amylase on amylopectin as substrate indicated that the enzyme was an α-amylase from its *endo*-lytic activity, but displayed properties unlike the typical α-amylase isolated from endosperm tissue. A time-sequence analysis of the starch digestion pattern in germinating rice showed that activity in the initial stage, was legalized in the epithelium septum between the scutellum and the endosperm.

The breakdown of starch in endosperm tissue began thereafter. Amylase activity could only be detected in the aleurone layer after two days.

α-D-Glucopyranose 1,2-cyclic phosphate is a potent inhibitor of phosphorylase-catalysed starch elongation in potato. The affinity of the enzyme for the cyclic ester is *ca.* thirty times that of D-glucose 1-phosphate. Under conditions where elongation has been retarded by a factor of three, UDP-D-glucose, ADP-D-glucose, D-glucose 6-phosphate, and D-glucose 2-phosphate cause a rate reduction of less than 10%.

FRUCTANS

Methylation analysis of the D-fructose polymers from cocksfoot *(Dactylis glomerata)* indicated that both (2→1)-β-D-(inulin) and (2→6)-β-D-(levan) types of linkages were present. Chemical analysis failed to detect any low molecular weight oligosaccharides during fructan synthesis in the leaves of cocksfoot *(D. glomerata)*. These results suggest that direct transfer of D-fructose to existing fructan polymers takes place. As the percentage dry weight of individual onion bulbs increased, the D-fructan content also increased, with a higher proportion of molecules with higher molecular weight being present. The percentage dry matter and fructan content increased from outer to inner layer of the bulbs.

CYCLOAMYLOSES

Cycloamyloses have been separated by h.p.l.c. on a μ-Bondapak-carbohydrate column using acetonitrile-water mixtures as eluant. The molecular dynamics of the inclusion complexes formed between cyclohexa-amylose and some aromatic amino-acids and dipeptides have been studied by ^{13}C n.m.r. spectroscopy. The forces binding the complexes were found to be weak. The c.d. spectra of cyclohepta-amylose which had been complexed with 2-substituted naphthalenes were measured at various concentrations of cyclohepta-amylase and temperatures between 10-70 °C.

The complex with 2-naphthoxyacetic acid showed 1 : 1 stoicheiometry. The molar ellipticity and thermodynamic parameters were determined and enthalpy and entropy ranges calculated. The correlation was explained by a cyclohepta-amylose guest molecule interaction where the guest molecule was highly solvated. The induced c.d. spectra of cyclohepta-amylose complexes with substituted benzenes confirmed that an axial inclusion was favoured, whereby the long axis of the substituted axis of the substituted benzenes is parallel to the axis of the cyclohepta-amylose cavity.

The binding forces contributing to the association of cyclohexa- and cyclohepta-amyloses with alcohols in aqueous solution have been determined from a spectrophotometric examination of the inhibitory effect of the various alcohols on the association of the cycloamyloses with Azo dyes.

Hydrophobic and van der Waals interactions were found to be of primary importance in complex formation. As the bulkiness of the alcohol increased, the stability of the cyclohexa-amylose adduct decreased, whereas the stability of the cyclohepta-amylose adduct was enhanced owing to the larger cavity in the cyclohepta-amylose molecule allowing better contact of the alcohol within the cavity. Inclusion complexes of poly-(cyclohepta-amylose) with two separate fluorescent probes have been used as models to measure the pressure effects upon ligand-protein complexes. The data permitted the calculation of compressibility curves, which were in broad agreement with those of liquid aliphatic and aromatic hydrocarbons in the low-pressure range, but indicated a reduced compressibility at higher pressures. Metastable crystals of cyclohepta-amylose complexes with propan-1-ol have been prepared and their structure has been discussed in terms of the membrane diffusion model proposed previously by these authors.

CELLULOSE

Single crystals of cellulose II have been grown from dilute aqueous solutions of cellulose acetate by deacetylation and consequent precipitation. The crystals had a ribbon-like lamellar appearance and the best results were obtained using methylamine, at 90 °C, as the deacetylating agent. It has been suggested that the transformation from cellulose I to cellulose II during mercerization is the result of a progressive shift of the sheets of cellulose chains within the crystallites of a microfibril from the quarter-staggered relationship in cellulose I to the complete correspondence found in cellulose II. This idea agrees with the observed changes in lateral disorder, cell dimensions, swelling, and X-ray diffraction reflections of cellulose fibres.

Such a shift may take place in the transformation of native celluloses with antiparallel structures as well as those with parallel polarity. The dielectric constant and dielectric loss for moist cellulose fibres was measured over a frequency band, a temperature range, and relative humidity range. The dielectric constant increased with increasing frequency and temperature owing to an increase in the rotation and polarization of the flexible part of the fibre. As the relative humidity was increased, the dielectric constant and dielectric loss both increased. This may have been due to the presence of the polar water molecules; the freeing of polar groups; or the freeing of ions in the fibre molecule. The ^{13}C n.m.r. spectra of peracetylated cello-oligosaccharides have been recorded and compared with that of cellulose acetate. All of the signals from cellulose acetate were only observed in the spectra of cellotetraose and cellopentaose peracetates.

Some signals were not observed in the spectra of cellobiose and cellotriose peracetates. A bioengineering approach has been used for the separation of cellodextrins. Cellulose was hydrolysed with HCl and a combination of columns of Sephedex G15 and Dowex AG50WX × 4 allowed a one step desalting and separation of the hydrolysate using water as the eluant. The column was stable and did not require to be regenerated after every run. Up to 3 g of cellodextrins could be produced per day. A method for the gel permeation chromatographic analysis of the molecular weight distribution of wood pulp holocellulose as the carbanilate derivative has been applied to red maple *(Acer rubrum)* and loblolly pine *(Pinus taeda).* Either the chlorine-ethanolamine or acid-chlorite 'method could be used to prepare the holocellulose and the derivative was obtained by heating at 80 °C with phenylisocyanate in pyridine.

Higher temperatures caused depolymerization. Scanning electron microscopy of cellulose xerogels indicated that a membraneous phase existed with individual pores being present which were separated by thin walls. These pores had a short diameter of 50–90 µm and a long diameter of 80–300 µm. The cellulose xerogels were distinct from agarose xerogels but were similar to chitin xerogels which had been chemically prepared from chitosan by *N*-acetylation. Dye fixation on modified celluloses depended on the pH, temperature, time of thermofixation, type and concentration of the dye used.

Diffusion data have been obtained for helium, nitrogen, oxygen, and carbon dioxide gases through membranes of the acetate derivative of cellulose. The thermodynamic data have been evaluated and the mechanism of diffusion was discussed. To assist with the statistics of crosslinking, the relationship between relative reactivity and accessibility of the hydroxy groups in cellulose have been determined. When epichlorhydrin was used as the cross-linking agent, the reactivity was in the order OH-3 > OH- 6 > OH-2. Considerable interest has been shown in the use of anthraquinone in wood pulping operations. The optimal yield of pulp and pulp viscosity at a given Kappa number was obtained with 0.25 M-NaOH. The rate of delignification was increased by the addition of anthraquinone (38 mg l^{-1}). Reducing end groups of hydrocelluloses were stabilized more effectively against degradation during alkaline cooking by the presence of anthraquinone 2-sulphate than by anthraquinone itself.

This is probably due both to the better solubility and oxidation potential of the sulphate. Neither anthraquinone or naphthaquinone had any effect on the cooking of wood meal with sodium hydrogen carbonate in the presence of oxygen, but the presence of amines did retard the rate. Pretreatment of wood meal in the presence of oxygen with anthraquinone gave an increased yield of pulp on NaOH cooking. The rates of delignification and depolymerization were increased as well as the stabilization of the carbohydrate towards an endwise β-alkoxy elimination reaction. The presence of magnesium hydroxide retarded the rate of depolymerization and delignification.

During the pulping of the tracheid cell wall from the sapwood of *Pinus sylvestris* by alkali in the presence of anthraquinone, the rate of delignification of the secondary wall and middle lamella was increased, with the middle lamella being degraded more rapidly than the secondary wall. Low molecular weight products from the alkaline degradation of cellulose have been identified by g.c.–m.s. of their trifluoroacetate derivatives. Both the residual aqueous phase and the floating oily product were examined and found to contain essentially the same products. These included unsaturated, aliphatic and alicyclic hydrocarbons, aldehydes, ketones, alcohols, and furans.

The synthesis of molecules with more than six carbon atoms also occurred. Reductic acid, separated as its lead salt, has been isolated from the pyrolysis products of cellulose by reaction at 450 °C in the presence of oxygen. Amylose, maltose, and D-glucose also gave reductones. A mass spectrometer with a data system was required to study the products of the direct pyrolysis of cellulose and cellulose which had been treated with a fire-retardant.

The amount of products evolved at any point during, or throughout, the entire pyrolysis could be analysed. A philosophical discussion of cellulose as a vast and renewable resource and potential source of food has been published. The properties of cellulose and the uses to which it can be put, as well as a discussion on cellulases, which can convert the potential value of the cellulose into reality are presented. An immunochemical approach has been used to study the cellulolytic complex of *Trichoderma reesei* QM91414. *Endo*-(1 → 4)-β-D-glucanase (CM-cellulase) and cellobiohydrolase (Avicelase) activities were separated on diethylaminoethyl-Sepharose CL6B. The *endo*-(1 → 4)-β-D-glucanase activity was located in a single peak, but the cellobiohydrolase activity was associated with two peaks. Each peak showed several components on isoelectric focusing. The digestion of cellulose by anaerobic micro-organisms from a variety of habitats has been investigated. Of the rumen organisms, *Ruminococcus albus* was the most effective, being able to digest 10 g cellulose per litre producing 1.4 g dry cells, 2.8 g acetic acid, and 2.6 g ethanol. A strain from a compost, similar to *Clostridium thermocellum,* digested 4 g cellulose per day giving 0.8 g dry cells, 0.6 g acetic acid, and 0.9 g ethanol. In studies on the biosynthesis of cell-wall polysaccharides in cultured carrot cells, D-glucose was only incorporated into the neutral sugar residues, particularly cellulose, whereas *myo*-inositol appeared in the uronic acid and pentose residues.

An increase in cellulose content was observed during the thickening of the cell walls of parenchymous tissue of *Discorea dumetorum* tubers during storage. The ^{18}O : ^{16}O ratio of the cellulose from two sets of wheat plants, grown under conditions similar in all respects except for a large difference in the ^{18}O : ^{16}O ratio of the carbon dioxide supplied to them, only differed by a small amount. The difference in ratio of the cellulose was similar to the ^{18}O:^{16}O ratio of the water present in the plants. These results indicate that the oxygen of the carbon dioxide undergoes complete exchange with the oxygen of the water in the plant during cellulose synthesis and that the ^{18}O: ^{16}O ratio of the water within the plant is the primary influence on the ^{18}O : ^{16}O ratio of cellulose in terrestrial plants.

HEMICELLULOSES

Two different L-arabinans have been isolated from azuki beans *(Phaseolus radiatus)* by alkaline extraction and separated by gel chromatography. No other sugar residues were present and, from structural analysis, the polymers were composed of chains of 1,5-linked L-arabinofuranose residues with branch points at O-2 and/or O-3. These branches were composed of non-reducing terminal L-arabinofuranose residues and there were more (1 → 2) linkages than (1 → 3)-linkages.

The polysaccharides differed in their molecular weights and solubilities in aqueous ethanol. An *endo*-L-arabinanase and two α-L-arabinofuranosidases are produced when *Bacillus subtilis* is grown in a medium containing sugar beet arabinan. The *endo*-L-arabinanase was homogeneous by sodium dodecyl sulphate-polyacryl-amide gel electrophoresis, had a molecular weight of 3.2×10^4, and a pH optimum of 6.0. It is able to release residues of L-arabinose from suspension-cultured sycamore cell walls. α-L-Arabinofuranosidase activity was determined in 306 species of *Actinomycetales*.

Streptomyces species produced these enzymes at a faster rate than *Ascomycetes* or *Basidiomycetes* species. Activity appeared in the *Streptomyces* species after 2-4 days and, although the pH of the medium was *ca.* 8.5. the pH optimum of the enzyme was *ca* 6.0. An L-arabino-D-galactan, in 2% yield, has been isolated from the fruits of *Dillenia indica.* From the results of partial acid hydrolysis and methylation analyses, when 2,6-di-*O*-methyl-D-galactose, 2,3,6-tri-*O*-methyl-D-galactose, 2,3,4,6-tetra-*O*-methyl-D-galactose, and 2,3,5-tri-*O*-methyl-L-arabinose residues in the molar ratio 7 : 3 : 1 : 8 were identified, it was concluded that the poly-saccharide was a (1 → 4)-β-D-galactan that contained L-arabinofuranose residues linked at O-3 of some of the D-galactose residues in the main chain. The occurrence, isolation, chemistry, and physico-chemistry of the plant L-arabino-D-galactans and their protein conjugates have been reviewed.

The structural relationship between those found in woods, gum exudates, and plant callus cells, and whole tissue is discussed and the nature of the glycoproteins compared with L-arabinose- and D-galactose-containing cell-wall glycoproteins. The possible biological role of these materials is also considered. Structural studies have been carried out on the L-arabino-D-galactan-protein which is the major component of the mucilage of *Gladiolus* style. It was isolated by affinity chromatography on the D-galactose-binding lectin from the clam coupled to macroporous agarose.

The preparation contained 3% protein and a D-galactose: L-arabinose molar ratio of 6 : 1. Carbohydrate and protein were co-eluted even on chromatography in urea buffers. Ultracentrifugation showed the material to be polydisperse in the molecular weight range $1.5 - 4.0 \times 10^5$. From the results of methylation, partial acid and enzymatic hydrolysis analyses, the structure was confirmed as that of a (1 → 3)-D-galactan with (1 → 6)-D-galactose side chains, some of which contain terminal non-reducing α-L-arabinofuranosyl residues. Removal of these residues reduced the solubility of the complex. A new water-soluble polysaccharide has been isolated from the seeds of *Cassia multifuga,* and the polysaccharide, on methylation analysis, gave residues of 2,3-di-*O*-methyl-D-galactose, 2,3,6-tri-*O*-methyl-D-galactose, 2,3,4,6-tetra-*O*-methyl-D-galactose, 2,3-di-*O*-methyl-D-mannose, 2-*O*-methyl-D-xylose, 2,3-di-*O*-methyl-D-xylose, and 2,3,4-tri-*O*-methyl-D-xylose in the molar ratio 2 : 4 : 4 : 2 : 1 : 2 : 1, respectively.

Partial acid hydrolysis gave the oligosaccharides (4) – (7) and periodate oxidation studies indicated that 32.4% of the monosaccharide residues were end groups. The polysaccharide was therefore considered to have a highly branched structure. The alkali-extracted gum from oats contained a β-D-glucan as the major component. Some endogenous D-glucanases survived the alkaline treatment so deactivation of these enzymes increased the yield of the β-D-glucan with a corresponding increase in the viscosity of the extract. The high viscosity gums had a a high limiting viscosity number (17-18 dL g^{-1}) in both dimethylsulphoxide and 7 M urea. Low concentrations of the dyes Calcofluor White M2R, Tinopol CBS-X, or Congo Red caused preferential precipitation of the β-D-glucan from an aqueous solution of oat gum in the presence of protein, pentosan, and starch.

α-D-Gal*p*-(1 → 6)-D-GaI
(4)

α-D-Gal*p*-(l → 6)-D-Man
(5)

β-D-Gal*p*-(1 → 4)-D-Xyl
(6)

β-D-Xyl*p*-(1 → 3)-D-Xyl
(7)

The presence of salt suppressed the precipitation while higher temperatures (>25 °C) dissolved the precipitate. The precipitation seemed to be specific for (1 → 4)-β-D-glucose residues. A (1 → 3)-β-D-glucan occurs in cotton fibres. Its solubility varies with fibre age, a proportion being soluble in cold water, while some cannot even be extracted with alkali. Enzymatic digestion studies indicated that all of the linkages are of the β-configuration, whereas methylation analysis revealed that the predominant linkage is (1 → 3) although a small proportion of (1 → 6)-linkages art also present.

The timing of the deposition of the (1 → 3)-β-D-glucan during fibre development coincides closely with the onset of secondary wall-cellulose synthesis. In the hemicellulosic fraction from rice endosperm cell walls, methylation and Smith degradation analyses as well as cellulase fragmentation studies have confirmed the presence of a mixed (1 → 3), (1 → 4)-β-D-glucan in which blocks of (1 → 3)-β-D-glucose residues occur. D-Glucose-containing polysaccharides have been isolated from *Colletotrichum trifoli, C. destructivuum,* and *C. lindemuthianum,* which elicit browning and phytoalexin production in *Phaseolus vulgaris.*

A high molecular weight D-glucan was isolated from each organism and this polysaccharide initiated the browning symptoms at a concentration of 10^{-7} g D-glucose equivalents per hypocotyl. The molecular weight distribution of sinistrin, a polysaccharide isolated from the bulbs of the red squill and which has been used for renal clearance, covered the range 8.0 × 10^2 to 1.6 × 10^4.

Viscosity measurements indicated a highly branched molecule in which residues of D-glucose were present at the reducing end of most molecules. The fine structures of curdlan and paramylon, the (1 → 3)-β-D-glucans from *Alcaligenes faecalis* and *Euglena gracilis,* respectively, have been determined, and it was found that they resembled the native-regenerated cellulose system with curdlan being the regenerated polysaccharide and paramylon the native one. From the negative birefringence of annealed fibres and positive birefringence of paramylon granules, a tangential disposition of these chains in the granules may be concluded.

A new α-D-glucan, elaborated by *Elsinoe leucospila,* has a structure based on the repeating sequence (8).

...[→ 3)-α-D-Glc*p*-(1 → 4)-α-D-Glc*p*-(1 → 3)-α-D-Glc*p*-(1 →]$_n$
(8)

The D-glucan produced by *Botrytis cinerea,* when grown on grape berries, has a structure based on a backbone of (1 → 3)-β-D-glucopyranose residues with (1 → 6)-β-D-glucopyranose side chains. The D-glucan is not produced from cellulose, but only from simple hexoses.

Two novel tetrasaccharides, (9) and (10), have been isolated and characterized from cotton seed.

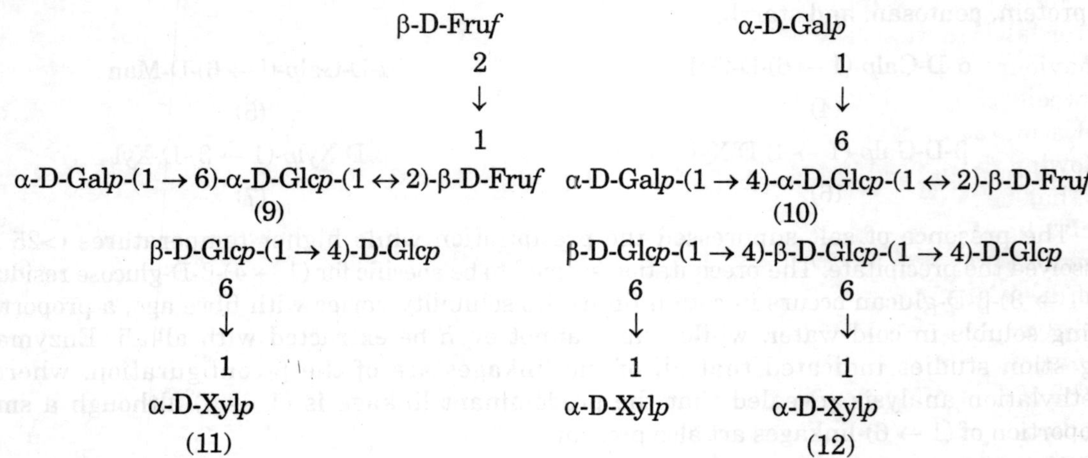

The oligosaccharides (11) and (12) have been isolated from the degradation products obtained from cellulase treatment of the L-arabino-D-xylo-D-glucan from tobacco leaves. Their structures were determined from the results of methylation analysis both before and after reduction with borohydride ion as well as by ^{1}H and ^{13}C n.m.r. spectroscopies. Present in the hemicellulose fraction from the endosperm cell walls of rice was a D-galacto-D-xylo-D-glucan.

This polysaccharide was based on a (1 → 4)-β-D-glucan backbone to which were attached residues of D-xylose and D-galactose at O-6. The modes of action of β-D-glucanases of *Penicillium notatum* on galactoxylo-D-glucans from various sources have provided a means of establishing their structure. The main chains are based on (1 → 4)-β-D-glucans although an occasional (1 → 3)-linkage may also be present. The side chains are nearly equimolar amounts of β-D-xylopyranosyl and *O*-β-D-galactopyranosyl-(l → 2)-β-D-xylopyranosyl units linked at O-6 of some of the main chain D-glucose residues.

The regions with disaccharide substitution were inaccessible to the enzyme, supposedly confirming the irregularity of this substitution. Auxin only induced growth in segments of rice coleoptile when the cell wall was rich in non-cellulosic D-glucose residues. The effect decreased with age and with the total amount of non-cellulosic polysaccharide present. A decrease in the relative amount of non-cellulosic D-glucose and an increase in the content of D-xylose residues were found in the growth cycle of cell walls of rice coleoptiles irrespective of the culture conditions. D-Galacto-D-marmans, which differed their average molecular weight, poly-dispersity, and D-mannose: D-galactose molar ratio, have been isolated from seven legumes.

The Mark-Houwink relationship between intrinsic viscosity and average molecular weights held for these D-galacto-D-mannans despite the variation in the D-mannose: D-galactose molar ratio. Estimates were made of the polydispersity of the polysaccharides from both the sedimentation coefficient distribution and also from the diffusion coefficient ratios, D_M/D_A, and the two methods gave similar results. The apparent viscosities of aqueous solutions of guar and locust bean D-galacto-D-mannans were measured over a range of shear rates both alone and in the presence of D-glucose, sucrose, or D-glucose syrups.

For all of the solutions, the variation of the relative viscosity with shear rate fitted a power law equation. The addition of the sugar solutions had no effect on the non-Newtonian behaviour of the gum solutions. The presence of casein had no effect on the relationship between relative viscosity and shear rate for the two D-galacto-D-mannans but its presence had a marked effect on κ-carrageenan. The location of inter-residue cyclic acetals, formed following the oxidation of guar D-galacto-D-mannan and other heteroglycans by the peridate ion, has been determined by methylation analysis of the oxidized glycans. The formation of cyclic hemiacetals led to the protection of hydroxy-groups during methylation in dimethylsulphoxide and their positions were located by the analysis of the resulting mono- and di-methyl ethers. These derivatives were not observed in either the native glycans or the glycans treated sequentially with periodate and borohydride ions. Inter-residue hemiacetals were identified in all glycans and were formed between aldehydo-groups at C-2 or C-3 of the oxidized residues and the hydroxy-groups at C-3 or C-2 of the adjacent unoxidized residues. The selective removal of six *O*-substituents from the oxidized residues resulted in a decreased ability of the latter to form inter-residue hemiacetals. An analysis of the types and proportions of the methyl ethers resulting from inter-residue hemiacetal formation could yield further structural information on the glycan. A new D-galacto-D-mannan, isolated from the seeds of *Ipomoea fistulosa* possessed a D-galactose: D-mannose molar ratio of 3 : 10.

The results of methylation and periodate-ion oxidation analyses confirmed that the polysaccharide was based on a main chain of (1 → 4)-D-mannose residues to which were attached single residues of D-galactose at O-6. Five oligosaccharides were found in the hydrolysis products of the polysaccharide from the seeds of *Caisas-Grandis*. There were three disaccharides, a trisaccharide, and a tetrasaccharide, whose structure was confirmed by methylation, periodate ion oxidation, and Smith degradation analyses. The tri- and tetra-saccharides had the unusual structures (13) and (14).

β-D-Gal*p*-(1 → 4)-β-D-Gal*p*-(1 → 4)-D-Man

(13)

β-D-Man*p*-(1 → 4)-β-D-Gal*p*-(1 → 4)-β-D-Gal*p*-(1 → 4)-D-Man

(14)

A number of enzymic complexes containing j3-D-mannanase activity were purified by affinity chromatography on D-gluco-D-mannan-derivatized AH-Sepharose. Isoelectric focusing gave multiple protein bands all of which showed β-D-mannanase activity, while a single band was obtained on sodium dodecyl sulphate-polyacrylamide gel electrophoresis. The purified enzymes differed in their physical properties and had wide variations in their ability to hydrolyse highly substituted D-galacto-D-mannans. A series of papers has appeared on the gelation of Konjac D-mannan.

On reaction with either sodium hydroxide or sodium carbonate, an induction period occurred before gelation due to the requirement for deacetylation. The rate of the induced reaction was of the form $V = 4.8[OH^-]^{0.61}$ where V is the reaction rate (min^{-1}). The power value of 0.61 suggested that the rate of deacetylation was proportional to the concentration of hydroxide ion and that gelation of the deacetylated molecules was inhibited by the hydroxide ion. The degree of deactylation, calculated from the Arrhenius equation, increased with a decrease in the concentration of D-mannan and an increase in the concentration of hydroxide ion, but was not influenced by temperature or the nature of the cation (Na^+ or K^+).

Deacetylation was an induced reaction which was essential for gelation and occurred after the equilibrium of the deacetylation and peptization reactions by hydroxide ion. To clarify the macromolecular and chemical properties of Konjac D-mannan, light scattering and viscosity measurements have been carried out on partially methylated derivatives. The weight average molecular weight was found to be $1.0 - 1.2 \times 10^6$ and it was suggested that a random coil configuration applied. The intrinsic viscosity was related to the molecular weight by a simple mathematical relationship, which meant that the molecular weight could easily be determined from viscosity measurements. The ^{13}C n.m.r. spectra of D-gluco-D-manno-oligosaccharides have been recorded. Considerable structural heterogeneity was shown by this method, mainly with respect to the sequence of linkages in the oligosaccharides. Four partially acetylated D-gluco-D-mannans have been isolated from the leaves of *Aloe vera.*

The polysaccharides differed in the ratio of D-glucose to D-mannose–as–well as in their acetyl content. The results of methylation and periodateion oxidation analyses showed that the polymers were all linear with only $(1 \rightarrow 4)$ linkages present. The hemicellulosic fraction of the seed endosperm of groundnuts *(Arachis hypogea)* contains a $(1 \rightarrow 4)$-β-D-gluco-D-mannan. The homogeneity of the polymer was confirmed by electrophoresis, sedimentation, and sugar analysis.

Hydrolysis of the methylated polysaccharide yielded 2,3,4,6-tetra-*O*-methyl-D-glucose and -D-mannose, 2,3,6-tri-*O*-methyl-D-glucose, and 2,3,6-tri-*O*-methyl-D-mannose in the molar ratio 1.6 : 21.2 : 5.6 respectively. High-resolution electron microscopy of fractions from potassium hydroxide extracts of sprucewood holocellulose showed the presence of D-galacto-D-gluco-D-mannan and 4-*O*-methyl-D-glucurono-D-xylan in the form of microfibrils with different diameters and flexibility.

The suprarmolecular elements of the D-mannans from the 5% and 24% KOH extracts were thin fibrils. The longer D-mannan fibril obtained from the 24% KOH extract had a higher molecular weight. Lignin was sometimes present in the fractions. Methyl glycosides of D-xylo-oligosacchandes (linear and branched) and 4-*O*-methyl-D-glucuronic acid-containing oligosaccharides, closely reflecting the main structural features of native D-xylans, were studied by thermal analysis and pyrolysis-g.c. The number of monomeric units in the oligomers markedly affected the course of active thermal decomposition.

Although the thermal stability increased with increasing numbers of monomeric residues, the ratio of 2-furaldehyde to 2-hydroxy-2-penteno-l,5-lactone produced remained constant. Considerable differences between reducing sugars and glycosides were observed during the course of thermal decomposition and other major differences were also observed when an ionic dehydrating catalyst was added to the pyrolysis sample. The results suggested that the 4-*O*-methyl-D-glucopyranosyl uronic acid linkage was the most thermally stable in the acetylated 4-*O*-methyl-D-glucurono-D-xylan, and acetyl groups do not significantly accelerate the thermal decomposition of the polysaccharide. The kinetics of the hydrolysis of D-xylo-oligosaccharides, ranging from di-to penta-saccharide, in dilute sulphuric acid have been studied.

The rate of hydrolysis of the glycosidic bond in 4-*O*-β-D-xylopyranosyl-D-xylose was similar to that of the same bond at the reducing end of the other oligosaccharides and was 1.8 times the rate of each internal bond. The yield of D-xylose from each oligosaccharide could be calculated from this ratio and from the empirical equation which describes the rate constant for the above xylobiose. The hemicellulose component of the endosperm cell walls of rice is mainly composed of L-arabino-D-xylan. This L-arabino-D-xylan is based on the usual (1 → 4)-β-D-xylan backbone to which are attached L-arabinose residues at O-3 of 78% of the D-xylose residues. Alkalme extraction of the husks of *Ispaghula* has yielded a highly branched acidic L-arabino-D-xylan.

The polysaccharide is unusual in containing both (1 → 3)- and (1 → 4)-D-xylose residues in the main chain with the majority of residues further substituted at O-2 or O-3 with residues of L-arabinose, D-xylose, or D-galactopyranosyluronic acid-(1 → 2)-L-rhamnose. Another acidic polysaccharide, based on a (1 → 4)-D-xylan main chain has been isolated from the seeds of *Ocimum basilicum*. The polysaccharide contained D-xylose, L-arabinose, L-rhamnose, and D-galacturonic acid in the molar ratio 15 : 9 : 7 : 12 and the branch points were present at both O-2 and O-3 of the D-xylose residues in the main chain. Two oligosaccharide sequences, (15) and (16), were isolated from the side chains.

D-Gal*p*A-1 → L-Rha*p*-(1 → 4 or 5)-L-Ara-1 → D-Xyl*p*-(1 → 3)-L-Rha
(15)

α-D-Gal*p*A 1
1
↓
3 or 2
D-Gal*p*A-(1 → 4)-D-Xyl*p*-(1 → 2 or 3)-L-Ara
(16)

Exhaustive extraction of the milled-wood lignin fraction from beechwood was unable to remove small amounts of carbohydrate. N.m.r. spectroscopy of model compounds as well as the acetylated milled-wood lignin confirmed that the carbohydrate was a xylan and was covalently bonded to the lignin. Both milled-wood lignin and lignin obtained after exhaustive cellulase treatment of Douglas fir and red alder sapwood contained covalently bound lignin. The method of pulping, duration of pulping, type of polysaccharide, and method of isolation all affected the lignin-carbohydrate bond. During studies on the etherification reactions of vanillyl and veratryl alcohols, it was observed that lignin-carbohydrate bonds could be formed, subsequent to lignification, through esterification of the *p*-hydroxybenzyl alcohol groups in the lignin polymer. The carbohydrates present in groundnuts have been reviewed.

As well as starch and cellulose, the more abundant polysaccharides were D-gluco-D-mannans and highly-branched D-xylans. Higher levels of non-starchy polysaccharides were present in legume flours than in wheat flour. The distribution of monosaccharide residues and consequently the polysaccharides present varied considerably between the different types of legume although D-xylans were present in each species examined except field bean.

PECTINS

The effect of the degree of polymerization of D-galacturonic acid oligosaccharides (DP 2-5) and D-galacturonans on their c.d. spectra has been examined in aqueous solutionr. The chiroptic properties were correlated with the degree of polymerization in terms of the optical superposition of monomeric unit increments. The interpretation of the results led to the conclusion that the conformation of D-galacturonans in solution is close to that of a helical structure. Above the critical concentration of pectin (0.02 g l^{-1}) in 0.1 M-NaCl, the addition of an increased amount of Ca^{2+} ion resulted in an apparent increase in molecular size owing to the formation of intermolecular cross-linkages.

Below the critical concentration, any increase in the apparent molecular weight was slight and due to a stretching of the polymer chains with expansion to a rodlike shape although the amount of cross-linking did increase with time. When the concentration was as low as 0.004 g l^1, no increase in molecular weight was observed and the chains contracted owing to intramolecular cross-linking. Grapefruit pectic polysaccharides have been shown to contain D-galacturonic acid residues (78%), and amongst the neutral sugar residues found were D-galactose, D-glucose, D-xylose, L-arabinose, and D-mannose of which D-galactose accounted for about half.

Oxidation of the complex with D-galactose oxidase followed by reduction with potassium borotritide gave a radio-labelled product in which about 90% of the incorporated tritium was located in D-galactose residues. D-Galacturonic acid oligosaccharides, with degrees of polymerization from 2 to 9 have been isolated in 54.8% yield by ion-exchange chromatography from an enzymic digest of D-galacturonan. The trisaccharide was the major product. Improved purification could be gained by t.l.c. on cellulose-coated plates. The recovery and degradation of sugars from potato pectin fractions by hydrolysis with 2M trifluoroacetic acid has indicated that between 61 and 96% of each sugar residue was recovered, depending on the monosaccharide residue.

Lower recoveries were obtained from intact onion cell walls. Various commercial pectins and pectinesterase-treated pectins have been fractionated according to their degree of esterification in the range 6–70%. These fractions were tested for their ability to destabilize the 'cloud' in orange juice. The juice was clarified by fractions with degrees of esterification between 6 and 14%, but not at all by those with degrees of esterification greater than 21%. A new dialdose dianhydride derivative has been obtained from an acid hydrolysate of the water-soluble pectic fraction of wobaku wood by successive treatment with methanolic hydrogen chloride and acetic anhydride-pyridine.

The compound was the 1,2 : 1′, 2′ dianhydride of 3,4-di-*O*-acetyl-β-L-rhamno-pyranose and methyl 3,4-di-*O*-acetyl-α-D-galactopyranosyluronate. This is the first report of a dianhydride containing two different sugar residues. The role of secondary alcoholic groups in the degradation of D-galacturonan by the *endo*-D-galacturonanase of *Aspergillus niger* has been investigated using 0-acetyl derivatives of pectic acid. The initial rate depended on the numbers of free hydroxy-groups on C-2 and C-3, with substitution giving an enhanced value for the Michaelis constant. The rate of degradation decreased in the presence of acetyl groups according to the relationship for competitive inhibition, and it was concluded that the lowered rate of reaction and extent of degradation was due to the lowered affinity of the enzyme for the substrate caused by modifying the groups which are required for enzyme-substrate binding. From the rates of hydrolysis of D-galacturonic acid oligosaccharides from pure heptamer to trimer, as well as mixed higher oligomers, it was suggested that the active site of the *endo*-D-galacturonanase from *Aspergillus niger* could accommodate at least seven units.

The release of end units from larger oligomers also suggested that the interaction was with four sugar units from the non-reducing side of the bond to be hydrolysed and three sugar units from the reducing side. The content of D-galacturonic acid, D-galactose, and L-arabinose residues of tomato cell walls declined during ripening. A water-soluble D-galacturonan found in isolated walls probably arises from the action of D-galacturonanase *in vivo.* This polysaccharide and that from the walls of mature green fruit, after incubation with galacturonanase, contained fewer neutral sugar residues than intact walls.

The loss of D-galactose and L-arabinose residues on ripening was not related to the loss of D-galacturonan. The predominant change that took place in the cell walls of apples and pears on ripening was a dissolution of the middle lamella. This change could be brought about in unripe pears by application of solutions of D-galacturonanase and cellulase, while the change in unripe apples could be initiated by D-galacturonanase alone. Ripe strawberries, artificially inoculated with various fungi, contained *endo*-D-galacturonanase and pectinesterase activities in their culture filtrates, whereas uninfected strawberries had *exo*-D-galacturonanase and pectinesterase activities.

No other pectin-degrading enzymes were detected in either sound or infected berries. Since the disintegration could be substantially reduced by inhibition of the D-galacturonanase activity, it was concluded that the D-galacturonanases secreted by the fungi were primarily responsible for the breakdown in sulphited strawberries. Sodium pectate can be cross-linked with epichlorhydrin in an alkaline medium. Different degrees of cross-linking can be obtained by varying the concentration of epicholrhydrin.

These materials could be used for the separation of the two pectinesterase isoenzymes from oranges. Both components had a 7-8 fold increase in activity and could saponify 93% of the pectin. The D-galacturonanase activity secreted by an aggressive strain of *Colletotrichum lagenarium* can be separated into two proteins. These proteins differ in their molecular weight and in their mode of action on the hydrolysis of D-galacturonans. The anti-inflammatory drugs benzydamine and ketoprofen formed co-precipitates with pectin. A Langmuir-type of binding isotherm was obtained on equilibrium dialysis and this slow dissolution of the precipitate suggested that binding may help to reduce the adverse effects of these drugs to the stomach on oral administration.

ACIDIC GUMS

The acidic polysaccharide from peach gum and macromolecular products obtained from it by mild acidic hydrolysis have been characterized by their molecular weight and composition. The single ion activity coefficient of Ca^{2+} ions bound to the polysaccharide and to the corresponding aldobiouronic acids was measured and the c.d. spectra of both K^+ and Ca^{2+} salts were recorded. Ca^{2+} ions were bound to the carboxy-groups by electrostatic bonds in disperse solutions. There is a close arrangement of uronic acid residues which does not undergo alteration with a stepwise degradation of the macromolecules. It was concluded that the main chain of the polysaccharide is substituted in long segments with uronic acid residues at least on each second D-galactose unit.

The enzymic hydrolysis of peach gum polysaccharide has yielded the oligosaccharides (17)-(20), whose structures were confirmed by g.c.-m.s. of their partially methylated alditol acetates.

β-D-gal$p$(1 → 6)-D-Gal (17)

α-D-Man$p$(1 → 3)-D-Gal (18)

β-D-Gal$p$-(1 → 3)-β-D-Gal$p$(1 → 6)-D-Gal

(19)

β-D-Gal$p$-(1 → 6)-β-D-Gal$p$(1 → 3)-β-D-Gal$p$(1 → 6)-D-Gal

(20)

The Hofman degradation has been applied to mangle gum, and a variety of fragments were obtained on borohydride ion reduction and acidic hydrolysis of the product. The same fragments were obtained on the Barry degradation of the Hofman-degraded product.

The results supported the earlier conclusions 'about the structure of the gum. An analytical study of eight *Acacia* gums (four each from the *Gummiferae* and *Vulgares* series) has confirmed the main chemotaxonomic differences between the species that have been previously reported. During the methylation analysis of *Combretum nigricans* gum by the Hakomori procedure, losses of D-galacturonic acid and L-rhamnose residues were observed, which did not occur in the Haworth or a modified-Haworth procedure.

However, during the Haworth procedure, *C. nigricans* gum was degraded to a greater extent than the gum from *A cacia seyal*. A new acidic polysaccharide, isolated from the bark of *Pterospermum acerifoli,* contains D-galacturonic acid, D-galactose, and L-rhamnose in the molar ratio 5 : 3 : 3. The oligosaccharide (21) was isolated from a partial acid hydrolysate and its structure was confirmed by methylation and periodate oxidation analyses. The polysaccharide of Junsai mucilage (*Brasenia schreberi*) contained D-galactose, D-glucuronic acid, L-fucose, D-mannose, L-rhamnose, D-xylose, and L-arabinose in decreasing order of magnitude. The actual composition varied between different samples. From the results of methylation

α-D-Gal$p$ A-(1 → 2)-L-Rha$p$-(1 → 4)-D-Gal

(21)

analysis, the precise structure was difficult to establish but it was found to be highly branched. Partial acid hydrolysis of paniculatan, the mucous polysaccharide from the inner bark of *Hydrangea paniculata,* led to the isolation of five acidic oligosaccharides.

The structure of each oligosaccharide was established by methylation analysis although each has previously been isolated from other sources. A possible structure of the polysaccharide was proposed. The polysaccharide had a molecular weight of 5.46×10^5. There were *O*-acetyl groups present on 40% of the L-rhamnose residues. Complete deacetylation of the polysaccharide in 0.1 M alkali led to a reduced molecular weight of 4.2×10^4. A mucilage, Abelmoschus-mucilage M, isolated from the roots of *Abelmoschus maniliot,* appeared homogeneous. The polysaccharide contained L-rhamnose, D-galacturonic acid, and D-glucuronic acid in the molar ratio 1.2 : 1.0 : 1.0 and had a molecular weight of 2.53×10^4.

Reduction of the carboxy-groups followed by methylation analysis suggested that the main chain comprised (1 → 2)-L-rhamnose and (1 → 4)-D-galacturonic acid residues. These latter residues were substituted at O-3 by D-glucuronic acid residues. These results were confirmed by the isolation, *inter alia,* of the oligosaccharide (22) from a partial acid hydrolysate of the poly saccharide.

$$
\begin{array}{ccc}
\alpha\text{-D-Gal}p\text{A-}(1 \rightarrow 2)\text{-}\alpha\text{-L-Rha}p\text{-}(1 \rightarrow 4)\text{-}\alpha\text{-D-Gal}p\text{A-}(1 \rightarrow 2)\text{-L-Rha} \\
3 \qquad\qquad\qquad\qquad\qquad\qquad 3 \\
\uparrow \qquad\qquad\qquad\qquad\qquad\qquad \uparrow \\
1 \qquad\qquad\qquad\qquad\qquad\qquad 1 \\
\beta\text{-D-Glc}p\text{A} \qquad\qquad\qquad\qquad \beta\text{-D-Glc}p\text{A}
\end{array}
$$

(22)

The mucous polysaccharide from *Lilium longiflorum* was homogeneous, contained D-mannose and D-glucose residues in the molar ratio 5 : 2 and had a molecular weight of 2.63×10^5. From the results of methylation and periodate ion oxidation analyses and the products of partial acid hydrolysis studies, it was concluded that the main chain was composed of a (1 → 4)-β-D-glycan with ten residues per non-reducing terminal group. The branch points were located on O-2 or O-3 of the D-mannose residues. The polysaccharide isolated from the mucous of *L. maculatum* was qualitatively similar to that from *L. longiflorum*. The water-soluble extract from *Opuntia ficus-indica* contained a number of polysaccharides. Four acidic components were present, each of which contained residues of D-galacturonic acid, L-arabinose, L-rhamnose, D-galactose, and D-xylose in a variety of proportions. Other studies on the same mucilage suggested that the core contained D-galacturonic acid, L-rhamnose, and D-galactose residues with the pentose sugars being present on the periphery. The mucilages in callus cultures of higher plants have been investigated particularly with regard to their monosaccharide composition. In studies on the effect of polysaccharides as icecream stabilizers, the gum from *Khaya grandifoliola* was most effective. The viscosity effect was low but the polysaccharide had a wide range of desirable effects.

ALGAL POLYSACCHARIDES

The vacuum-u.v. c.d. spectra of solid films of agarose have been measured and show a positive band near 180 nm and a larger negative band about 152 nm. The positive band remains accessible in aqueous solution and has been used to characterize the changes in molecular conformation and interactions during sol-gel transitions. The temperature profile of the vacuum-u.v. c.d. spectra shows sharp discontinuous changes around melting and setting points of the gel. These changes were interpreted in terms of co-operative intermolecular associations through double helices and also show pronounced hysteresis, which was considered as evidence of helix-helix aggregation. The rates of tumbling and exchange of free nitroxides in aqueous solutions are unaffected by high concentrations of agarose gels. Low-level covalent attachment of this spin-label to the agarose gels by means of a stable acetamido-ether linkage (23) caused considerable diminution of their rate of reorientation. Dissolution of this material to form a gel, and melting and setting of the gel, 'caused further changes in the e.s.r. spectrum.

$$
\left|\begin{array}{l}-OH \\ -OH\end{array}\right. + ClCH_2CONH\text{—}\langle\rangle N\text{—}O \xrightarrow{OH^-} \left|\begin{array}{l}-OCH_2CONH\text{—}\langle\rangle N\text{—}O \\ -OH\end{array}\right.
$$

(23)

Studies on the interaction between agarose and the *endo*-D-galacturonanase from *Aspergillus niger* has shown it to be of an ion-exchange nature which did not involve biospecific interactions. The sequence and composition of uronate residues in intact alginate samples have been determined by high resolution ^{1}H n.m.r. spectroscopy. Problems encountered owing to the viscosity of the samples were overcome by controlled and slight depolymerization before n.m.r. at 90 °C. The D-mannuronic acid: L-guluronic acid molar ratio was obtained from the intensities of the signals from the anomeric protons.

A sequence-dependent deshielding of H-5 of the L-guluronic acid residues made it possible to determine the fractions of the four possible doublets of nearest neighbours along the chain. The results were consistent with ^{13}C n.m.r. spectroscopic data. Significant deviation from results obtained by chemical analysis appeared only in samples containing a large fraction of mixed doublets. An enzyme preparation from the marine brown alga *Pelvetia canaliculata* contained both alginate 5-epimerase and alginate lyase activities. Ca^{2+} ions activated both enzymes, but Mn^{2+} ions only inhibited lyase activity.

Increasing concentrations of substrate D-mannuronanate removed the inhibition of lyase activity by Mn^{2+} ions. The storage D-glucans of the algae *Chlorella* and *Prototheca* species are identical in that they consist of a linear polymer, akin to amylose, and a branched amylopectin component. The branched D-glucans from the above two species are markedly different from that formed by the alga *Cyanidium caldarium* which exists in hot springs. The amylopectin component of the hot springs alga was more highly branched. The osmotic coefficients of Na^{+}, K^{+}, and Ca^{2+} counter-ions have been determined in aqueous solutions of *κ*-, ι- and λ-carrageenans and the results were correlated with those calculated from the limiting law of Manning.

An ordered secondary structure exists in κ- and ι-carrageenans, whose stability can be discussed as a function of temperature, ionic strength, and nature of the counter-ion. The viscosity and other physical properties of two carrageenan-water gels have been compared with those of an agar gel. The carrageenan gels had a smaller viscosity gradient and a higher viscosity at a given concentration. The structure of the gels were also different as revealed by microscopic analysis; those from the carrageenans were coarse and spongy, while that from agar contained closely intertwined fibrous molecules. Carrageenans from male and female gametophytic plants of *Rhodoglossum calif ornicum, Chondrus crispus, Gigartina pistillata, Irideae cordata,* and another unidentified *Gigartina* species have been fractionated into KCl-soluble and insoluble components.

An anti-(κ-carrageenan; antibody was used to analyse the components immunochemically. The KCl-insoluble fractions were highly reactive κ-types whereas the soluble fractions were less reactive. These latter fractions showed anti-(λ-reactivity). These data suggest that the KCl-soluble carrageenans contain either λ- or μ-carrageenan as no increase in immunological activity to anti-(κ-carrageenan) antibody was observed on treatment with alkali. A χ–carrageenanase from *Pseudomonas carrageenovora* has been isolated and purified and its identity has been confirmed by an inspection of the products obtained by its enzymic action. The limit digest was neo-carrabiose 4-sulphate. Addition of carrageenan to flours from soft wheats gave a product that was more akin to the flour from hard wheats in its bread-making quality. The non-Newtonian behaviour of the vlscosity of κ-carrageenan solutions at various shear rates increased on addition of D-glucose or sucrose. The results were explained in terms of changes in the solute-solute interactions. The structure of the laminaran from *Eisenia bicyclis* has been determined by methylation and periodate ion oxidation analyses as well as by ^{1}H and ^{13}C n.m.r. spectroscopies.

The polysaccharide possessed a branched structure with both (1 → 3)- and (1 → 6)-β-D-glucopyranose residues being present in the ratio of 3 : 2. (1 → 3)-β-D-Glucans were specific activators of the prophenoloxidase from crayfish serum. Laminaran chains, terminated at the reducing end with a D-glucose residue (G-type chains) were more potent activators than the laminaran chains terminated at the reducing end with a D-mannitol residue (M-type chain) The extracellular polysaccharide of *Porphyridium cruentum* contained D-xylose, D-glucose, D- and L-galactose, 3-*O*-methyl-D-xylose, 3(or 4)-*O*-methyl-D-galactose, D-glucuronic acid, 2-*O*-methyl-D-glucuronic acid, and a 2-*O*-methyl hexose in the molar ratio 3 : 1 : 2.5 : 0.13 : 0.13 : 0.8 : 0.2 : 0.13, whereas that from *P. aerugineum* contained the first six sugar residues in the molar ratio1.7 ; 1.0 : 1.1 : 0.3 : 0.6 : 0.5, respectively, as well as 2,4-di-*O*-methyl-D-galactose.

Each polysaccharide contained, in addition, 10% half-ester sulphate and *ca.* 0.5% protein. Structural studies showed, *inter alia,* that the D-glucuronic acid residues were linked to O-3 of the D-galactose residues and that 2-*O*-methyl-D-glucuronic acid residues were linked to O-4 of the L-galactose residues. The polysaccharide from *P. cruentum* had a molecular weight of 4 x 10^6, and that from *P. aerugineum* had a molecular weight of 5 × 10^6. The sulphated heteropolysaccharide from the brown alga, *Dyctyota dichotoma,* contains residues of D-glucuronic acid, D-galactose, D-mannose, D-xylose, and L-fucose.

Evidence, from chemical analyses, was provided that the polysaccharide is partially sulphated and that the sugar residues are (1 → 4)-linked except L-fucose which is (1 →2)-linked. The carbohydrates of the brown seaweed *Padina pavonia* contained a sulphated heteropolysaccharide that contained residues of D-glucuronic acid, L-fucose, D-glucose, D-mannose, and D-xylose as well as a D-mannan-protein complex. The red alga *Laurencia spectabilis* contains a glycoproteineous species which, on purification, was found to contain 92% carbohydrate and 8% protein. The carbohydrate portion mainly consisted of D-galactose and uronic acid residues and there was no ester sulphate polysaccharides from the heterocyst and spore envelopes of the blue-green alga *Anabaena cylindrica* have a backbone, which consists of the repeating units (24) to which are attached residues of D-glucose, D-xylose, D-galactose, and D-mannose as side chains.

The main chain contained between 128 and 150 sugar residues. After treatment of the polymer with various glycoside hydrolases, the degraded product gave, on treatment with *endo*-(1 → 3)-β-D-glucanase, the trisaccharide (25) and the pentasaccharide (26).

–β.D-Man*p*-(1 → 3)-β-D-Glc*p*-(1 → 3)-β-D-Glc*p*-(1 → 3)–1–β-D-Glc*p*-(1 → 3)

(24)

```
β-D-Manp-(1 → 3)-β-D-Glc
                     4
                     ↑
                     1
```

(25)

```
β-D-Glcp-(1 → 3)-β-D-Manp-(1 → 3)-β-D-Glc
    4                                4
    ↑                                ↑
    1                                1
 D-xylp                            D-Glcp
```

(26)

The use of ^{13}C n.m.r. spectroscopy, and methylation and Smith degradation analyses have confirmed the structure of the (1 → 2)-β-D-mannopyranan from *Crithidia deanei.* It was assumed that the polymer arises from the protozoon and not a bacterial endosymbiont, since a related organism *C. fasciculata* also contains a D-mannan.

The protozoon *Herpetomonas samuelpessoai* also contains a (1 → 2)-β-D-mannopyranan as well as a branched chain D-glucurono-D-xylan. Since injection of cells or flagellar suspensions from *H. samuelpessoai* cause protection against infection with *Trypanosoma cruzi,* the immunological protection may be due to these polysaccharides or glycoproteins. The surface coat of *T. congolense* has been characterized as a glycoprotein. At high N : P ratios in the growth medium, the marine diatom *Chactoceros affinis* produces large amounts of extracellular polysaccharide, whereas *Skeletonema costatum,* grown under similar conditions, produces *only* very small amounts of extracellular polysaccharide.

CHAPTER

3 Microbial Polysaccharides

TEICHOIC ACIDS

The ^{13}C n.m.r. spectra of a number of ribitol teichoic acids substituted with glycosyl residues, and of their dephosphorylated repeating units, have been recorded. Identical spectra are obtained for teichoic acids substituted with 2-*O*-and 4-*O*-glycosyl residues, so it is not possible to differentiate between these positional isomers using this technique. Investigation of the cell-wall composition and associated properties of strains of *Staphylococcus aureus,* which are resistant to methicillin, has not revealed evidence for any unusual wall polymers.

The ribitol teichoic acid isolated from *Staphylococcus hyicus* contains 2-acetamido-2-deoxy-D-glucosyl residues. Interaction of the teichoic acid with concanavalin A, and its susceptibility to α-, but not to β-D-2-acetamido-2-deoxyglucosidase, showed that the aminosugar is in the α-configuration. The cell walls of *Streptococcus pneumoniae* contain a teichoic acid (C polysaccharide) composed of 2-acetamido-2-deoxy-D-galactose, 2-acetamido-4-amino-2,4,6-trideoxy-D-hexose, ribitol phosphate, choline phosphate, and D-glucose. From the results of periodate oxidation and methylation analysis, together with an examination of the ^{13}C n.m.r. spectra, the structure (1) has been proposed. It is speculated that the C-3 hydroxy-group of the chain terminal diaminotrideoxy-D-hexose unit is glycosylated with isomaltose. Pneumococcal cell-wall teichoic acid, rather than peptidoglycan, is considered to be responsible for the activation of the alternative complement pathway.

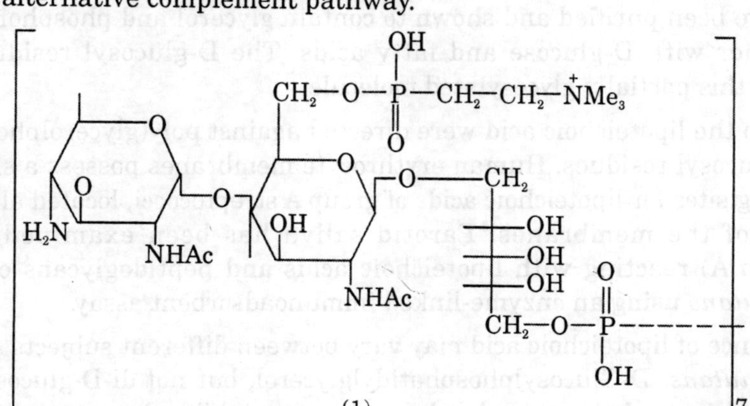

(1)

$_{7-8}$

This contrasts with group A streptococci in which the greatest activity was found associated with a heat labile protein of the cell membrane. The addition of a pulse of phosphate to a phosphate-limited chemostat culture of *Bacillus subtilis* has been found to lead to the synthesis of teichoic acid and the consequent development of bacterial ability to bind phage SP50 within about one generation line after the addition of the pulse. The time taken for newly synthesized wall material to become maximally exposed at the surface of the cells was directly related to their growth rate and corresponded to about three-quarters of a generation time.

The newly synthesized material does not become exposed exclusively at any single highly localized zone, but appears near the middle and along the cylindrical length of the bacteria at newly formed polar caps. The regulation of teichoic acid synthesis during phosphate limitation in *B. subtilis* has been shown to be due to the inactivation of the enzyme catalysing the formation of P^1-2-acetamido-2-deoxy-D-glucopyranosyl P^2-polyisoprenyl pyrophosphate. Conditions have been described for the continuous culture of a derivative of *Staphylococcus aureus* H, with L-cysteine as the sole amino-acid and under various nutrient limitations.

The proportion of ribitol teichoic acid present in the wall and the extent to which it is substituted with 2-acetamido-2-deoxy-D-glucose varies on growth of the organism under different conditions, as does the extent of cross-linking of the peptidoglycan. Neither the derivative nor the original strain H produced teichuronic acid when grown under phosphate limitation. Phosphate-repressible phosphodiesterases, isolated from *Bacillus subtilis,* have been reported to be responsible for the depolymerization of teichoic acids. Lipoteichoic acid from *Streptococcus rhutans* has been purified by a combination of gel filtration, hydrophobic chromatography, and adsorption on to phospholipid vesicles. Plasma membranes from *Staphylococcus aureus,* purified on columns of immobilized human immunoglobulin G, have been characterized as lipoteichoic acids, composed of D-glucosylglycerol teichoic acid containing ester-linked L- 7 alanine with pentadecanoic acid as the major fatty acid.

Growth conditions for the production of extracellular lipoteichoic acid by *Streptococcus mutans* have been investigated. The greatest levels were produced by slow-growing organisms. A lipoteichoic acid fraction and a presumed deacylated form were separated by column chromatography. The effect of growth rate on lipid and lipoteichoic acid composition in *Streptococcus faecium* has been reported. Lipoteichoic acids from *Bacillus subtilis* 168 and its *gta* mutants have been purified and shown to contain glycerol and phosphorus in equimolar amounts, together with D-glucose and fatty acids. The D-glucosyl residues have the α-configuration on this partially glycosylated molecule.

Antibodies to the lipoteichoic acid were directed against poly(glycerolphosphate) and not against the D-glucosyl residues. Human erythrocyte membranes possess a single population of specific binding sites for lipoteichoic acids of group A streptococci, located almost exclusively at the surface of the membranes. Parotid saliva has been examined for antibodies (immunoglobulin A) reacting with lipoteichoic acids and peptidoglycans of whole cells of *Streptococcus mutans* using an enzyme-linked immunoadsorbent assay.

The significance of lipoteichoic acid may vary between different subjects and for different serotypes of *S. mutans.* D-Glucosylphosphatidylglycerol, but not di-D-glucosyldiacylglycerol (the head group of lipoteichoic acids), has been shown to inhibit the extracellular autolysin of *Staphy locoecus aureus.* This suggests a requirement for the phosphate esters of the lipoteichoic acid molecule, together with the fatty acid ester group, for inhibition. Scanning electron microscopy has revealed concentric circular structures on newly exposed surfaces of the walls and isolated cross-walls of *Staphylococcus epidermidis*. Extraction of the walls with trichloracetic acid removes most of 'the phosphorus and about half of the 2-acetamido-2-deoxy-D-glucosyl residues, and destroys this type of structure. Hence, teichoic acid and possibly peptidoglycan (since these polymers are covalently linked) may be arranged circularly on the surface of the wall.

$$\rightarrow 4)\text{-}\beta\text{-D-Glc}p\text{-}(1\rightarrow 3)\text{-}\alpha\text{-L-Rha}p\text{-}(1\rightarrow 4)\text{-}\alpha\text{-L-Rha}p\text{-}(1\rightarrow$$

$$\begin{array}{c} 3 \\ \uparrow \\ 1 \\ \beta\text{-D-Glc}p\text{A} \end{array}$$

(2)

The primary structure (2) of the teichuronic acid of *Bacillus megaterium* M46 has been elucidated, using a combination of partial acidic hydrolysis, enzymic hydrolysis, periodate oxidation, and Smith degradation studies. The teichuronic acids of *B. megaterium* are unusual in that they resemble capsular polysaccharides in their size (mol. wt. 5×10^5) and composition, yet are structurally and functionally integral components of the cell wall. The proton-decoupled ^{15}N. n. m. r. spectra of ^{15}N-labelled intact cells, isolated cell walls, and cell-wall digests have been described. Each of five Gram-positive bacteria studied displayed a unique set of cell wall ^{15}N resonances, which reflected variations in the primary structure of peptidoglycans and in the amounts of teichoic acid and teichuronic acid in the cell wall.

Specific assignments of cell wall ^{15}N resonances assigned to teichoic acid D-alanyl residues, teichuronic acid and acetamido groups, and to peptidoglycan acetamido, peptide amido and free amino-groups, have been made on the basis of a combination of specific isotope labelling and dilution experiments, and from the results of cell wall fractionation. In *Micrococcus various* all three glycerol phosphate residues in the linkage unit are derived from CDP-glycerol and are therefore not transferred from a lipoteichoic acid carried.

PEPTIDOGLYCANS

A range of helical conformations for the glycan chains of bacterial peptidoglycans has been proposed on the basis of X-ray diffraction studies of cell walls and peptidoglycans from several bacterial species. These studies support the view that the peptide conformation has proportions of the residues in the helical and hydrogen-bonded β-sheet conformations. The balance between the two conformations is presumably related to the mean separation of the glycan chains, which varies with the water content by a factor of approximately two. The pentapeptide chains of peptidoglycans have a sequence of amino-acids terminated with D-Ala-D-Ala at positions 4 and 5.

Possible conformations of D-Ala-D-Ala and its analogues L-Ala-D-Ala, D-Ala-L-Ala, D-butyryl-L-Ala, D-Ala-D-butyric acid, D-Val-D-Ala, and D-Ala-D-Val have been analysed by theoretical methods. From theoretical studies it is predicted that L-Ala or D-Val at the 4 or 5 position of the pentapeptide group of the peptidoglycan will reduce the cross-linking in peptidoglycan biosynthesis, whereas the effect of D-butyric acid will be marked at the 4 position and moderate at the 5 position. This is in agreement with experimental results. Possible conformations of the disaccharide-peptide sub-units of the peptido-glycans of *Staphylococcus aureus* and *Micrococcus luteus* have been studied using an energy-minimization procedure.

Contrary to earlier reports the favoured conformation of the disaccharide 2-acetamido-2-deoxy-β-D-glucosyl-(1 → 4)-*N*-acetylmuramic acid is different from that of cellulose or chitin. Three types of conformation, two compact and one extended, are postulated with all three being stabilized by intramolecular hydrogen bonds.

Two different models were proposed for the three-dimensional arrangement of peptidoglycan in the cell wall- X–Ray crystallographic studies of the binding of the trisaccharide β-MurNAc-(1 → 4)-β-D-Glc*p*NAc-(1 → 4)-MurNAc to sub-sites B, C, and D of lysozyme indicate no distortion of sugar residues at sub-site D. The specificity of antisera to oligoglycylalbumin and to L-alanyl-tetraglycyl-albumin has been used to demonstrate serological differentiation of some bacterial strains on the basis of the primary structure of their cell-wall peptidoglvcan. Numerous synthetic glycopeptide analogues of peptidoglycans have been analysed by m.s. The immunoadjuvant activity of a number of synthetic glycopeptide analogues of peptidoglycans appears to be dependent on the type of substitution of the γ-carboxy-group of the D-isoglutamine residue. The peptidoglycan of *Acetobacterium woodii* contains D-ornithyl residues, which function as interpeptide bridges between the γ-carboxy-group of D-isoglutamic acid and the carboxy-group of the terminal D-alanyl residue of an adjacent peptide sub-unit.

The usual D-alanyl residue in position 1 of the peptide sub-unit is replaced by an L-seryl residue. The primary site of action of the antibiotic amphomycin has been demonstrated to involve inhibition of the phospho-*N*-acetylmuramylpentapeptide translocase, an enzyme involved in the first step of a lipid cycle of peptidoglycan synthesis. Although *Bacillus cereus* 569 is known to be resistant to lysis by lysozyme, because of the presence in the peptidoglycan of residues of deacetylated 2-amino-2-deoxy-D-glucose, the action of lysozyme on isolated cell walls has been reported to release some free reducing groups, indicating a limited breakage of the polysaccharide chains of the peptidoglycan.

The action of the enzyme modifies the peptidoglycan and makes it more susceptible to autolysin (s) and also enhances the rate of septum separation. A carboxypeptidase-transpeptidase, which incorporates free diaminopimelic acid into previously formed cell walls, has been isolated from *Bacillus megaterium*. Membranes from *B. megaterium* have been solubilized and then reconstituted in a form capable of synthesizing peptidoglycan. Using this reconstituted system, an assay has been developed for a factor necessary for peptidoglycan synthesis. The factor may be a polymerase, which assembles the disaccharide-pentapeptide sub-units of the peptidoglycan into polymers. A peptidoglycan complex, possessing a linear, non-crosslinked structure composed of disaccharide-pentapeptide units, has been isolated from a penicillin-treated mutant of *Brevibacterium divaricatum* requiring biotin for growth.

The peptidoglycan fragments appear to be synthesized after incubation of the cells with penicillin. A peptidoglycan, of different type from that of other species of *Cellulo-monas,* has been isolated from *C. cartalyticum. L-Lysine* replaces ornithine and the interpeptide bridge consists of D-aspartyl-D-serine. The same peptido-glycan type has been identified in *Arthrobacter luteus, Brevibacterium liticum,* and *Corynebacterium manihot.* The peptidoglycan isolated from the photosynthetic organelles of *Cyanaphora paradoxa* contains 2-acetamido-2-deoxy-D-glucose, *N*-acetylmuramic acid, L-alanine, L-glutamic acid, and diaminopimelic acid (1 : 1 : 1.6 : 1 : 1) and may be present as a lipoprotein-peptidoglycan complex.

Close similarities to the structures of other Gram-negative bacteria and cyanobacteria were observed preferential orientation of the glycan chains perpendicular to the axis of the cell has been observed in an electron microscopic study of the sacculus of *Escherichia coli*. Selective partial hydrolysis of inter-peptide bridges was achieved with an *endo*-peptidase. The regulation of peptidoglycan synthesis in *rel* A^+ and *rel* A^- strains of *Escherichia coli* during diauxic growth on D-glucose and lactose has been reported. In both strains, the biosynthesis of peptidoglycan, lipid intermediates, and nucleotide precursors abruptly halted at the onset of diauxic lag from D-glucose to lactose, with the concomitant accumulation of guanosine 5'-diphosphate and guanosine 3'-diphosphate.

This is consistent with the proposal that the accumulation is involved in the inhibition of the incorporation of disaccharide-pentapeptide into peptidoglycan and in regulating nucleotide precursor synthesis. Lanthionine, a monosulphated analogue of diaminopimelic acid, has been isolated from the peptidoglycan of *Fusobacterium nucleatum,* which contained all the other normal constituents, with the exception of diaminopimelic acid. Glycopeptides have been isolated from the cell walls of *Lacto bacillus plantarum.* Arthritogenic and immunoadjuvant activities are associated with peptidoglycan fractions having an average glycan chain length of more than five Uisaccharide units.

The peptidoglycan of *L. vaccinostercus* contains *meso*diaminopimelic acid, an unusual component in heterofermentative lactobacilli. The turnover of cell walls during myxospore formation in *Myxococcus xanthus* has been studied by measuring the incorporation and release of *meso*-diamino[^{14}C]pimelic acid.

Following addition of glycerol to induce myxospore formation, there is a greatly increased rate of incorporation, resulting in substantial replacement of existing wall with newly synthesized wall. None of the representatives of four genera of methanogenic bacteria grown in pure culture was found to contain typical peptidoglycan as a cell-wall polymer. However, each of the four genera was found to contain different cell-wall polymers, indicating a very early divergence, not only of the methanogens from all other prokaryotes, but also of the various genera of the methanogens from one another. The appearance of soluble peptidoglycan fragments and concurrent turnover of the glycan and peptide region of peptidoglycan from growing Gonococci indicate that both glycan-splitting and peptidoglycan amidase activities are involved.

If released during gonococcal infections, these or similar soluble peptidoglycan fragments might influence the consequences of host-gonococcus interactions. Muramic acid has been detected in the cell walls of the anomalous alga *Prochloron* species, which although prokaryotic, shares several factors with the Chlorophyta. Lipoprotein, covalently linked to peptidoglycan, has been found absent from exponential phase cultures, but present in stationary phase cultures of *Proteus mirabilis*. The overall peptidoglycan structure did not change during transition from logarithmic to stationary growth. The covalently linked lipoprotein is not as essential for cell-wall stability in this organism as it apparently is in *Escherichia coli*.

It is possible that there is a higher degree of hydrophobicity in the rigid layer, owing to the presence of O-acetyl groups on some of the muramic acid residues, and that this suffices for the association of the outer-membrane components to the peptidoglycan later in exponentially growing cells. Compearison of the structural features in normalpeptidoglycan and the peptidoglycan of the unstable sphaeroplast L-form of *Proteus mirabilis* confirms earlier reports of a continuation of the peptide cross-linking reaction during L-form growth and peptidoglycan synthesis in the presence of benzylpenicillin.

However, the peptidoglycans formed, in the presence or absence of the antibiotic, were found to differ in the degree of *O*-acetylation of their muramic acid residues. The mechanism of peptidoglycan synthesis in *Pseudomonas aeruginosa* appears to be similar to that of *Escherichia coli,* although evidence has suggested that in the former the DD-carboxypeptidase and transpeptidase systems may have somewhat different specificity restrictions from those in *E. coli.* Some similarities in the effects of Bactam antibiotics on the biosynthesis of peptidoglycan in the two bacteria were also observed. A combined ^{15}N- and ^{13}C-n.m.r. spectroscopic study of the chain segmental motion of the peptidoglycan-pentaglycine chain of [^{15}N]glycine- and [2-^{13}C]-glycine-labelled *Staphylococcus aureus* cells and isolated cell walls has been reported.

Main-chain motions of the pentaglycine bridge have been interpreted in terms of flexible pentaglycine chains attached to immobile glycan strands, whose packing arrangements can vary as a function of temperature and growth conditions. Although cell surface components of *Staphylocoecus aureus,* including teichoic acid, peptidoglycan, and protein A, have all been shown to be capable of activating the classical complement pathway, only peptidoglycan consumed complement *via* the alternative pathway. Using a quantitative immunofluorescence assay, the peptidoglycan was shown to bind to the C-3 component of complement in both pathways and in the absence of immunoglobulins G and A antibodies. Peptidoglycan synthesis in *Staphylococcus aureus* is inhibited by 3-amino-3-deoxy-D-glucose. The primary mode of action appears to be the prevention of formation of 2-amino-2-deoxy-D-glucose 6-phosphate from D-fructose 6-phosphate. The physical state of the membrane lipid in *Staphylococcus aureus,* as measured by fluorescence and e.s.r. spectroscopy, has been shown to affect the activity of the phospho-*N*-acetylmuramylpentapeptide transferase (E.G. 2.7.8.13). A correction to conclusions from earlier studies on the biosynthesis of the peptidoglycan of *Staphylococcus aureus* has been occurrence of 2-amino-2-deoxy-D-glucosyl residues in the cell-wall peptidoglycan of group A *Streptococcus pyogenes* type 4 has been confirmed? The ultrastructural localization of peptidoglycan in streptococcal cell walls has been explored using an indirect immunoferritin technique.

The ferritin particles were bound predominantly to filamentous structures, which were produced both from surfaces of isolated peptidoglycan fragments and from isolated walls. The results contradict proposed models of the streptococcal wall in which the peptidoglycan is thought to form the innermost layer of the wall. The alternative view is of a mosaic structure in which peptidoglycan forms a network of peptidoglycan-polysaccharide complexes.

LIPOPOLYSACCHARIDES

A radioimmunoassay for bacterial lipopolysaccharide has been developed. The lipopolysaccharide could be derivatized and radiolabelled with. ^{125}I without apparent loss in its biophysical, immunological, or biological activities. From studies of the interactions between lipopolysaccharide and phosphatidyl ethanolamine in molecular monolayers at airwater interfaces, it was concluded that each lipopolysaccharide molecule is surrounded by approximately sixteen phosphatidyl ethanolamine molecules.

The hydrocarbon chains of lipopolysaccharides can undergo a reversible thermal order-disorder transition, as shown by thermal phase-transition studies. The ordered conformation of lipopolysaccharides interpreted from wide angle *X*-ray studies is considered less developed than that in normal phospholipid bilayers. Pulse-chase studies with D-galactose on suspensions of *Alteromonas haloplanktis* have been used to follow the progress of newly synthesized lipopolysaccharide through the three outermost layers of the cell wall. The lipopolysaccharide moves through the periplasmic space, through the outer membrane, and spreads over the outer surface of the organism; its presence in each of the three layers was confirmed by polyacrylamide gel electrophoresis. The lipopolysaccharide of *Fusobacterium nucleatum* has been coupled to horse radish peroxidase and used for locating the lipopolysaccharide binding site on erythrocytes.

Fluorescence analysis of dansylated derivatives of various lipopolysaccharides has been used to define the nature of metal binding sites of lipopolysaccharides. A high affinity binding site for Ca^{2+} and Mg^{2+}, which appears to be formed by the 3-deoxy-D-*manno*-octulosonate trisaccharide unit, is involved in the assembly and maintenance of the normal structural organization of the outer membrane.

Lipopolysaccharide-protein interactions, as well as protein-protein interactions, have been shown to be significant in maintaining ghost structures in strains of *Escherichia coli.* The recognition of unopsonized bacteria by mouse peritoneal macrophages is thought to involve capsular and lipo-polysaccharides.

Bacterial lipopolysaccharides have been degraded by enzymic preparations from *Helix pomatia,* resulting in an extensive loss of anticomplementary activity. The lipid A moiety of the lipopolysaccharide appeared to be the main site of attack with little evidence for the degradation of the polysaccharide component. Two fatty acyl amidases, amidases I and II, purified from *Dictyostelium discoideum,* act upon the disaccharide (3, R = R^1 = H, n = 1), which has been isolated from the lipopolysaccharide of a heptose-less mutant of *Escherichia coli* K-12.

Amidase I removes the 3-hydroxymyristoyl unit, which is esterified to the amino-group of the 2-amino-2-deoxy-D-glucosyl residue adjacent to the C-1 phosphate of the disaccharide. The product is then acted upon by amidase II, which removes the remaining 3-hydroxymyristoyl unit from the amino-group on the distal 2-amino-2-deoxy-D-glucosyl residaue. Amino-sugars derived from bacterial lipopolysaccharides have been identified by g.l.c.–electron impact-m.s. and g.l.c.-chemical ionization-m.s. The absence of amino-sugars from the lipopolysaccharide (mol. wt. 8.5×10^4) isolated from *Acholeplasma oculi* has been reported.

Neutral sugars (L-fucose, D-galactose, D-glucose, 1 : 12 : 3) were detected but the lipopolysaccharide elicited no antibody response in rabbits. Treatment of the lipopolysaccharide of a mutant of *Escherichia coli,* containing no heptose, with alkali released two phosphorylated components, LPS I and II, both consisting of a substituted 2-amino-2-deoxy-D-glucosyl disaccharide, but differing in the nature of the phosphate groups present at the glycosyl position. Their structures (3) were determined. LPS I contained two phospho monoester groups, one

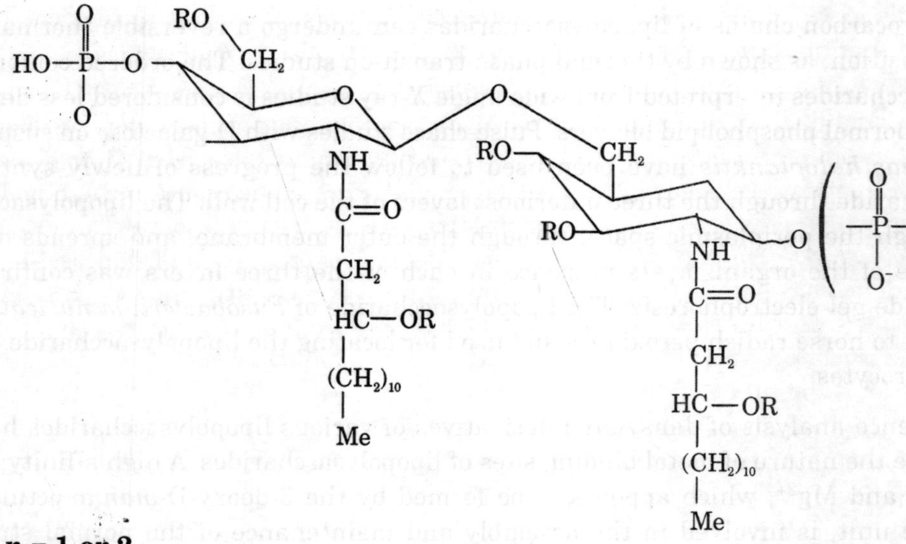

***n* = 1 or 2**

R = fatty acyl residues or H (acyl groups are myristic, lauric, and β-hydroxymyristic acids)

R′ = 3 moles of 3-deoxy-D-Hmanno-2-octulosonic acid

(3)

at the 4′-hydroxy-position and one at the glycosyl position. The remaining hydroxy-groups were substituted with lauryl, myristoyl, and 3-hydroxymyristoyl groups although the exact assignment of the different fatty acyl groups to specific hydroxy-groups in the amino-sugar rinks was not determined.

Further information on the nature of the phosphate linkages has been obtained from P n.m.r. spectroscopic studies on the original lipopolysaccharide and on two products of alkaline treatment. The structure of the K-13 antigenic polysaccharide of *Escherichia coli* consists of a repeating sequence of 3-linked D-ribofuranosyl residues and 7-linked 3-deoxy-D-manno-octulosonic acid (KDO) residues, approximately 50% of which are O-acetylated at C-4 or C-5. The serological specificity of this polysaccharide is expressed through KDO and its O-acetylated derivative. The polysaccharide chain of the cell wall lipopolysaccharide of *Escherichia coli* O58 contains D-mannose, 2-acetamido-2-deoxy-D-mannose, 3-*O*-[(1′ *R*)-carboxyethyl)]-L-rhamnose (rhamnolytic acid), and *O*-acetyl groups in the molar ratios of 2 : 1 : 1 : 1.

Structural investigations show that this polysaccharide, which contains a substituted trisaccharide repeating unit, has an identical structure (4) to that of *Shigella dysenteriae* type 5.

→3)-β-D-Glc*p*NAc-(1 → 4)-α-D-Man*p*-(1 → 4)-(α-D-Man*p*-2/3-OAc)-(1 →

3

↑

1

L-RhaLA

L-RhaLA = rhamnolytic acid

(4)

The O8- and O9-specific lipopolysaccharides of *Escherichia coli* lose their serological activity upon mild acid treatment to liberate their polysaccharide (mannan) moieties.

Restoration of the activity was achieved by substitution of one or two stearoyl groups per polysaccharide chain. The D-mannans obtained by *in vitro* biosynthesis were serologically active only when bound to the membrane-associated hydrophobic carrier. After removal of lipid A from the lipopolysaccharide of *Fusobacterium nucleatum,* the resultant high molecular weight fraction, which was devoid of phosphate and 3-deoxy-D-*manno*-octulosonic acid, was composed of *L-glycero-D-manno*-heptosyl-, 2-amino-2-deoxy-D-glucosyl-, and D-glucosyl-residues. Serological evidence has shown that this fraction contains the O-antigenic side-chains of the lipopolysaccharide. Alterations in the cell envelope of *Pseudomonas aeruginosa* resistant to polymyxin are associated with the alteration of the outer membrane through a diminution of lipopolysaccharide and outer membrane proteins.

Analysis of the lipopolysaccharides of *P. aeruginosa* PAO and three mutants selected for resistance to a virulent lipopolysaccharide-specific bacteriophage shows alterations in both O-specific side-chains and core regions. In a typical lipopoly-saccharide, isolated from the walls of *P. cepacia,* phosphate, L-rhamnose, D-glucose, heptose, and amino-sugars were found but there was no deoxy-octulosonic acid. The lipid A component of the lipopolysaccharide of *P. diminuta* has been reported to contain 2,3-diamino-2,3-dideoxy-D-glucose instead of 2-amino-2-deoxy-D-glucose. Some *Rhodopseudomonas* species also contain this unusual amino-sugar. The lipopolysaccharides of three strains each of *Rhizobium leguminosarum, R. phaseoli,* and *R. trifolii* have been purified and partially characterized.

The composition and immunodominant structures of the lipopolysaccharides were found to vary as much among strains of a single *Rhizobium* species as among the different species of *Rhizobium.* No obvious correlation between the nodulation group to which a *Rhizobium* species belongs and the chemical composition or immunochemistry of the lipopolysaccharides was observed. A component of the lipopolysaccharide of *R. trifolii,* previously designated as 'heptose I' is not *D-glycero-D-manno*-heptose but a 3-*O*-methylheptose. Heterogeneity of the lipopolysaccharide moiety of the intact rhizobium of free-living and bacterial forms of *R. leguminosarum* has been reported. A low molecular weight polypeptide exhibiting high affinity for lipopolysaccharides has been isolated from the cell surface of a heptose-less mutant of *Salmonella minnesota.*

Antiserum adsorption studies reveal that this protein also represents a common antigen of the Enterobacteriaceae. Antibodies against the synthetic disaccharide 3-*O*-α-abequosyl-L-rhamnose, representative of *Salmonella* O-antigen 8, have been used in the diagnosis of *Salmonella* bacteria by indirect immunofluorescence and co-agglutination using sensitized protein A-containing staphylococci. The octasaccharide (5), the synthesized disaccharide methyl

$$
\begin{array}{l}
\alpha\text{-D-Gal}p\text{-}(1 \rightarrow 2)\text{-}\alpha\text{-D-Man}p\text{-}(1 \rightarrow 4)\text{-}\alpha\text{-L-Rha}p\text{-}(1 \rightarrow 3)\text{-}\alpha\text{-D-Gal}p\text{-}(1 \rightarrow 2)\text{-}\alpha\text{-D-Man}p\text{-}(1 \rightarrow 4)\text{-D-Rha} \\
\qquad\quad 3 \qquad\qquad\qquad\qquad\qquad\qquad\qquad\qquad\qquad\qquad 3 \\
\qquad\quad \uparrow \qquad\qquad\qquad\qquad\qquad\qquad\qquad\qquad\qquad\qquad \uparrow \\
\qquad\quad 1 \qquad\qquad\qquad\qquad\qquad\qquad\qquad\qquad\qquad\qquad 1 \\
\quad \alpha\text{-Tyv}p \qquad\qquad\qquad\qquad\qquad\qquad\qquad\qquad\qquad \alpha\text{-Tyv}p
\end{array}
$$

Tyv*p* = tyvelopyranosyl residue.

(5)

3-*O*-α-tyvelopyranosyl-β-D-mannopyranoside, and methyl α-tyveloside inhibit the precipitation of *S. typhi* T2 alkali-treated lipopolysaccharide by O-factor 9 antibodies.

From the relative effectiveness of the oligosaccharides in inhibiting the precipitation reaction it appears that the 3-*O*-α-tyvelopyranosyl-D-mannopyranosyl structure is immunodominant in the *Salmonella* O9 antigen. 3-*O*-α-Tyvelopyranosyl-D-mannose coupled to bovine serum albumin elicited O-antibodies of higher specificity than those obtained by adsorption of antibacterial immune serum. The polysaccharide moiety of the lipopolysaccharide of *Schizothrix calcicola* contains 2-amino-2-deoxy-D-glucose, D-galactose, D-glucose, D-mannose, D-xylose, and L-rhamnose. In contrast to many enterobacterial lipopolysaccharides, the major fatty acid component has been identified as 3-hydroxy-palmitic acid and not 3-hydroxymyristic acid. A lipopolysaccharide has been isolated from *Shigella flexneri,* type 2, and some of its structural aspects have been elucidated. The specific polysaccharide isolated from the lipopolysaccharide of *Shigella newcastle (S. flexneri,* type 6) contains L-rhamnose, 2-acetamido-2-deoxy-D-galactose, D-galacturonic acid, and *O*-acetyl groups (in the molar ratio 2 : 1 : 1 : 1).

On the basis of ^{1}H and ^{13}C n.m.r. spectroscopy, methylation analysis, partial acid hydrolysis, and chromium oxide oxidation, the structure (6) has been assigned. Both structural and immunochemical evidence indicate that the lipopolysaccharide from *Shigella newcastle* belongs to the 'non-classical' type of somatic antigens with acidic O-specific polysaccharide chains. The hexose region of the core polysaccharides of lipopolysaccharides isolated from *Shigella flexneri* and some *Escherichia coli* species have a common structure (7).

[→ 4]-α-Gal*p*A-(1 → 3)-α-Gal*p*NAc-(1 → 2)-α-L-Rha*p* 3Ac-(1 → 2)-α-L-Rha*p*-(1 →]$_n$

(6)

α-D-Glc*p*-(1 → 2)-α-D-Glc*p*-(1 → 2)-α-D-Gal*p*-(1 → 3)-α-D-Glc*p*-(1 →

3
↑
1
α-D-Glc*p*NAc

(7)

A lipopolysaccharide isolated from *Vibrio cholerae* 4715 (NAG) has been reported to be devoid of 2-amino-2,6-dideoxy-D-glucose (quinovosamine), a characteristic sugar of cholera vibrios, but to contain significant amounts of glycine in the lipid A moiety.

The occurrence of odd-numbered fatty acids in the lipopolysaccharides of a number of *Vibrio cholerae* strains has been reported. A high molecular weight heptose-free fraction representing the O-specific side-chain polysaccharide, and a low molecular-weight fraction containing heptose and representing the core polysaccharide region have been isolated from the lipopolysaccharides from *Vibrio cholerae* and *V. el-tor* (Indaba). The core polysaccharide is composed of residues of D-glucose, D-mannose, heptose *(D-glycero-L-manno*-heptose and *D-glycero-L-gluco*-heptose), 2-amino-2-deoxy-D-glucose, and D-glucuronic acid (in molar ratio 9 : 4 : 5 : 1 : 2 : 5).

A branched structure for the polysaccharide has been proposed, with (1 → 2)-D-mannopyranosyl-, (1 → 4)-D-glucopyranosyluronic acid-, heptopyranosyl-. and 2-amino-2-deoxy-D-glucosyl-residues in the interior part of the molecule with D-glucopyranosyl- and heptopyranosyl-residues as the non-reducing end groups. Mild acid hydrolysis of the lipopolysaccharide released an oligosaccharide, representing the O-antigen of the lipopolysaccharide and composed of (1 → 2)-D-perosamine (4-amino-4,6-dideoxy-D-mannose) residues. The amino-groups are acylated with 3-hydroxypropionyl residues. Some of the toxic and immunological properties of the lipopolysaccharide and lipid A moiety of *V. el-tor* have been described. By use of different extraction procedures, two types of lipopolysaccharides have been isolated from a slightly virulent strain of *Yersinia pestis.*

Some structural information was obtained from methylation and periodate oxidation studies. The immunodominant moiety of lipid A has been shown to consist of a residue of 2-amino-2-deoxy-D-glucose, *N*-acylated with 3-hydroxytetradecanoic acid. A lipopolysaccharide isolated from *Y. enterolitica* exhibits serological cross-reactions with lipopolysaccharides from *Brucella abortus* and *Vibrio cholerae.* Common antigenic relationships between the lipopolysaccharides of *Escherichia coli, Kloechera africana,* and *Salmonella aberdeen* have been demonstrated. A survey of the composition of lipopolysaccharides and exopolysaccharides produced by *Cystobacter, Archangium, Sorangium,* and *Stigmatella* species has been reported. 3-*O*-Methyl-D-xylose was detected in several lipopolysaccharides. UDP-D-glucose, in addition to being required for the synthesis of the core oligosaccharide region of the lipopolysaccharide of *Escherichia coli* and for the synthesis of UDP-D-galactose, is now recognized to be required as a precursor for the D-glucosyl residues of the membrane-derived oligosaccharides. Mutant strains of *E. coli* specifically blocked in the synthesis of UDP-D-glucose were unable to synthesize membrane-derived oligosaccharides. A technique has been described for the selection of hybrid plasmids from *E. coli* and *Salmonella typhimurium,* containing genes for the glycosyl transferases involved in lipo-polysaccharide biosynthesis. Plasmids carrying the *rfa* G^+ gene, responsible for the synthesis of UDP-D-glucose: (heptosyl) lipopolysaccharide D-glucosyl-transferase, and the *rfa* M^+ gene, responsible for the synthesis of UDP-D-glucose: (D-glucosyl) lipopolysaccharide 1,3-D-glucosyltransferase have been identified. In addition, the identity of a plasmid carrying a gene capable of restoring the activity of UDP-D-galactose :(D-glucosyl)lipopolysaccharide α-1,3-D-galactosyl transferase in an *rfa* H mutant of *Salmonella typhimurium* has been reported.

The D-galactosyl transferase was also identified in *E. coli* K12 cells despite the apparent absence of the product of the enzyme-catalysed reaction in the cellular lipopolysaccharide of that organism. A glucolipid is considered to be an intermediate in the sequence of reactions involved in the biosynthesis of the O-9-D-mannan of *Escherichia coli.* The nature of the lipid, the characterization of the glucosylation reaction, and the sequence of the D-mannosyl transfer reaction have still to be clarified. The capacity of membranes derived from *E. coli* O8 and O9, when grown in the absence of D-glucose, to synthesize D-mannan, is stimulated by lipid-deriyed material from cells of either strain that had been grown on D-glucose.

It is possible that the first stage of the pathways of D-mannan synthesis may be the same, namely the synthesis of a hydrophobic D-glucosyl-D-mannose acceptor group. The incorporation of 2-acetamido-2-deoxy-D-glucose into the core region of the lipopolysaccharide of *E. coli* involves a 2-acetamido-2-deoxy-D-glucosyl-transferase that is localized in the inner membrane of the organism. The endogenous acceptor for the enzyme, which is also localized at the inner membrane, appears to be the basic polysaccharide structure of the organism. Subsequent to the incorporation of the amino-sugar, the polymer is assembled into a protein-lipopolysaccharide complex. Lactose, covalently attached to bovine serum albumin, has been used to elicit antilactose antibodies that are active against *Neisseria gonorrhoeae* lipopoly-saccharide.

Gonococcal cell-wall polysaccharide is immunogenic in BALB/c mice, but the kinetics of the response differ from those of *Salmonella typhosa* antigen and pneumococcal type III polysaccharide. Resistance against ascites tumour development and interferon-inducing activity has been demonstrated in the lipopolysaccharide derived from the protein-lipopolysaccharide complex of *Pseudomonas aeruginosa.* Both the lipid. A portion and the covalently linked polysaccharide-protein complex appear to be necessary for inhibition of tumour development, whereas the lipid A with amide-linked fatty acid is suficient to induce the *in vitro* formation of interferon. The lipopolysaccharide of *P. aeruginosa* has a more pronounced effect as immunogen, mitogen, and interferon inducer than the lipopoly-saccharide of *Brucella melitensis,* although they have similar toxicity levels. Since lipid A is responsible for most of these activities, the variations may be partially explained by structural differences in the lipopolysacchartdis. Further evidence has supported the theory that the interaction between lectins of host plants and *Rhizobium* is the key to the mechanism of host specificity in *Rhizobium*-legume symbiosis and that the lectin receptor of Rhizobial cells exists in the cell surface lipopolysaccharide.

No evidence was obtained to show interaction of extracellular polysaccharides with the lectins. It is possible that a specific recognition system may involve the specific binding of, for instance, pea *(Pisum sativum)* root hairs to an antigen that is situated on the lipopolysaccharide of infective strains of *R. leguminosarum.* Some aspects of the mechanism of O-specific polysaccharide biosynthesis in *Salmonella newport* and *S. kentucky* have been reported. Cell envelope and soluble glycosyltransferase preparations from both organisms catalysed the formation of polyprenyl pyrophosphate oligosaccharides with carbohydrate units corresponding to the structures of the main-chain polysaccharides (Scheme 1). The D-glucosylation reaction occurs before the polymerization of the repeating units to the polysaccharide. The bindmg sites for lipopolysaccharides on mouse lymphocytes have been identified as membrane proteins.

Active fractions isolated by affinity chromatography from B cell membranes were identified as immunoglobulins, possibly IgM and IgD, whilst the histocompatibility-2 complex proteins (H-2D, H-2K, and la antigens) were found to be binding sites for lipopolysaccharides on both B and T cells. Some aspects of the chemistry and biology of bacterial lipopolysaccharides have been reviewed.

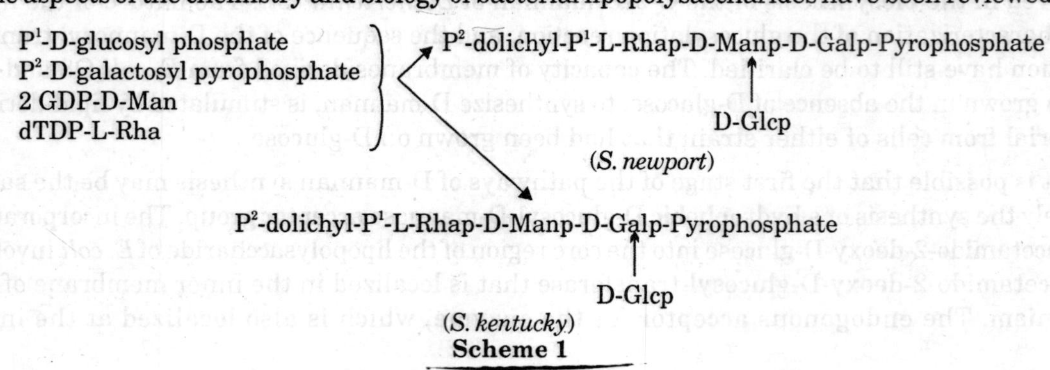

Scheme 1

CAPSULAR POLYSACCHARIDES

Colominic acid, a homopolymer of (2 → 8)-α-*N*-acetylneuraminic acid residues, is the *Escherichia coli* K1 capsular polysaccharide. Random *O*-acetylation at C-7 and C-9 of the polysaccharide, as determined by ^{13}C n.m.r. spectroscopy, has been shown to be responsible for the alternating antigenic change (form variation) associated with the organism. A serological method has been established for the identification and isolation of the form variants. *X*-Ray diffraction patterns for the capsular polysaccharides of *Escherichia coli* K29 and its M13 and M41 mutants have indicated that the molecular conformation and packing are the same for all the polysaccharides, suggesting that their primary structures are essentially unchanged.

Similar phage specificity and antibody reactions, together with the similarity of the diffraction patterns, indicated an overall chemical identity of the polysaccharides elaborated by the mutant strains with that of the parent strain. A model with anti-parallel two-fold helices at the corner and centre of the unit cell was proposed. The interaction of phage K29 with its host cell, *E. coli* K29, has been studied by use of combined virological and immunological methods.

The virion penetrates the capsule by a mechanism which involves receptor destruction and results in the release of DNA over a membrane-adhesion site. Fully methylated and fully acetylated disaccharides isolated from the capsular polysaccharide of *Escherichia coli* K100 and *Haemophilus influenzae* type b have been identified by m.s. as 2-*O*- and 1-0-D-ribofuranosyl ribitol, respectively. The capsular antigen of *H. influenzae* type c has been purified by ion-exchange chromatography. Structural investigations have identified the disaccharide (8) as the repeating unit of the polysaccharide. The highly branched capsular polysaccharide from *Cryptococcus neoformans* serotype c consists of a backbone of (1 → 3)-α-D-mannopyranosyl residues with single chain non-reducing (1 → 2)- and (1 → 4)-β-D-xylopyranosyl- and (1 → 2)-β–D-glucopyranosyl uronic acid residues.

```
                                                        O
                                                        ‖
→ 4)-β-D-GlcpNAc 3Ac-(1→3)-α-D-Galp-(1—O—P—O—
                                                        |
                                                        O⁻
                         (8)
```

```
 β-D-Xylp                          β-D-GlcpA
    1                                  1
    ↓                                  ↓
    2                                  2
→ 3)-α-D-Manp(1 → 3)-α-D-Manp-(1 → 3)-α-D-Manp-(1 →
    4              2                   4
    ↑              ↑                   ↑
    2              1                   2
 β-D-Xylp       β-D-Xylp            β-D-Xylp
                    (9)
```

$$\begin{array}{ccc} \rightarrow 3)\text{-}\alpha\text{-D-Man}p(1 \rightarrow & 3)\text{-}\alpha\text{-D-Man}p\text{-}(1 \rightarrow & 3)\text{-}\alpha\text{-D-Man}p\text{-}(1 \rightarrow \\ 2 & & 2 \\ \downarrow & & \downarrow \\ 1 & & 1 \\ \beta\text{-D-Xyl}p & & \beta\text{-D-Glc}p\text{A} \end{array}$$

(10)

Further examination of the polysaccharide (9) by alkaline degradation, followed by methylation, revealed that the D-glucuronic acid residues are located on D-mannosyl residues already substituted with a D-xylosyl residue. The capsular polysaccharide of *C. neoformans* serotype D is less highly branched, but with many structural similarities to (10). When high molecular weight pneumococcal capsular polysaccharide is injected intraperitoneally into rats, the antigen is excreted in the urine.

Examination of the molecular weight profile of the excreted product indicates that significant degradation has taken place. Pneumococcal type 19 polysaccharide is composed of residues of D-glucose, L-rhamnose, 2-acetamido-2-deoxy-D-mannose, and phosphate. Strong acidic hydrolysis has been found to be accompanied by migration of the phosphate groups. The structure of the capsular antigen of *Streptococcus pneumoniae* type 26 has been established (11). The only difference between the type 26 and the type 6 antigen is that the α-L-rhamnopyranosyl residue is linked to O-4 of the D-ribitol moiety in the former but to *O*–3 in

$$\rightarrow 2)\text{-}\alpha\text{-D-Glc}p\text{-}(1 \rightarrow 3)\text{-}\alpha\text{-D-Gal}p\text{-}(1 \rightarrow 3)\text{-}\alpha\text{-L-Rha}p\text{-}(1 \rightarrow 4)\text{-D-ribitol-5—O—}\overset{\overset{\displaystyle O}{\|}}{\underset{\underset{\displaystyle O^-}{|}}{P}}\text{—O—}$$

(11)

the latter. The configuration of the 4,6-0-pyruvic acid acetal in *S. pneumoniae* type 27 has been established from ^{1}H and ^{13}C n.m.r. spectral data of the natural product and model compounds.

Inhibition of the homologous precipitin reaction by synthetic diastereoisomeric pyruvic acid acetal showed that the configuration of the pyruvic acid acetal contributes to the serological specificity of the type 27 polysaccharide. As a result of the elucidation of the chemical structures of many *Pneumococci* and *Klebsiella* species, several previously uninterpretable serological cross-reactions have now been rationalized. The acidic capsular polysaccharides of *Klebsiella* species have been visualized by application of polycationic derivatives of ferritin which bind electrostatically to anionic sites on the surface of other glutaraldehy de-fixed or unfixed cells. Pyruvic acid has been identified in approximately half of seventy-seven *Klebsiella* species investigated.

On the basis of o.r.d. measurements and their interpretation from Moffitt-Yang plots, a helical superstructure is proposed for *Klebsiella* capsular polysaccharides of serotypes K1, K5, K6, K8, K11, K56, and K57. Measurements of the temperature dependence of the specific optical rotation reveal, in all cases, co-operative order-disorder transitions at temperatures between 25 and 40°C. Structures have been established for the capsular polysaccharides of *Klebsiella* K30 (12), K31 (13), K33 (14), K61 (15), K63 (16), and K64 (17). Limited *X*-ray studies on the *Klebsiella* K30 capsular polysaccharide suggest the existence of a trisaccharide-repeating structure with all the glycosidic link-ages disposed 1,4-diequatorial and producing a two-fold helix with a ribbon-like appearance. The antigenic determinant of hyporesponsiveness in BALB/c mice to *K. pneumoniae* type 47 has been reported to be L-rhamnose.

α-D-Glc*p*A
1
↓
3
→ 4)-β-D-Glc*p*-(1 → 4)-β-D-Man*p*-(1 → 4)-β-D-Man*p*-(1 →
6
↑
1
β-D-Gal*p*
3 4
C
Me CO_2^-
(12)

→ 3)-β-D-Glc*p*-(1 → 3)-α-D-Glc*p*A-(1 → 3)-β-D-Gal*p*-(1 →
4
↑
1
α-D-Man*p*
2
↑
1
4 6
C
Me CO_2^-
(13)

α-D-Glc*p*A
1
↓
3
→ 4)-β-D-Glc*p*-(1 → 4)-β-D-Man*p*-(1 → 4)-β-D-Man*p*6 Ac-(1 →
6
↑
1
β-D-Gal*p*
3 4
C
Me CO_2^-
(14)

$$\rightarrow 4)\text{-}\beta\text{-D-Glc}p\text{A-}(1\rightarrow 2)\text{-}\alpha\text{-D-Man}p\text{-}(1\rightarrow 3)\text{-}\beta\text{-D-Glc}p\text{-}(1\rightarrow 6)\text{-}\alpha\text{-D-Glc}p\text{-}(1\rightarrow$$

$$\begin{array}{c} 3 \\ \uparrow \\ 1 \\ \alpha\text{-D-Gal}p \end{array}$$

(15)

$$\rightarrow 3)\text{-}\alpha\text{-D-Glc}p\text{-}(1\rightarrow 3)\text{-}\alpha\text{-D-Gal}p\text{A-}(1\rightarrow 3)\text{-}\alpha\text{-L-Fuc}p\text{-}(1\rightarrow$$

(16)

$$\begin{array}{c} \alpha\text{-L-Rha}p \\ 1 \\ \uparrow \\ 3 \end{array}$$

$$\rightarrow 4)\text{-}\alpha\text{-D-Glc}p\text{A-}(1\rightarrow 3)\text{-}\alpha\text{-D-Man}p\text{-}(1\rightarrow 3)\text{-}\beta\text{-D-Glc}p\text{-}(1\rightarrow 4)\text{-}\alpha\text{-D-Man}p\text{-}(1\rightarrow$$

$$\begin{array}{c} 6 \\ \downarrow \\ 1 \\ \beta\text{-D-Gal}p \\ 4 \quad 6 \\ C \\ Me \quad CO_2^- \end{array}$$

(17)

Physical parameters, such as molecular weight, radius of gyration, and translational diffusional coefficient have been reported for group C meningococcal polysaccharide. Viscosity, light scattering, and sedimentation results all showed that the polysaccharide forms aggregates in solution, the aggregates behaving like random-coiled polymers with no measurable shape factors.

EXTRACELLULAR AND INTRACELLULAR POLYSACCHARIDES

Several review articles on extracellular polysaccharides have dealt with such topics as the structures of acidic and neutral extracellular microbial polysaccharides, the production of xanthan gums and other bacterial polysaccharides, and the chemistry and biochemistry of microbial extracellular polysaccharides and polysaccharidases. Water-soluble and water-insoluble dextrans, synthesized *by Streptocorpus mutans,* have been fractionated from solutions of alkali.

The insoluble fraction contained a greater proportion of (1 → 3)-linkages and was less susceptible to dextranase than was the soluble fraction. The water-soluble dextrans produced by a strain of *Streptococcus mutans* have been fractionated by gel chromatography into three types having different molecular sizes, chemical structures, and serological properties.

One of the fractions and a water-insoluble dextran, also synthesized by the organism, appear to represent different physical states of an inherently identical D-glucan. The molecular compression of dextrans has been examined using viscosity measurements. Marked osmotic compression of the flexible molecular structures occurs within the semi-dilute concentration range. Decreases in molecular volumes may have important effects on transport processes such as the multi-component systems found in living organisms. Chemically defined, highly branched dextrans have been studied by a ^{13}C spin-lattice relaxation method.

Correlations were made to the positions of carbon atoms associated with branching residues, permitting the values of *O*-substitution to be established for D-glucopyranosyl residues. Fourier-transform-i.r. difference spectra of dextrans have been correlated to the type and degree of branching, which had been established previously by permethylation for these polysaccharides. A combination of ^{1}H n.m.r.- and Fourier-transform-i.r.-spectroscopy has been used to give complementary information on the structure of dextrans. The results of the high-temperature enhancement of ^{13}C n.m.r. chemical shifts of dextrans have been correlated with those of methylation analysis. The diagnostic nature of the 70 – 75 p.p.m. spectral region with regard to the type of dextran branching, and an increase in resolution of the polysaccharide spectra at higher temperatures, are reported.

A combination of chemical-ionization m.s., ammonia m.s., and ^{13}C n.m.r. spectroscopy has been used for the unambiguous identification of peracetylated aldononitrile derivatives of methylated dextrans. The method has been used to distinguish α-D-glucosyl residues substituted at O-3 in the linear chain and branch-point positions, or only in the branch-point positions. Several ^{13}C n.m.r. signals that are diagnostic for 4,6-di-O-substituted α-D-glucopyranosyl residues have been identified. Analytical data for g.l.e.-m.s. methylation fragmentation and periodate oxidation for numerous dextrans have shown that results obtained using the two techniques are in general agreement for slightly branched dextrans, but are divergent for highly branched dextrans.

Hitherto unrecognized effects of (1 → 2)-, (1 → 3)-, (1 → 4)-linkages in the dextrans were observed:. A water-insoluble α-D-glucan has been isolated from growing cultures of *Streptococcus salivarius*. The results of methylation analysis and periodate oxidation studies indicated the presence of about 80% (1 → 3)-α-D-glucosidic linkages, together with short side-chains attached to the backbone by (1 → 4)-, (1 → 3)-, and (1 → 2)-α-D-glucosidic linkages. A linear (1 → 3)-α-D-glucan was recovered after periodate oxidation of the parent polysaccharide. Further studies have confirmed that immunoglobulin A (W3129) binds only to the non-reducing end of dextrans and not to any intercatenary segments of five D-glucopyranosyl residues. Extended, *X*-ray absorption fine structure analysis of ferritin and imferon (a synthetic complex of iron oxide coated with dextran) show that the near-neighbour environment around the iron atom in the two macromolecules is identical. The action patterns of an intracellular dextranase and three α-D-glucosidases (oligo-D-glucanases) from *Pseudomonas* UQM733 have been used to determine the structure of two branched isomalto-oligosaccharides obtained by enzymic degradation of dextran.

The adherence of heat-killed *Streptococcus mutans* cells to glass, induced by incubation with sucrose and cell-free dextransucrase (EC 2.4.1.5), is markedly inhibited by low concentrations of dextranases. The dextranases repress the production of water-insoluble α-D-glucan and reduce adhesiveness and cell-agglutinating activity of the D-glycans, by lowering the content of (1 → 6)-linkages and molecular weights, thus inhibiting the formation of D-glucan film and adherence of *S. mutans* cells on glass. Three extracellular proteins having affinities for binding to *Streptococcus mutans* serotype *c* cells have been isolated by affinity chromatography on immobilized α-D-glucans containing (1→ 3)- and (1→ 6)-linkages.

Dextransucrase activity was found to be associated with two of the proteins. Neither of the proteins had a unique binding specificity for (1 → 3)-α-D-glucosyl residues, and they were able to bind to (1 → 6)-α-D-glucans, other dextrans, and also to the (1 → 3)-α-D-glucosyl regions in pseudonigeran. The role of antibodies against cell-wall antigens of *Streptococcus mutans* in the protection against dental caries in humans has been reviewed. Soluble antigen preparations containing dextransucrase and from one serotype of *S. mutans* have been shown to elicit a protection immune response against infection with cariogenic *S. mutans* from many or possibly all serotypes.

Significant protection from experimental dental caries has been achieved by oral administration of soluble dextransucrase in animal feeding experiments. A survey of anti-dextransucrase activities in rabbit antisera against whole cells of *S. mutans* in terms of inhibition of cell-free and cell-associated dextran-sucrases has been reported. Inhibition of antiserum against serotype *c* was restricted to dextransucrases of serotypes *c, e,* and *f* but the dextransucrases of serotypes *a, d,* and *g* were not affected. The production, purification, and properties of dextransucrase from *Leuco nostoc mesenteroides* NRRL B-512F have been reported. The glycosyltransferase complexes [dextransucrase and levansucrase (EC 2.4.1.10)] from two strains of *Streptococcus mutans* have been resolved by polyacrylamide gel electrophoresis.

An improved method for the purification of dextransucrase involves the use of the non-ionic detergent Tween 80 to prevent aggregation of the enzyme, and affinity chromatographic absorption of the enzyme on insoluble dextran. Ammonium sulphate precipitation of dextran-sucrase appears to favour aggregation of the enzyme to a form which catalyses the synthesis of insoluble α-D-glucan containing mainly (1 → 3)-linkages. Gel filtration (hence dilution of the enzyme) promotes disaggregation to a low molecular weight form that catalyses only the synthesis of water-soluble a-D-glucans, which contain predominantly (1 → 3)-linkages. An intracellular polysaccharide, mol. wt. 3×10^5, isolated from *Bacteroides fragilis,* has been characterized as being of the glycogen-type, as it stains with iodine and is degraded to D-glucose by glucoamylase?

The intracellular glycogen-like polysaccharide present in the cells of two strains *of Streptococcus mutans* at various stages of growth has been measured by quantitative electron microscopy. The nature of the binding sites for the substrates, activators, and inhibitors located on each of the four sub-units of adenosine diphosphate-D-glucose pyro-phosphorylase (EC 2.7.7.27) from *Escherichia coli* have been investigated. The binding of adenosine triphosphate and D-fructose 1,6-diphosphate to the enzyme is synergistic, showing heterotrophic co-operativity, and the two phosphorylated compounds must also be present together to inhibit effectively the binding of adenosine monophosphate.

A mechanism has been proposed to explain some of the kinetic and binding properties in terms of asymmetry in the distribution of the conformational states of the four identical sub-units. Quantitative correlation of the changes in the rates of synthesis of glycogen and the utilization of D-glucose with simultaneous changes in the cellular levels of both D-glucose 6-phosphate and D-fructose 1,6-diphosphate have been reported for *E. coli.* An equation has been developed which describes the *in vivo* velocity of ADP-D-glucose pyrophosphorylase, and the rate of glycogen synthesis. Conditions have been described which cause significant changes in glycogen accumulation during periods of nitrogen starvation.

Differences have been observed in the values of velocity of ADP-D-glucose synthetase in the intact *E. coli* cell, when the carbon source is varied from D-glucose to D-fructose, glycerol, or succinate.

These differences are a result of the different cellular levels of adenosine 3',5'-phosphate in the cells using different carbon sources. A kinetic mechanism describing the action of the D-glucose 6-phosphate-dependent form of glycogen synthase (EC 2.4.1.11) of *Neurospora crassa* has been derived from initial velocity experiments and inhibition studies and has proved to be consistent with a rapid equilibrium random Bi-Bi mechanism. *X*-Ray diffraction studies on the (1 → 3)-β-D-glucan, curdlan, from *Alcaligenes faecalis* have shown the existence of microfibril formation, and a triple helical structure is proposed for the crystalline form of this polysaccharide. The maximum production of the D-fructan, levan, has been shown to occur during the mid-exponential to the late exponential phase of growth of *Actino-myces viscosus* and is followed by a rapid decline as a result of the production of levan hydrolase activity.

The levan produced by *Streptococcus mutans* OMZ 176 appears to have a structure similar to other known D-fructans and inulin. On the basis of chemical shift displacements of the ^{13}C n.m.r. spectra, resulting from *O*-substitution at specific carbon atoms, resonances have been assigned to the carbon atoms of the β-D-fructofuranosyl residues in levans. A variety of different levans show almost identical ^{13}C n.m.r. spectra in contrast to a wide divergence in the corresponding spectra of dextrans. The extracellular polysaccharide isolated from *Arthrobacter stabilis* NRRL B3225 contains residues of D-glucose, D-galactose, pyruvate, succinate, and acetate (in molar ratio 6 : 3 : 1 : 1 : 1.5).

The viscosity of aqueous salt-free solutions is relatively low, but atypical of anionic polysaccharides, increasing rapidly in the presence of salts, acids, and alkali. The improvements of the rheological properties of microbial exopolysaccharides through mutagenesis and mutant selection has been demonstrated for *Escherichia coli* and *Enterobacter (Klebsiella aerogenes).* The chemical composition of each group of polysaccharides produced by the mutants did not show any significant differences from those originating from the parent strain. Antigenic polysaccharides from *Escherichia coli* 29M and *E. coli* 72M have been isolated from chicken intestine and the structures of their repeating units, (18) and (19) respectively, have been established using the standard methodology. When grown on methanol an obligate methylotroph, *Methylocystis parvus,* originally isolated as a methane-utilizing bacterium, elaborates an extracellular viscous heteropolysaccharide composed of residues of D-glucose and L-rhamnose (6 : 1).

The extracellular polysaccharide of a methylotrophic culture contains residues of D-glucose, D-galactose, D-mannose, L-fucose, and L-rhamnose when grown on methane as the sole carbon source. Transfer of the culture from methane to methanol results in the loss of its ability to utilise methane as the sole carbon source and is accompanied by a loss of its ability to synthesize 6-deoxyhexoses in the extracellular polysaccharide. An acidic polysaccharide that exhibits unusual properties, such as formation of a brittle water-insoluble gel in the presence of Ca^{2+}, has been isolated from a *Pseudomonas* species when grown on methanol.

The polysaccharide is composed of a main chain of (1 → 3)-linked β-D-galactopyranosyl residues, with additional residues of D-glucose, D-mannose, D-glucuronic acid, and D-allose in undetermined positions. The slime polysaccharide synthesized by *Pseudomonas* PB1 contains

```
                      D-GlcpA
                        1
                        ↓
                        6
→4)-D-Manp-(1 → 4)-D-Manp-(1 → 4)-D-Manp-(1→
                        2
                        ↑
                        1
                      D-Galp
                        4
                        ↑
                        1
                      D-Manp
                        3
                        ↑
                        1
                      D-Glcp
                        4
                        ↑
                        1
                      D-Manp
                        3
                        ↑
                        1
           D-GlcpA-(1→ 3)-D-Galp
                        6
                        ↑
                        1
                      D-Galcp
                      3     4
                        \  /
                         C
                        /  \
                      Me    CO2-
                       (18)
```

```
                      D-GlcpA
                        1
                        ↓
                        6
→4)-D-Manp-(1 → 4)-D-Manp-(1 → 4)-D-Manp-(1→
                        2
                        ↑
                        1
                      D-Manp
                        4
                        ↑
                        1
                      D-Galp
                        3
                        ↑
                        1
                      D-Glcp
                        4
                        ↑
                        1
                      D-Manp
                        3
                        ↑
                        1
           D-GlcpA-(1→ 3)-D-Galp
                        6
                        ↑
                        1
                      D-Galcp
                      4     6
                        \  /
                         C
                        /  \
                      Me    CO2-
                       (19)
```

D-glucose, D-galactose, acetate, and pyruvate and therefore differs from many exopolysaccharides formed by other pseudomonads in not containing residues of hexuronic acid. Solutions of the polysaccharide were pseudo-plastic, in that the measured viscosity was shear rate dependent, and formed gels in the presence of Fe^{3+} and Al^{3+}. D-Fructose favours the production of exopolysaccharides by *Rhizobium meliloti*. In contrast, D-glucose is metabolized to 2-ketogluconic acid, an inhibitor of the exopolysaccharide biosynthesis.

$$\begin{array}{l}\rightarrow 4)\text{-}\beta\text{-D-Glc}p\text{A-}(1\rightarrow 4)\text{-}\alpha\text{-D-Glc}p\text{-}(1\rightarrow 4)\text{-}\beta\text{-D-Glc}p\text{-}(1\rightarrow \\ \qquad\qquad\qquad\qquad\qquad 6 \\ \qquad\qquad\qquad\qquad\qquad \uparrow \\ \qquad\qquad\qquad\qquad\qquad 1 \\ \beta\text{-D-Gal}p\text{-}(1\rightarrow 3)\text{-}\beta\text{-D-Glc}p\text{-}(1\rightarrow 4)\text{-}\beta\text{-D-Glc}p\text{-}(1\rightarrow 4)\text{-}\beta\text{-D-Glc}p \\ \quad 4,6\text{-}C(Me)(CO_2^-) \qquad 4,6\text{-}C(Me)(CO_2^-)\end{array}$$

(20)

$$\begin{array}{l}\rightarrow 4)\text{-}\beta\text{-D-Glc}p\text{-}(1\rightarrow 4)\text{-}\beta\text{-D-Glc}p\text{-}(1\rightarrow 3)\text{-}\beta\text{-D-Glc}p\text{-}(1\rightarrow 4)\text{-}\beta\text{-D-Glc}p\text{-}(1\rightarrow \\ \qquad\qquad\qquad\qquad\qquad 6 \\ \qquad\qquad\qquad\qquad\qquad \uparrow \\ \qquad\qquad\qquad\qquad\qquad 1 \\ \beta\text{-D-Glc}p\text{-}(1\rightarrow 3)\text{-}\beta\text{-D-Glc}p\text{-}(1\rightarrow 4)\text{-}\beta\text{-D-Glc}p\text{-}(1\rightarrow 4)\text{-}\beta\text{-D-Glc}p \\ \quad 4,6\text{-}C(Me)(CO_2^-)\end{array}$$

(21)

Growth yields, polysaccharide production, and energy conservation in chemostat cultures of *R. trifolii* have been reported. The extracellular polysaccharides of different strains of *R. meliloti* have been found to induce the production of polygalacturonase by roots of *Medicago sativa* seedlings. A comparative study of polysaccharide production and monosaccharide composition of partially purified fractions capable of inducing the enzyme demonstrated differences according to their origins. The structure (20) of the heptasaccharide repeating unit of an extracellular polysaccharide from *Rhizobium trifolii* has been established by a combination of methylation analysis, sequential oxidation, and elimination of oxidized residues. This is not a repeating unit in the true sense as approximately one third of the units lack the terminal β-D-galactopyranosyl residues. Comparative studies on the polysaccharides elaborated by *R. meliloti, Alcaligenes faecalis,* and *Agrobacterium radiobacter* indicate that they have a common structure (21) except for the position of acylation. A D-gluco-D-mannan (mol.wt. $1.4 \times O^6$) has been isolated from *Rhodococcus erythropolis*. The nutritional requirements for the production of xanthan by *Xanthomonas campestris* have indicated that D-glucose and sucrose are the best sources of carbon for the production of the polysaccharide.

Xanthan has been shown to possess a rigid rod-like conformation, 6000 Å in length and independent of ionic strength $>10^{-2}$ M . The conformation has been studied as a function of temperature, ionic strength, and polymer concentration. C.d. measurements indicate that the polysaccharide exists in a combination of random and ordered forms regardless of these conditions. The virulence of the plant pathogen *Erwinia amylovora* has been correlated with the ability of the organism to produce extracellular polysaccharide. A soluble carrier protein (HPr), one of the common components of all phosphoenolpyruvate-dependent phosphotransferase systems, has been purified from the phosphoenolpyruvate-dependent phosphotransferase system of *Escherichia coli*. The protein has been co-purified with a (1 → 6)-α→-D-glucan (mol.wt. 2.5 × 10^3). Results of steady-state kinetic measurements of the phosphotransferase activity demonstrate that the polysaccharide works as an activator of the enzymic system.

MISCELLANEOUS CELL-WALL POLYSACCHARIDES

The polysaccharide antigen from *Eubacterium saburreum,* which has been isolated by gel- and ion exchange-chromatographic separation of proteolytic digestion products, contains residues of *D-glycero-D-galacto*-heptose and a 6-deoxyheptose. Cross-reactivity with a related polysaccharide from strain L49 indicates that deoxyheptose is 6-deoxy-D-*altro*-heptose. The structure (22) of the corresponding antigen from *E. saburreum* O2 contains (1 → 6)-β-*D-glycero*-D-*galacto*-D-heptopyranosyl residues, all of which are substituted with 6-deoxy-α-D-*altro*-heptofuranosyl residues at O-3. This differs from the previously identified polysaccharide from the L49 strain, which contains chains of alternating (1 → 3)- and (1 → 6)-β-*D-glycero-D-galacto*-heptopyranosyl residues, the latter being substituted with 6-deoxy-α-D-*altro*-heptofuranosyl residues at O-3. In addition this polysaccharide contains *O*-acetyl groups at *O*-7 of some heptosyl residues and at *O*-2 of some 6-deoxy-α-D-heptosyl residues. The antigenic determinant on the L49 polysaccharide may be the 2-O-acetyl-6-deoxy-α-D-*altro*-heptofuranosyl residues.

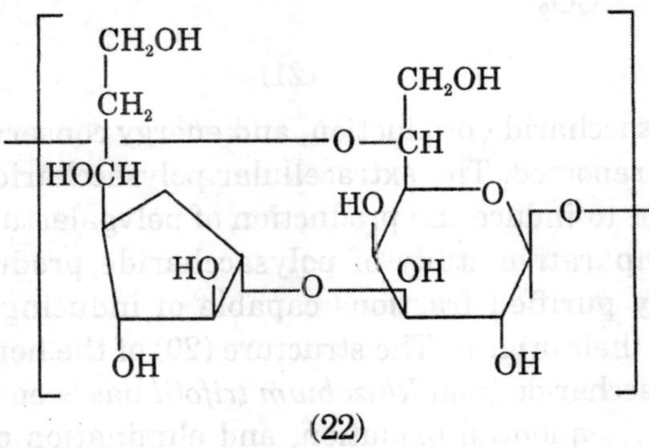

(22)

One common characteristic of the cell walls of halococci so far investigated is the high content of glycine. A 2,4-dinitrophenyl derivative of 2-deoxy-2-glycylamido-D-glucose has been isolated after hydrolysis of 2,4-dinitrophenylated cell walls.

It is postulated that glycine may play a role in connecting glycan strands through peptide linkages between the amino-group of 2-amino-2-deoxy-D-glucose and the carboxy-group of a uronic acid residue. When incubated under appropriate conditions with tritiated L-methionine, *Mycobacterium smegmatis* accumulates significant amounts of labelled oligosaccharides that are related to known 3-*O*-methyl-D-mannose-containing polysaccharides. A homologous series of oligosaccharides has been isolated, consisting predominantly of (1 → 4)-3-*O*-methyl-α-D-mannopyranosyl residues, terminated at the non-reducing end by D-mannopyranosyl residues or 3-*O*-methyl-α-D-mannopyranosyl residues and terminated at the reducing end of the -chain by a methyl aglycone.

These oligosaccharides are probably biosynthetic precursors of the larger 3-*O*-methyl-D-mannan homologues. ^{1}H N.m.r. spectroscopic studies suggest that the 3-*O*-methyl-D-mannan undergoes a major change in conformation on complexation with palmitic acid; the change is suggested to result from the tightening of a loosely coiled polymer. The fatty acid is orientated, at least part of the time, with its carboxy-group near the *O*-methyl groups of the polysaccharide. The phylogenetically closely related bacterium, *Streptomyces griseus,* also produces an *O*-methyl-D-mannose-containing polysaccharide, which in addition contains approximately four esterified acetyl groups.

These *O*-acetyl groups in the polysaccharide are located at position 6 of approximately half of the ten contiguous (1 → 4)-3-*O*-methyl-α-D-mannosyl residues. The affinity of this polysaccharide for palmitic acid suggests that it plays a role similar to the 3-*O*-methyl-D-mannose-containing polysaccharide of *M. smegmatis* in its regulation of the bacterium's fatty acid synthase. An acidic arabino-D-mannan isolated from *M. smegmatis* also contains two phosphate, six monoesterified succinate, and four ether-linked lactate groups per molecule.

Saponification of the polysaccharide releases a phosphorylated and a non-phosphorylated polysaccharide, the main features of which are the presence of chains of contiguous (1 → 5)-arabinofuranosyl residues attached to *O*-4 of L-arabinopyranosyl residues. A serologically active D-arabino-D-mannan, isolated from the cells of *M. tuberculosis,* is composed of (1 → 5)-α-D-arabino-furanosyl residues and (1 → 6)- and (1 → 2)-D-mannopyranosyl residues. Methylation analysis and enzymic degradation studies established the presence of a structure (23) containing short side-chains of (1 → 2)-α-D-mannosyl residues attached to a main backbone of (1 → 6)-α-D-mannopyranosyl residues.

In addition, a serologically inactive α-D-mannan whose structure (24) resembles that of the core L-arabino-D-mannan has been isolated. In addition to a high molecular weight lipopolysaccharide, the cell surface of *Rhizobium* contains a low molecular weight (1 → 2)-β-D-glucan. Curie point m.s. of a polysaccharide isolated from the cells walls of the cyanobacterium *Spirulina platensis* resembles that of the (1 → 2)-β-D-glucan of *Agrobacterium tumefaciens.* It is speculated that this polysaccharide may be the innermost of four layers of the cell wall.

→ [5-α-D-Ara*f*-1]$_m$ →– α-D-Ara*f*-(1 → 6)-α-D-Man*p*-(1 → 6)-α-D-Man*p*-(1 → 6)-α-D-Man*p*-(1 →

	3	2	2
	↑	↑	↑
	1	1	1
	α-D-Ara*f*	α-D-Man*p*	α-D-Man*p*
	2	2	
	↑	↑	
	1	1	

$[\alpha\text{-D-Ara}f]_n$ α-D-Man*p*

5
↑
1

α-D-Ara*f* $m + n = 8 - 9; n > 5.$

(23)

→ 6)-α-D-Man*p*-(1 → 6)-α-D-Man*p*-(1 → 6)-α-D-Man*p*-(1 → 6)-α-D-Man*p*-(1 → 6)-α-D-Man*p*-(1 →

2 2 2 2
↑ ↑ ↑ ↑
1 1 1 1

α-D-Man*p* α-D-Man*p* α-D-Man*p* α-D-Man*p*

2 2
↑ ↑
1 1

α-D-Man*p* α-D-Man*p*

(24)

Chemical and immunochemical studies have been used in the assignment of the structures (25)-(27) of the group-specific carbohydrates of Groups A, A-variant, and C streptococci, respectively. A close relationship among the L-rhamnan structures is revealed. Human antibodies to Group A and Group C carbohydrate have been measured in a radioimmunoassay system. The Group-specific antibodies arise after infection or persistent colonization of the pharynx with either Group A or Group C streptococci.

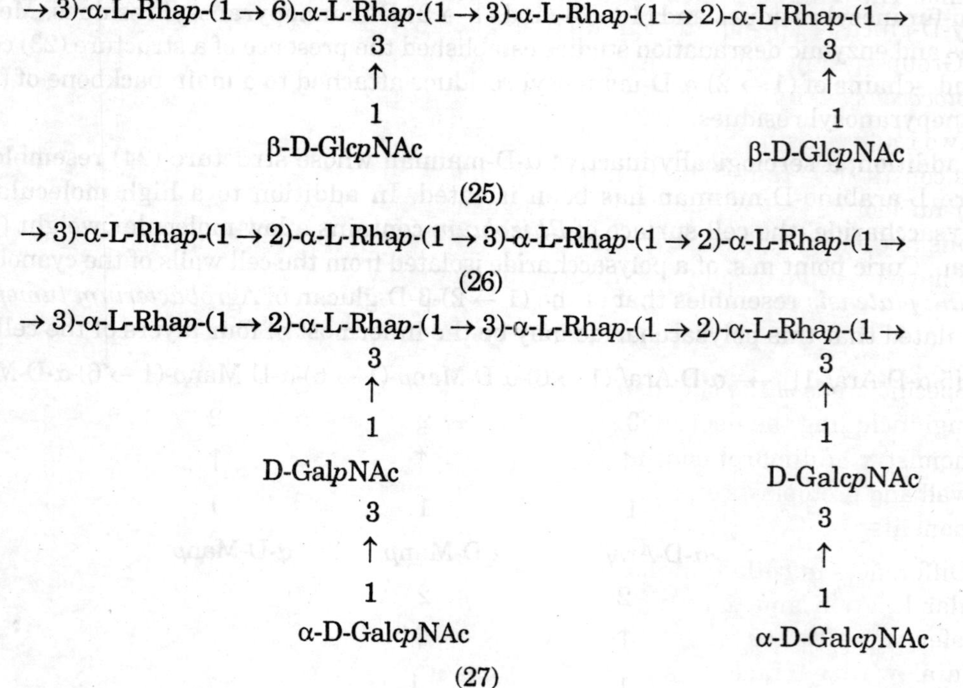

Several characteristic differences in the chemical composition of the cell walls among some serological types of *Streptococcus mutans* have been investigated through their susceptibility to *Flavobacterium* L-11 enzyme.

Generally, non-peptidoglycan components of the walls are responsible for serological differences in bacterial strains. The glycan isolated from *Streptococcus faecalis,* the structure of which (28) has been previously established, behaves abnormally on periodate oxidation.

Cyclic acetals are formed following periodate oxidation and this leads to the protection of hydroxy-groups during methylation. Inter-residue hemiacetals were identified between the aldehyde groups at C-2 or C-3 of oxidized residues and hydroxy-groups at C-3 or C-2 of an adjacent unoxidized residue.

→ 4)-D-Glc*p*-(1 → 4)-D-Glc*p*-(1 → 4)-D-Gal*p*-(1 →
6
↑
1
D-Glcp
4
↑
1
β-D-Galp

(28)

The purification of b-antigen of *Streptococcus sanguis* has been achieved by separation from teichoic acid by affinity chromatography on immobilized anti-poly(glycerol phosphate) columns. The antigen, which contains residues of D-glucose, L-rhamnose, and 2-acetamido-2-deoxy-D-glucose, appears to contain a single antigenic determinant related to isomaltose. A new Group-specific antigen, designated W-antigen, has been characterized for five strains *of Streptococcus sanguis.*

Two routes exist in *Streptococcus faecalis* for the synthesis of TUP-D-galactose, an inducible and a constitutive route. Incorporation studies with TDP-D-[^{14}C]hexoses have shown that the sugar nucleotide serves as a donor for D-galactosyl moieties for the synthesis of immunogenic glycans in the cell wall of *Streptococcus faecalis.* Antibodies to a heteroglycan of *Streptococcus bovis* have been purified by an affinity chromatographic method.

FUNGAL POLYSACCHARIDES

Specific areas of the chemistry and biochemistry of fungi have been reviewed: the autolysis of fungi including the degradation of carbohydrates, the biosynthesis of fungal cell walls, the biochemistry of dimorphism in *Mucor,* and cell wall and hyphal growth. Fractions from the cell wall and protoplasts of *Aspergillus fumigatus* have been shown to contain common antigenic components.

Differences in both immunological reactivity and chemical structure may be linked to their cellular location and may reflect incomplete structures in the process of biosynthesis and transfer to wall sites where they become immobilized. Electron microscopic studies of the conidial wall surface of the deutero-mycetes *Cladobotrytum varium, Cladosporium macrocarpum, Aspergillus niger,* and *Penicillium* species, and of the sporangiophore surface of the zygomycete *Mycotypha microspora* have been reportedf. The wall layers, isolated from *A. niger,* exhibited distinct differences in total carbohydrate, monosaccharide. amino-sugar, peptide, and lipid composition.

In *Phycomyces blakesleeanus,* during the formation of the germ-tube cell-wall, the composition shifts from that of a D-glucan, in the formant spore wall, to that of a chitin-like polymer in the mycelial wall. Heat-induced germination of the sporangiophore results in an increase in the content of chitin and a decrease in the D-glucan contents. In the presence of 2-deoxy-D-arabinohexose, *Rhodosponridiumtoruloides* synthesizes higher proportions of D-glucan and chitin, while the relative contents of D-mannan and D-galactan decrease drastically.

D-Glucans. — A study designed to ascertain whether nigeran might serve in determining phylogenetic relationships within the large and complex genera *Aspergillus* and *Penicillium* has shown that the nigeran content, used in conjunction with traditional morphological characteristics, can provide a useful and convenient tool for the determination of natural groupings of these two species. An alkali-soluble D-glucan which contains minor proportions of D-xylose, D-mannose, 2-amino-2-deoxy-D-glucose, and amino-acids has a large proportion of (1 → 4)-α-D-glucopyranosyl residues and a small proportion of (1 → 3)- and (1 → 6)-β-D-glucosyl residues.

On the basis of the identification of 2-*O*-α-*D*-glucopyranosylerythritol, it appears that segments of the (1 → 4)-α-D-glucopyranosyl residues are joined by single (1 → 3)-linkages. The relationship between the chemical structure and anti-tumour activity of D-glucans prepared from *Grifora umbellata* has been investigated. The anti-tumour D-glucans appear to contain the basic structural unit (29) although the length of the branches, branching frequency, molecular size, and molecular

R → 6)-β-D-Glc*p*
1
↓
6
→ 3)-β-D-Glc*p*-(1 → 3)-β-D-Glc*p*-(1 →
R = other D-glucosyl residues

(29)

conformation are also important. An extracellular (1 → 3)-β-D-glucan (mol. wt. 7×10^4) with single (1 → 6)-β-D-glucosyl side chains on every second D-glucosyl residue of the main chain has been isolated from *Monilinia fructigena.* Alkali-soluble and alkali-insoluble polysaccharides from the fruiting bodies of the basidiomycetes *Laetiporus sulphureus* and *Piptoporus betulinus* have been examined by *X*-ray diffraction, u.v. spectroscopy, and chemical methods.

The presence of (1 → 3)-α-D-glucan, (1 → 3)-β-D-glucan, and chitin was established. Variations in the levels of these polysaccharides were noted in different parts of the fungal structure. Potato, red kidney bean, and soybean plants respond to highly purified β-D-glucan from *Phytophthora megasperma* by accumulating phytoalexinst. The plants appear to respond to the fungus following recognition of the same mycelial wall component and it is possible that plants may respond to a variety of fungi by recognizing the same or similar mycelial wall components. Cell-wall formation in *Saccharomyces cerevisiae* appears to be the result of two main patterns of deposition of wall material, *viz.,* around the whole periphery of the non-budding cells and mainly at the tip of the daughter cell, or at the cross-wall that separates dividing cells.

Synthesis and secretion of β-D-glucan is unaffected by inhibition of RNA and protein synthesis. Growth of *S. cerevisiae* under conditions of phosphate limitations results in a decreased content of D-glucan and increased protein content as compared with control cells. Papulacandin B, an antibiotic produced by the deuteromycete *Papularia sphaerosperma,* inhibits the growth of yeasts, but has no effect on other fungi, bacteria, or protozoa. The mode of action involves the inhibition of synthesis of alkali-insoluble β-D-glucan of *Saccharomyces cerevisiae* and *Candida albicans* in sphaeroplasts and causes lysis of cells by osmotic rupture.

Echinocandin B and a structurally related antibiotic aculeacin A also inhibit β-D-glucan synthesis in yeast, possibly in a similar manner. Papulacandin B has been shown to inhibit the transfer of D-glucose from UDP-D-glucose, but not D-mannose from GDP-D-mannose, into dolichyl phosphate. Some properties of the β-D-glucan synthetase from *Saccharomyces cerevisiae* have been described.

UDP-D-Glucose is a competitive inhibitor and D-glucono-1,4-lactone is a non-competitive inhibitor of this enzyme. High yields of laminaribiose (3-*O*-β-D-glucopyranosyl-D-glucose) have been achieved by treatment of the (1 → 3)-β-D-glucan *of Saccharomyces cerevisiae* with an *endo*-(1 → 3)-β-D-glucanase from *Kluyveromyces phaseolosporus*. Glycogen metabolism in resting and growing cells of *Saccharomyces carlbergensis* has been shown to be regulated by a variety of endogenous and environmental factors. A ^{13}C.n.m.r. study of the gel networks of the branched (1 → 3)-β-D-glucans lentinan (from *Lentinus edodes)* and schizophyllan (from *Schizophyllum commune*) has been reported.

Conformational changes of the helical to random-coil forms were induced by treatment with aqueous sodium hydroxide. The effect of low pH on the growth of *Aureobasidium pullulans* has been shown to prevent transformation of the cells into the yeast phase, with the elaboration of small amounts of pullulan. The main degradation product (75-90%) resulting from the action of selected α-amylases on the α-D-glucan of *Elsinoe leucospila* has been identified as the trisaccharide *O*-α-D-glucopyranosyl-(1 → 3)-α-D-glucopyranosyl-(1→ 4)-D-glucose. This confirms previous reports that the polysaccharide consists chiefly of maltotriose units and a small proportion of maltotetraose units joined by (1 → 3)-α-D-glucosyl linkages. Cell extracts of *Oyptococcus laurentii,* a yeast which synthesizes amylose when grown at low pH, contain an α-D-glucan phosphorylase (EC 2.4.1.1) which has been characterized. Evidence for a possible role of this enzyme in the biosynthesis of amylose was reported. The ability of yeast cells to liberate sphaeroplasts following treatment with (1 → 3)-β-D-glucanases has been proposed as a taxonomic criterion for differentiating basidiomycetous from ascomycetous yeasts and for classifying the yeasts of the fungi imperfecti. Purified preparations of extracellular β-D-glucans from *Sphacelia sorghi* and from the cell walls of *Claviceps purpurea* have been shown to contain enzymically active protein.

The binding of the protein, which exhibited β-D-fructo furanosidase and acid phosphatase activities, to the D-glucan appeared to be non-covalent. When the enzymes were liberated from the D-glucan they showed enhanced activity, although the K_m values remained the same, suggesting that the matrix acts in the same manner as a non-competitive inhibitor. The solubility characteristics of β-D-glucans in the fungal cell wall have been shown to undergo changes after specific removal of chitin. Two types of β-D-glucans were isolated, one water-soluble and highly branched, and the other alkali-soluble and with single residue branches.

D-Mannans. — Six (1 → 6)-α-D-manno-oligosaccharides have been isolated after partial acid hydrolysis of proteolytic digests of an extracellular glycan of *Absidia cylindrospora.* A pentasaccharide and higher molecular weight oligosaccharides inhibited significantly the precipitation reaction of intracellular glycans from *A. cylindrospora, Mucor hiemalis,* and *Rhizopus nigricans* with anti -A. *cylindrospora* serum. On the basis of the ^{1}Hn.m.r. spectra of the cell-wall D-mannans, and the base composition and sequence similarities of the deoxyribonucleic acids, no distinction could be made between strains of *Candida utilis* and *Hansenula jadinii* although differences could be detected between *H. jadinii* and *H. petersonii.* The catabolic and anabolic transformation of D-glucose by a *Candida* species, into cell wall D-glucan and D-mannan has been reported. Approximately 20-40% of the hexose of these cell wall polysaccharides has been derived from the resynthesis of hexoses *via* the pentose phosphate pathway. Methylation analysis and periodate oxidation studies have been used to characterize the extracellular β-D-mannan produced by *Rhodotorula rubra* as being composed of alternating (1 → 3)- and (1 → 4)-β-D-mannopyranosyl residues.

Of eight oligosaccharides isolated from the cell walls of *R. glutinis* after lysis with an enzyme from *Penicillium lilacinum,* two have been characterized as 3-O-β-D-mannopyranosyl-D-mannose and 4-*O*-β-D-mannopyranosyl-D-mannose. The remaining oligosaccharides, which contained both D-glucose and D-mannose, were composed of (1 → 3)- and (1 → 4)-β-D-mannopyranosyl residues with (1 → 6)-β-D-glucopyranosyl side chains. The transformation of D-glucose by a *Candida* species has been followed catabolically and anabolically.

Maximal incorporation into D-glucan and D-mannan is only achieved at a critical concentration of D-glucose. Dolichyl monophosphate-D-mannose is not an intermediate in the synthesis of the outer chains of the D-mannan-protein of *Saccharomyces cerevisiae* and dolichyl diphosghate oligosaccharides are synthesized by different D-mannosyl-transferases. An osmotically-sensitive mutant of *Saccharomyces cerevisiae* S288 contains more β-D-glucan, less α-D-mannan, and less alkali-soluble glycogen than the parent strain. The D-mannan has been characterized as containing more short side-chains and fewer long side-chains than does the parent D-mannan and also contains a different proportion of (1 → 2)- and (1 → 3)-α-D-mannosyl residues.

Chitin. — The ability of lytic β-D-glucanases of *Arthrobacter* GJM-1 to dissolve the cell wall of *Saccharomyces cerevisiae* with the exception of the chitin component has been reported. Electron microscopic examination of such wall residues isolated from various stages of the budding cycle shows a correspondence to the formation of an annular structure found as part of the bud scar after cell division. This annular chitin-rich structure could not be isolated at either cell-cycle stages preceding the bud emergence or at the earliest stages of bud development.

Observations confirmed that the annular structure (chitin ring) formed during bud growth represents a major part of the total chitin present in the bud scar after septum closure. Chitin synthetase (EC. 7.4.1.16) has been located, mostly in an inactive form, associated with membranous fractions of *Neurospora crassa,* and could be activated either endogenously or by addition of exogenous proteases. Plasma membranes from *Schizophyllum commune* protoplasts have been stabilized against fragmentation by coating the protoplasts with concanavalin A. Approximately half of the chitin synthase activity is associated with the plasma membranes in the form of a zymogen and could not be detached from the membranes. It is suggested that this inactive form represents the cyto-plasmic transport form of the enzyme.

The nature and location of proteases of *Aspergillus nidulans* and the interaction of one of these with the zymogen of chitin synthetase have been described. A potential industrial method for the production and isolation of chitosan from the hyphal walls of *Mucor rouxii* as an alternative to the product derived from shellfish chitin has been reported.

Miscellaneous Polysaccharides

An extracellular polysaccharide of *Aspergillus parasiticus* has been characterized as being a homopolymer of 2-amino-2-deoxy-D-galactosyl residues, of which approximately 50% are *N*-acetylated and the remainder unsubstituted. The occurrence of the free amino-groups may result from enzymic *N*-deacetylation, since an enzyme isolated from the organism is capable of *N*-deacetylation of oligosaccharides containing fourteen or more 2-acetamido-2-deoxy-D-galactosyl residues, at the same rate as the polysaccharide degraded. A D-galacto-D-mannan, isolated from the mycelia of *A. fumigatus,* reacts with antibodies specific for circulating antigen in invasive aspergillosis in umans and rabbits.

The polysaccharide is composed of a backbone of (1 → 6)-D-mannopyranosyl residues with trisaccharide side chains terminated with residues of D-galactofuranose. An enzyme-linked immunoadsorbent assay (ELISA) for immunoglobulin (IgG and IgE) antibodies to protein and antigens of *A. fumigatus* has been reported. D-Mannans, D-gluco-D-mannans and D-galacto-D-mannans have been identified in phytopathogenic strains of *Ceratocystis paradoxa* and *C. fimbriata* isolated from different hosts, including tropical fruits and sugar cane.

Isolates of *Ceratocystis* within the same species were found to synthesize different polysaccharides, depending on their original hosts. Different polysaccharides could also be synthesized by the same strain at different temperatures. No correlation was made between host specificity and synthesis of a specific polysaccharide. It is possible that adherence of the fungus to the plant cells involves either specific or non-specific affinity reactions, which in turn do not correlate with the capacity of the micro-organism to cause disease. In a study of the carbohydrate composition of *Geotrichum, Trichosporon,* and allied genera, the intact cells of *Geotrichum* were characterized as containing predominantly D-galacto-D-mannans. L-Iduronic acid and D-glucuronic acid have been identified by g.l.c.-m.s. of the *O*-TMS ethers of the corresponding hexuronolactones prepared from the glycuronan 'protuberic acid' from *Kobayasia nipponica.* A family of glycopeptides of disperse molecular weights, charge, and sugar composition have been isolated from the culture filtrates of *Fulvia fulvia.*

D-Galactose, D-glucose, and D-mannose were identified as the major sugar components, together with smaller amounts of 2-amino-2-deoxy-D-glucose, 2-amino-2-deoxy-D-galactose, D-glucuronic ácid, and diester phosphate. The molecular organization of these polymers was investigated using mild alkali- and acid-hydrolysis, and acetolysis of the residual polysaccharide. The proposed structure (30) is similar to that suggested for the peptidophosphogalactomannan isolated from *Penicillium charlesii.* The pathogenic fungus of rice blast disease *Piricularia oryzae,* contains a cell-wall heteroglycan-protein complex containing residues of D-mannose, D-glucose, and D-galactose in the molar ratio 6 : 2 : 1.

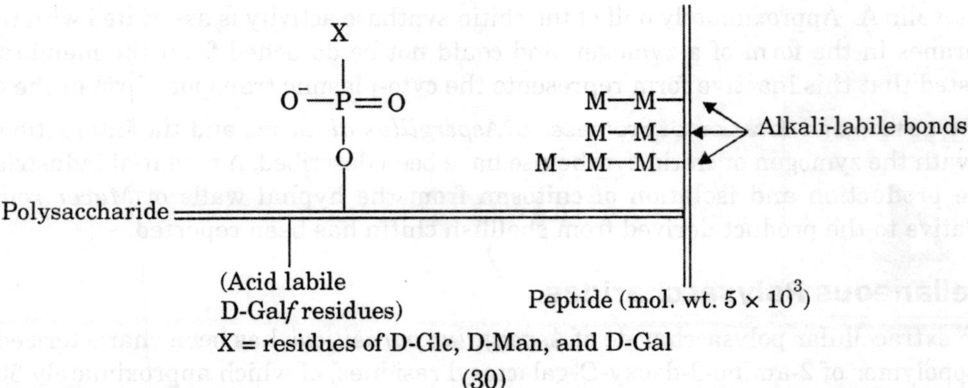

X = residues of D-Glc, D-Man, and D-Gal

(30)

Methylation analysis, partial acid hydrolysis, and digestion with an α-D-mannanase have shown the presence in the carbohydrate portion of a backbone of (1 → 6)-α-D-mannopyranosyl residues with side chains terminated by D-glucopyranosyl-, D-galactofuranosyl-, (1 → 2)- and (1 → 3)-α-D-mannopyranosyl residues. The heteroglycan is attached to the protein *via* O^3-D-mannopyranosyl-L-serine or O^3-D-mannopyranosyl-L-threonine linkages. Further studies have shown that the side chains comprise one to four D-mannosyl units, some of which are terminated by D-glucosyl-or D-galacto-furanosyl-residues. Histochemical staining at the cell surface of yeast-like forms, hyphae, and conidia of *Sporothrix schenckii* cell-types with iron(III) hydroxide and ferritin has been shown to correspond to the locations of concanavalin A-reactive peptido-L-rhamno-D-mannan complexes in the cells. Several acidic oligosaccharides, isolated from the acidic heteroglycan of the edible mushroom *Tremella fuciformis,* have been characterized by methylation analysis and ^{1}H n.m.r. spectroscopy and the structure (31) has been proposed for the parent polysaccharide.

A similar structure has been proposed for a cell-wall polysaccharide isolated from growing cultures of haploid cells of two strains of *T. fuciformis;* two extracellular polysaccharides showed only minor variations from this structure.

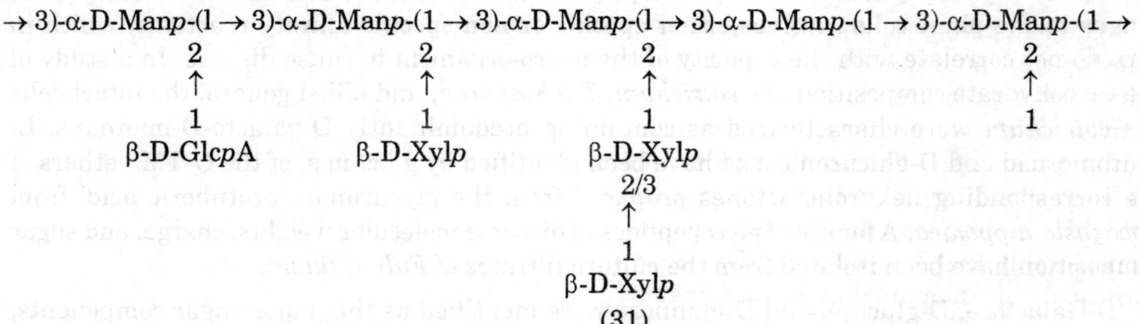

(31)

Water-soluble polysaccharides have been isolated from cultures of lichen mycobionts and phycobionts. The polysaccharides of the mycobionts show close similarities with those of the parent lichens, whereas the polysaccharides of the phycobionts possess different features. The mycobiont of *Parmelia caperta* contained an α-D-glucan having equal proportions of (1 → 3)- and (1 → 4)-linkages, and identical to a D-glucan of the parent lichen. The phycobiont of *Ramalina crassa* contains a major water-soluble D-galactan, whose properties are different from those of the intact parent lichen and its mycobiont.

CHAPTER 4 Energy Reserves

We now consider a group of polysaccharides responsible for storage of energy reserves: starch and its constituents amylose and amylopectin, the related storage polysaccharide glycogen from animals, and two furanose-based plant storage polysaecharides, inulin and levan.

STARCH

World production of isolated starch, in the form of grains and products made from them, is of the order of 20 million tonnes. The amount consumed as part of staple diets is very much greater than that and this is subject to attack by digestive enzymes in the gut. As will become evident below the extent to which it is attacked depends on the detailed structure of the starch and precisely which enzymes and enzyme inhibitors are present.

There is scope for biotechnology in improving starch which is simply consumed as part of a foodstuff, as well as in the numerous processes which depend on a more or less purified starch as a raw material. The most important sources of starch are maize, potatoes and wheat, with sorghum rice and barley as less significant localized sources. In some places minor crops such as cassava are also processed to yield starch.

Amylose and Amylopectin

Students who studied biology at school will be familiar with these molecules. Amylose, amylopectin and also the animal storage polysaccharide glycogen are all homopolymers of glucose: i.e. they are glucans. Amylose is a linear polymer involving $\alpha(1 \rightarrow 4)$ links whereas amylopectin (like glycogen) is a branched polymer containing an $\alpha(1 \rightarrow 4)$ linked backbone with regular $\alpha(1 \rightarrow 6)$ linked branches which can be up to 20 residues (each themselves linked $\rightarrow(1 \rightarrow 4)$) in length. Both polymers can be very large, particularly amylopectin: molecular weights for amylose have been reported between $0.2 - 2 \times 10^6$ Da and for amylopectin considerably higher ($\sim 50 \times 10^6$ Da). Amylose is soluble in aqueous solvents where it exists principally in an approximately random coil form (class D or C conformation type), with a Mark-Houwink a coefficient between 0.5 and 0.7 and a relatively small persistence length L_p between 20 – 40 Å. It can form under certain conditions a regular helical conformation, particularly when complexed with iodine. Amylopectin on the other hand is largely insoluble in aqueous media, but can be solubilized by solvents such as dimethyl sulphoxide.

Starch Biosynthesis and Variant Forms

Both amylose and amylopectin are branched and specific enzymes are responsible for forming the branch point linkages. In every case the glucose is activated by a reaction of the type

$$\text{Glc} + \text{ATP} = \text{ADP} - \text{Glc} + \text{Pi}$$

and the ADP-Glc is added to a primer chain of $(\text{Glc-Glc-Glc})_n$ with release of phosphate. Although ADP is the commonest activator, GDP and UDP are also used in what is the common synthetic mechanism for all the cell wall and most of the storage polysaccharides. There is evidence that higher plants contain two different starch synthase enzyme complexes, one using ADP-Glc and the other UDP-Glc, the latter possibly being an evolutionary relic. A similar situation seems to exist in lipid synthesizing enzymes.

The other common hexoses such as mannose, galactose, rhaninose and xylose and the pentose arabinose are all activated in this way, as are the uronic acids. With two exceptions it is the universal mechanism. This is on the whole not good news for biotechnology. Reactions involving nucleotides have proved very hard to use in that context. Biotechnology tends to do well with relatively simple hydrolysis reactions, or at least has done so far in its development. Fortunately the exception may offer a way out of the difficulty. Fructans (see below) are synthesized by using a disaccharide as the activated donor, in this case sucrose donating a fructosyl residue. Thus

Glc - Fru + Fru - Fru - Fru– = Glc + Fru - Fru - Fru - Fru

and

Glc – Fru + Glc – Glc – Glc– = Fru + Glc – Glc – Glc – Glc–

avoid the use of nucleotide activation.

Another system apparently able to donate glucosyl residues from sucrose has also been reported. These look much more exploitable and attempts have been made to use them, and we will consider this when we discuss synthetic polysaccharides. The other exception involves the synthesis of glycosubstances such as extensin where a dolichol –P–P-polysaccharide chain is the donor when the chain is attached to the protein part of the molecule.

The enzymes involved probably occur as a complex, within plastids in higher plants, and different enzymes are required for each sugar and for each specific reaction. Thus it will require at least two enzymes to synthesize amylose from ADP-Glc, as well as the presence of a primer site. One suggestion is that this may actually be a glycoprotein.

Amylopectin differs from amylose in having many more branches and the branching enzyme is relatively more involved. It is now clear that there are a number of branching enzymes, which broadly accounts for the differences between different species in the extent of branching in their amylose and amylopectin. This is not unreasonable. If one considers a chain in the process of extension, at some point a branching enzyme intervenes. This is likely to be determined by the length of the existing chain which probably reaches a point at which the ability of the branching enzyme to bind to it exceeds that of the extending enzyme.

However, as soon as a branch point is inserted the substrate becomes a quite different kind of molecule with a sharply different shape. A second branching enzyme could well be better adapted to this shape. The same argument can be pursued to a second branch point. At least three branching enzymes are known to exist in maize, which has an interesting series of mutants that have helped to elucidate the pathways. Knowledge about the enzymes themselves is fairly limited as yet. The maize enzymes have been partially isolated, and are similar to each other. They cross-react in immunological tests and have similar amino acid compositions.

The cDNA for these enzymes has not yet become available, so that the use, for example of anti-sense RNA methods is as yet limited. Antisense methods have been used on potato

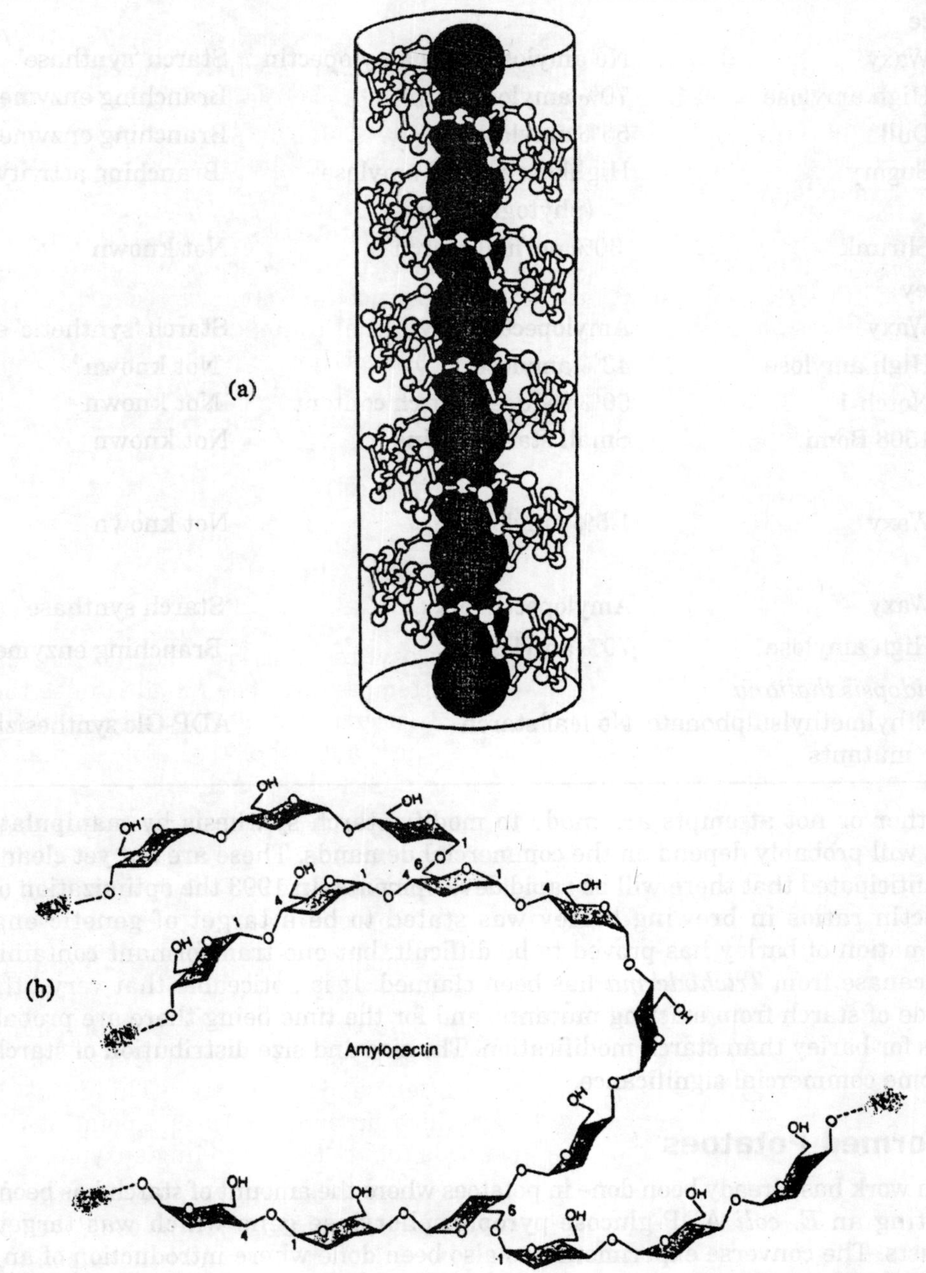

Fig. 4.1. Starch polymers, *(a)* The amylose-iodine complex: rows of iodine atoms (shown in grey) neatly fit into the core of the amylose helix. Unliganded amylose normally exists as an approximately random coil in solution (see text), *(b)* Part of the primary structure of amylopectin. The 'non-reducing' ends are indicated by an N and the reducing ends by an R.

Table 4.1. Mutations involving starch synthesis

Species	*Change*	*Enzyme involved*
Maize		
Waxy	No amylose, only amylopectin	Starch 'synthase'
High amylose	70% amylose	Branching enzyme 2b absent
Dull	55% amylose	Branching enzyme 2a absent
Sugary	Highly branched amylose (phytoglycogen)	Branching activity enhanced
Shrunk	30% normal starch	Not known
Barley		
Waxy	Amylopectin only	Starch 'synthetic' enzymes
High amylose	43% amylose	Not known?
Notch 1	50% reduced starch content	Not known
1508 Bomi	Small starch grains	Not known
Rice		
Waxy	1.5% amylose	Not known
Pea		
Waxy	Amylopectin only	'Starch synthase'
High amylose	70% amylose	Branching enzymes
Arabidopsis thaliana		
Ethylmethylsulphonate mutants	No leaf starch	ADP-Glc synthesizing system

Whether or not attempts are made to modify starch synthesis by manipulation of the enzymes will probably depend on the commercial demands. These are not yet clear cut, but it is to be anticipated that there will be rapid developments. In 1993 the optimization of amylose: amylopectin ratios in brewing barley was stated to be a target of genetic engineering. Transformation of barley has proved to be difficult, but one transformant containing a gene for β-glucanase from *Trichoderma* has been claimed. It is noticeable that very little use has been made of starch from existing mutants, and for the time being there are probably higher priorities for barley than starch modification. The size and size distribution of starch grains is also of some commercial significance.

Transformed Potatoes

Much work has already been done in potatoes where the amount of starch has been increased by inserting an *E. coli* ADP-glucose-pyrophosphorylase gene which was targeted to the amyloplasts. The converse experiment has also been done where introduction of an antisense gene to the same enzyme led to a reduction of starch accumulation and a corresponding increase in soluble sugars. The proportion of highly branched amylopectin-like starch has been increased by inserting an *E. coli* 'gene for glycogen synthase, though the amount of starch was greatly reduced.

A chimeric gene containing promoter sequences from potato patatin, and linkers from both soy and peas was constructed and inserted via *Agrobacterium* transformation of tuber discs. These were then cloned and mature tubers eventually harvested. Enough starch was obtained for amylograph analysis and differential scanning calorimetry from 2.5 kg of tubers, and showed distinct differences from the usual material. The presence of the active enzyme was demonstrated in the tubers. Potato starch unusually contains a number of phosphate groups which give it a charged character in suspension or 'solution' and can be as high as one group per 200 residues.

The transformed starch had less phosphate, but the main reason for altered properties was a much higher degree of branching. There was a lower content of amylose, and a higher level of soluble sugars. Perhaps significantly there was also an alteration in the morphology of the starch grains. This particular set of transformed potato plants probably do not represent the optimum result in terms of production of a highly branched amylopectin – the yields would probably have to be improved - but provides an impressive illustration of the potential for making novel starch in transformed plants. The use of antisense RNA has also lead to variants with reduced amylose content, and a variety completely free of amylose is likely to be available for cultivation shortly though this would, unlike the example given above, contain only normal amylopectin.

Shoot cultures of tetraploid *Solanum* cultivars were transformed with *Agrobacterium* carrying antisense gene constructs raised against the granule-bound starch synthase of the potato. This was incorporated into the genome, and 23 transgenic clones used in field trials. The actual degree of inhibition of the enzyme activity is related to the extent of expression of the antisense gene, as might be expected, and clones with a high degree of inhibition were chosen. Notably the amount of starch and the content of soluble sugars did not vary much from the untransformed plants. This result is obviously encouraging and suggests that the large-scale growth of transformed potatoes producing specific starches is feasible. The other possibility, a purely amylose containing variety, is currently being sought by using antisense RNA against the branching enzyme. So far the starch appears to be an amylopectin with longer linear chains but not an amylose. Similar results have been obtained in peas (*Pisum sativum*). Potato starch has a number of applications, and at least 60 per cent of it is used in non-food ways.

Paper coatings, where purity of the starch is crucial, may be one field where a genetically modified starch would have sufficient advantage to make its production viable. At present potato starch has to have its viscosity modified by hydrolysis before it can be used and it is thought that a high amylopectin could be used more directly with less preliminary treatment. High-amylopectin starches are less sensitive to the presence of ions such as calcium and sodium, which can adversely affect the viscosity of ordinary starch.

Starch Grain Size Distribution

Starch granules derived from waxy maize and a high-amylopectin maize, *amylomaize VII*, have been investigated for the sizes of the fragments of amylose and amylopectin produced when the grains are exploded by microwave heating. It is possible in these varieties to image the particles by electron microscopy and a representation of the results is given in Fig. 4.2. As expected amylose gives linear particles with some appearance of helicity, while amylopectin showed as either space filling patches or linear particles. About 30 per cent of the fragments were in the circular form with sizes up to 200 nm.

The average particle appeared to contain about six molecules and had an effective molecular weight of about 2 million. These results are comparable with those from gel filtration methods, and are relevant to the kind of particles important in food formulations. The formulation of quick-cook packet soups and similar products depends on having a very consistent range in the raw materials and the wrong size of grain can ruin the product.

Mutants are known where the grains do have a different size distribution, but the reasons for this must lie in complex interactions between the quantity and spatial distribution of the synthases as well as supply of substrates. It is rather remarkable that a single mutation can apparently have such an effect, and indicates the huge scope for manipulation potentially available in starch synthesis. If the same rules apply to heteropolymers, and they probably do, there are clearly possibilities of making novel materials. As well as all this, the presence of starches at high levels in the seeds of many species means that cheap and easy production should be possible. Demand for such materials remains much more problematic, and until and unless some uses are identified there does not seem to be much scope for production of modified starches by direct manipulation of the biosynthetic pathways.

Production of a range of materials from starch by biotechnology processes is a relatively mature industry and most of the niches have been filled by currently available materials. However, novel requirements may well be identified, for example in the cyclic dextran and related compounds and their applications, which if they are ever made on a large scale will almost certainly be made from starch. One raw material, xanthan gum, was originally conceived in 1959 as a way of using starch to produce more interesting materials, in this instance by using it as a medium for *Xanthomonas* growth in fermenters.

Specialist baking may also lead to new demands while the exploitation of unusual starches for brewing, such as sorghum in many parts of Africa, will certainly lead to new uses for enzyme breakdown, and might lead to a demand for novel forms of starch. There seem to be no uses for amylose or amylopectin, as such, which would lead to the rapid exploitation of suitable mutants other than the possibilities mentioned above for potato starch.

Starch Degradation

The enzymes that degrade starch are some of the most commercially important in current biotechnology both as crude isolates and *in vivo* in the various brewing processes based on cereals. These enzymes account for about 25 per cent of sales of enzymes by value, exceeded only by total proteinases, most of which would be due to enzyme detergents. For food use they are probably the most important in all the significant markets. In tonnages too they are prominent but tables of enzyme production are of little comparative value because the actual enzyme content of the dry powders usually sold is variable and low.

There is not much point in noting that 300 tonnes of *Bacillus* amylase was sold without also knowing what the actual enzyme content was. This is often very low, usually below 1 per cent of the protein, but it is possible that the 10 tonnes of fungal amylase also sold actually contained more enzyme. Using units of enzyme activity does not bring much improvement since these are quoted in units that are often arbitrary and not comparable between different preparations and different applications. The objective of enzyme digestion of isolated starch varies.

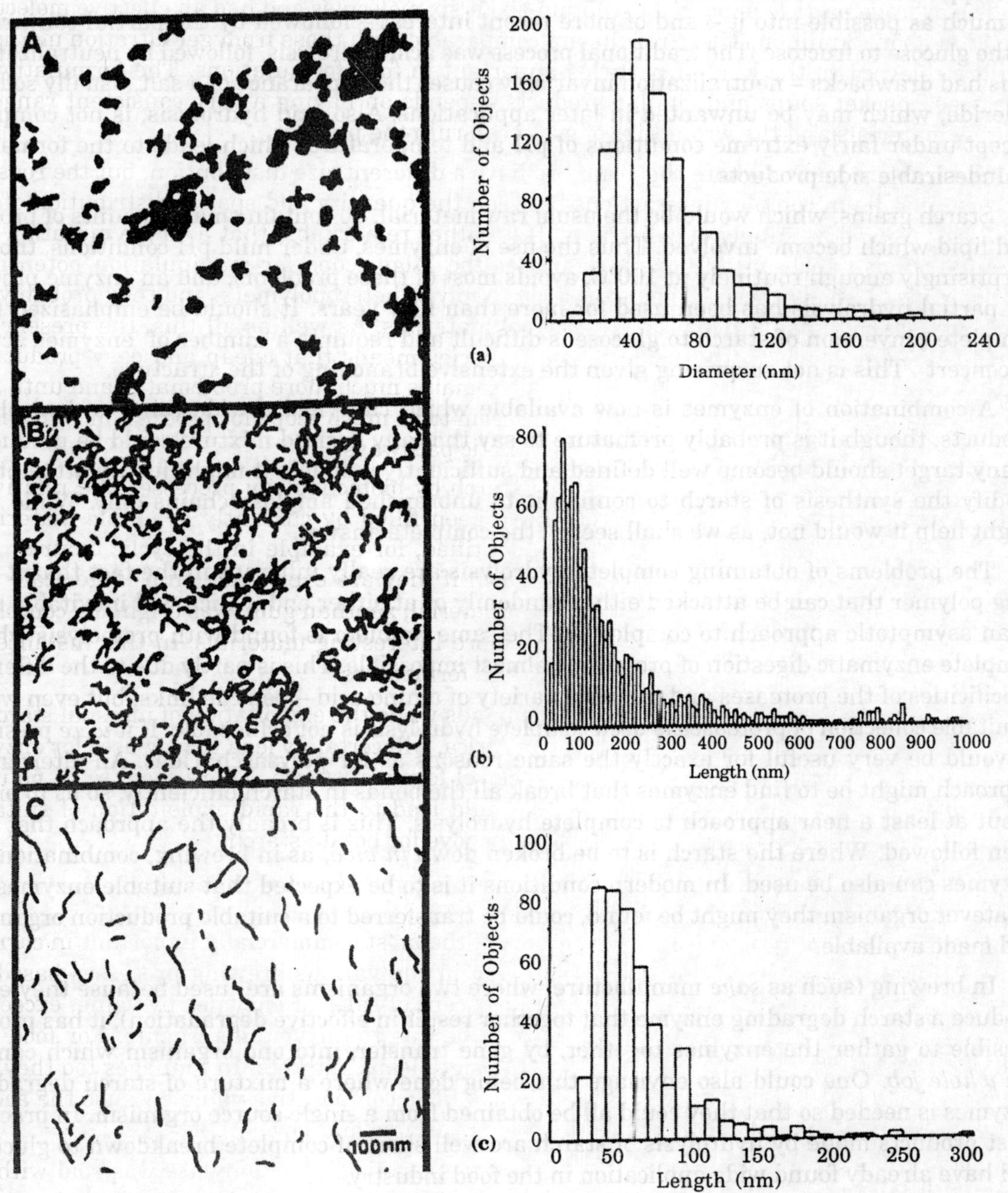

Fig. 4.2. Electron microscope images of amylose and amylopectin obtained with samples made by microwave heating of a suspension of maize starch in water. A: amylopectin. B: linear amylopectin. C: amylose pseudohelices. To the right is shown the corresponding size distributions of the particles. The amylose probably consists of an aggregate of six molecules on average.

Partial hydrolysates have found many applications and still constitute the most important products. One enzyme process leads to maltose, while a whole variety of dextrans are used in food applications. The ultimate product of hydrolysis is glucose, and another target is to convert as much as possible into it – and of more recent interest – followed by the partial conversion of the glucose to fructose. The traditional process was acid hydrolysis, followed by neutralization. This had drawbacks – neutralization invariably causes the appearance of a salt, usually sodium chloride, which may be unwanted in later applications. Also acid hydrolysis, is not complete except under fairly extreme conditions of pH and temperature, which leads to the formation of undesirable side products.

Starch grains, which would be the usual raw material, do contain small amounts of protein and lipid which become involved. Thus the use of enzymes, under mild pH conditions, though surprisingly enough routinely at 100°C, avoids most of these problems, and an enzyme process for partial hydrolysis has been used for more than fifty years. It should be emphasized that complete conversion of starch to glucose is difficult and requires a number of enzymes acting in concert . This is not surprising given the extensive branching of the structure.

A combination of enzymes is now available which can yield a wide range of hydrolysis products, though it is probably premature to say that any desired mixture could be produced. If any target should become well defined and sufficiently valuable it might justify attempts to modify the synthesis of starch to confine it to unbranched amylose chains only. While this might help it would not, as we shall see, be the complete answer.

The problems of obtaining complete hydrolysis are really inherent in the fact that it is a long polymer that can be attacked either randomly or at either end, which will inevitably lead to an asymptotic approach to completion. The same problem is found with proteolysis where complete enzymatic digestion of proteins is almost impossible. This is partly due to the differing specificities of the proteases and the wide variety of amino acid – peptide links, but even when a suitable collection of proteases is used complete hydrolysis is not achievable. If it *were* possible, it would be very useful for exactly the same reasons as for polysaccharides. An alternative approach might be to find enzymes that break all the bonds in starch efficiently, so as to bring about at least a near approach to complete hydrolysis. This is broadly the approach that has been followed. Where the starch is to be broken down *in vivo,* as in brewing, combinations of enzymes can also be used. In modern conditions it is to be expected that suitable enzymes, in whatever organism they might be found, could be transferred to a suitable production organism and made available.

In brewing (such as *sake* manufacture, where two organisms are "used because they each produce a starch degrading enzyme that together result in effective degradation), it has proved possible to gather the enzymes together, by gene transfer, into one organism which can do the *whole job.* One could also envisage this being done where a mixture of starch degrading enzymes is needed so that they could all be obtained from a single source organism. In practice most products made by hydrolysis of starch are well short of complete breakdown to glucose, and have already found wide application in the food industry.

Starch Grains and Heterogeneous Attack

In most organisms, and all those used as commercial sources, starch occurs in the form of insoluble grains. These vary in size and shape but usually have radial symmetry, which is the reason for their spectacular appearance in polarized light, and a highly ordered structure. On germination starch grains are used up and disappear along with the other storage materials of the seed. They may also appear and disappear during the maturation of the seed. The soybean for example has many starch grains in the cotyledon until just before maturation when all the starch is converted to lipid.

The mature seed contains none. This can happen quickly and there are clearly very active enzyme systems able to break down starch grains. Classical biochemistry was almost entirely concerned with enzymes acting on substrates in solution, and not much is known about the way in which enzymes attack lumps like starch grains. A similar situation in understanding the mode of action of lipases has only recently been improved, under the stimulation of a sharp increase in the uses of lipases in biotechnology. In the case of amylases one innovative study has used the technique of *fluorescence recovery after photobleaching* ('FRAP') to follow the movement of amylase molecules on a starch matrix.

Henis *et al.* (1988) found that the technique could be used to follow the lateral diffusion of labelled β-amylase molecules over the surface of a starch gel. There were two components to the motion. One depended on an exchange between bound and free enzyme molecules while the other was a simple lateral motion. The lateral diffusion disappeared when the enzyme was inactivated by reaction with iodoacetamide, suggesting that activity was linked in some way to motion, though not to simple binding. When starch grains are heated they swell and gelatinize. The temperature of gelatinization varies with the particular grain used, from 50°C for potato up to 68°C for sorghum, rice and high-amylose maize.

There is a correlation with the amylose content. The result is an amylose gel containing fragments of unexpanded parts of the original structure in an irregular way. Eventually a viscous mass called a starch paste is produced. The transition from an ordered structure to a random gelatinous one actually creates opportunities for interactions between the molecules which can then form either an insoluble white precipitate or in more dilute systems an elastic gel. This pnenomenon is called *retrogradation,* and probably makes access of degradative enzymes more difficult. In a different context it is part of the staling mechanism of bread. The hydrolytic enzymes are added to this mass and fortunately since they can operate at temperatures up to 110°C it is not necessary to cool the slurry first.

The substrate is physically heterogeneous and quite apart from the problems of enzymes in random attack on long polymers, access to retrograded particles must be different from those on gelatinous ones. Thus there appear to be at least three starch degrading enzyme systems. The first of these is that which is involved in the balance between lipid and starch content of the seed.

Seeds contain both lipid and starch as storage materials, though the relative amounts vary greatly. Some, such as groundnut contain both, while others such as peas may contain starch but little lipid, and mature soy has no starch but contains it earlier in the maturation process. The switch controlling this level is near to CoA carboxylase in the metabolic pathways, and in effect determines the fate of the acetyl CoA pool, but there is clearly a mechanism by which starch is broken down to acetyl CoA *in vivo.* It is likely that this is a simple reversal of the synthetic pathway. Secondly there are the enzymes responsible for breakdown on germination and mobilization of the reserves.

Dry mature seeds contain little enzyme activity and there is accumulating evidence that an early event in germination is the appearance of mRNA coding for the hydrolytic enzymes. These are clearly able to bring about hydrolysis, and are exploited in the malting process in brewing, and to a lesser extent in baking. Thirdly there are the enzymes associated with digestion. Saprophytic organisms such as fungi and bacteria all have them. They are effectively digestive enzymes on a par with the amylases of the mammalian gut.

There are many thousands of species producing such enzymes which probably all belong to a small number of families and exist in hundreds of variants. It is doubtful if they are specialized to produce complete hydrolysis. Because fungal and bacterial culture media are convenient sources for the production of enzymes most of those which have been used so far in biotechnology applications are of this type. The extent of degradation of starch in the human gut is of current interest because it is thought to be a factor influencing the fibre content of the diet. Although outside the scope of this book this is one influence which might lead to demands for a more controlled level of branching in dietary starch.

There is also some interest in non-ruminant animal feeds in the so-called resistant starch as a component of the diet. The enzymes used *in vivo* for germination of the seeds have not been exploited (other than indirectly in the malting process) and, since they presumably attack the highly ordered starch grain in the cell, the question of whether they could attack an isolated grain without preliminary gelatinization might be asked. Some studies on degradation *in vivo* have shown that the attack is heterogeneous and leads to holes in the starch grains, and can in some instances even lead to the development of a hollow shell.

Corn starch granules show similar effects when treated with bacterial α-amylase *in vitro*. It is believed that the potato starch granule is the most resistant, but even this is extensively digested by a *B. circulans* amylase which yielded maltohexose – though an amylase from *Chalara paradoxa* is claimed to yield only glucose. Many starch grains have small pits on the surface which contain protein, and there is some evidence that these are connected with the initial attack and hole formation. In some instances the attack seems to be by a general erosion. Recently a range of new products based on partial hydrolysis of starch grains that have not been gelatinized, or only partly gelatinized, has been introduced. Called 'starch hydrolysis products' to distinguish them from malto-dextrins and corn syrups, they have properties that appear to depend on the possession of higher molecular weights. During storage of potato tubers, some of the starch is converted to sucrose and reducing sugars when storage is at less than 5°C. This is reversible and starch is reformed when the temperature is raised, and although the question of whether this represents the reversible synthesis of a few starch grains, or the partial degradation of most of them, has been raised, it has not yet been answered.

Both phosphorylase which would be involved in any reversal of synthesis, and α-amylase are present at significant levels in the mature tuber, though little is known about the debranching enzymes that would be essential to hydrolytic breakdown. The amylase level does not increase much on budding, but little is known about starch degradation at this stage. In contrast the levels of amylase activity in fruits and seeds does increase substantially on germination, and potato tubers may be an exceptional situation.

Prolonged storage of tubers at ambient temperature also leads to a slow accumulation of reducing sugars, probably by amylase action. Starch breakdown in cereals is under hormonal control, and the giberellins are responsible for the production of a series of amylases that are able to break up the starch granules. However, phosphorylase is also prominent and it is also involved. The overall pattern which emerges is one where hydrolases such as α- and β-amylase break down the granules, and partially break up the amylose and amylopectin into fragments which are then available for phosphorolysis.

Thus the true function of amylases in the storage organs may be to break down the insoluble starch grains so as to render them accessible to the enzymes such as phosphorylase which would channel them into the metabolic pathways of the growing organism. It is possible that this is also the function of the 'digestive' amylases, which are intended to make the otherwise intractable starch break down to absorbable fragments, rather than to take them all the way to mono- or disaccharides. The ability of amylases to attack intact starch grains, of their own or other species, with apparently variable results is a factor which should always be taken into account when choosing a suitable enzyme. It is likely that different amylases will give widely different results, and a variety should be evaluated.

Enzymes of Starch Degradation

Some individual enzymes which are important in starch processing are:

α-Amylase

This enzyme, which is very widely distributed and is an important digestive enzyme of the human gut, breaks the (1 → 4) linkages in both amylose and amylopectin. It does this randomly, acting mostly as an endohydrolase, and cannot break the (1 → 6) branch point links. Also, presumably because of steric hindrance effects, it stops short of the branch points by one or two residues to yield the so-called limit dextrans. The α-amylases of *B. subtilis* and *B. licheniformis* are commonly used, have the useful property of high stability at 100°C and are used at this temperature in processing. This has the advantage of maintaining sterility, as well as a speedier reaction. It is known that proteins are stabilized against thermal unfolding in the presence of polyols such as hexoses, though the effect of soluble polysaccharides does not seem to have been studied and it is possible that this is a factor in the unusual stability of these enzymes. If so other enzymes should also show greater stability in these highly concentrated systems. A typical out-turn of extensive α-amylase activity would be about 5 per cent glucose, 50 per cent maltose and 30 per cent maltotriose, with significant amounts of higher polymers also present as well as the limit dextrans whose amount will vary with degree of branching of the particular starch chosen. They are typically around 20 per cent of the total.

Clearly using this enzyme alone is not the way to produce glucose. The presence of α-amylases has been reported in numerous organisms and the choice of the commercial producing organism is largely a matter of convenience. There will be many different sequences found amongst the enzymes though it does not follow that every species has a unique sequence, and the number of different enzymes is likely to be smaller than the total number of species producing them. From the viewpoint of the biotechnologist the main factor determining the choice of amylase is likely to be the ease of production and general availability rather than any specific property of an individual amylase.

There may be local specialization. For example, recently in Nigeria an amylase from *Aspergillus niger* has been described that is good at breaking down intact starch grains from sorghum, cassava and maize. Isolated from a cassava rotting strain it may well find local uses where it can outperform amylases adapted to rice or barley.

β-Amylase

This enzyme, which attacks only the α(1 → 4) linkages and like α-amylase cannot attack the branch links, is very widespread and again like α-amylase has the status of a digestive enzyme. It attacks only the ends of the chains to release maltose. That is, it attacks only the penultimate link. It occurs together with *a*-amylase and forms part of the digestive battery of enzymes. The two together produce almost entirely maltose and the limit dextrans. This combination is strongly reminiscent of the synergy between carboxypeptidases and endopeptidases in .digestion and is a neat solution to the problem of breaking down long polymers. The product however is maltose, and the combination is still not the answer to glucose production.

Glucoamylase

This enzyme, like amylases attacks the (1 → 4) linkages in the exomode, releasing glucose, but its main interest is that it can also break the branch point (1 → 6) linkages. Thus it increases the yield of glucose and is used after an initial treatment with α-amylase. They are not used together since glucoamylase cannot withstand temperatures above 60°C and the pH should be around 4.5 for optimum activity. A combination of amylase and glucoamylase activity can result in products with 95-98 per cent glucose which are widely used in food applications, as well as forming the feedstocks for high fructose syrup production.

Isoamylase

This (1 → 6) linked α-D-glucan maltohydrolase removes maltose from the non-reducing ends of amylose and amylopectin. It is found in bacteria, particularly *Pseudomonas* spp. It is effective in combination with β-amylase in raising the level of maltose.

Pullulanase : An endo-α(1 → 6) linked glucosidase, this enzyme is important because it splits the (1 → 6) links of amylopectin. It is obtained from *Bacillus* species and derives its name from its ability to hydrolyze pullulan, a linear polysaccharide made up of maltotriose units linked by α(1 → 6) linkages found in *Aureobasidium pullulans.* There are several hydrolytic enzymes found in this organism, and some thermostable pullulanases occur in *Clostridium thermohydrosulphuricwn.* These can break α(1 → 4) bonds in starch and could be used in the liquefaction step since they can withstand 90°C.

Protein Engineering of the Amylase Family

All the enzymes mentioned above belong to the same family, as judged from their sequences and nine crystallographic structures currently available. Table 4.2 lists the enzymes whose structures are already known. Study of these structures leads to the conclusion that all members of the family have a barrel-like arrangement of eight β sheets and eight α helices, shown as $(\beta/\alpha)_8$. The α-amylases have a domain between the third β strand and the third α helix.

The catalytic and substrate binding residues are at the C ends of the β strands, and in loops in the same regions binding sites have been identified in other locations as well. In particular the starch binding domain is found near the C terminus. Numerous sequences are now available representing eighteen different specificities in starch hydrolysis, and show conservation in four or five regions. These include the catalytic glutam ate, arginine and histidine, an important structural glycine, and a variety of substrate binding sites. Thus all the detail needed for protein engineering is available, and has been employed in studies on α-amylase, cyclodextrin glucotransferases and pullulanase. The α-amylase of *B. stearotfiermophilus* and a number of other organisms has had its detailed specificity altered by directed alteration of residues in the active centre and the binding site.

In most cases, as might be expected alteration of active centre residues led to loss of activity while alteration of binding sites led to changes in reaction rates. In one case, *Saccharomycopsis fibuligera* mutants showed a small change in preference for longer as against shorter chain substrates. The human pancreatic enzyme has enhanced maltase activity when the His210 residue is altered to asparagine. Other attempts to alter specificity from amylase to pullulanase type on a rational basis from sequence comparisons were rewarded with substantial loss of activity. The amylase of barley has been much studied because of its commercial importance. Cereal amylases bind b-cyclodextrin to a site that is probably the same as the starch grain

Table. 4.2. Starch hydrolases of known three-dimensional structure

Enzyme	Organism
Taka-amylase	*Aspergillus oryzae*
Pancreatic α-amylase	Pig
α-amylase	*Aspergillus niger*
α-amylase type 2	*Hordeum vulgare* (barley)
Cyclodextrin glucotransferase	*B. circulans*
Cyclodextrin glucotransferase	*B. stearothermophilus*
$(1 \rightarrow 6)$ glucosidase	*B. cereus*
β-Amylase	*Glycine max.* (soybean)
Glucoamylase	*Aspergillus awamori*

binding site, since it competes with it. It has a pair of tryptophan residues Tyr278–Tyr279, and changing 279 to alanine results in much lower cyclodextrin binding and an enhanced binding to starch. Residue 278 could not be altered with retention of structural integrity and is invariant in cereals. When barley amylase is expressed in yeast, one of the four cysteines is converted to a mixed disulphide with glutathione (so is about one third of human albumin in man). This results in inactivation, apparently because Cys95-glutathione prevents proper folding.

There is no reason to expect this to be a general effect of glutathione attachment, though it is interesting that yeast can do it in heterologous expression. There are two barley amylase isozymes, with only about 80 per cent sequence conservation, with distinctly different specificities. Hybrids have been made and expressed in yeast but the results are complicated and not fully interpreted as yet. A number of different hybrids can be made with varying results. There are so many possibilities that without some better idea of the targets for site directed mutagenisis and the use of mutants in general it is difficult to make use of the mass of detail that is beginning to accumulate. It is clearly going to be possible to obtain an array of enzymes with well-understood and defined specificities, but just how they might be used remains obscure.

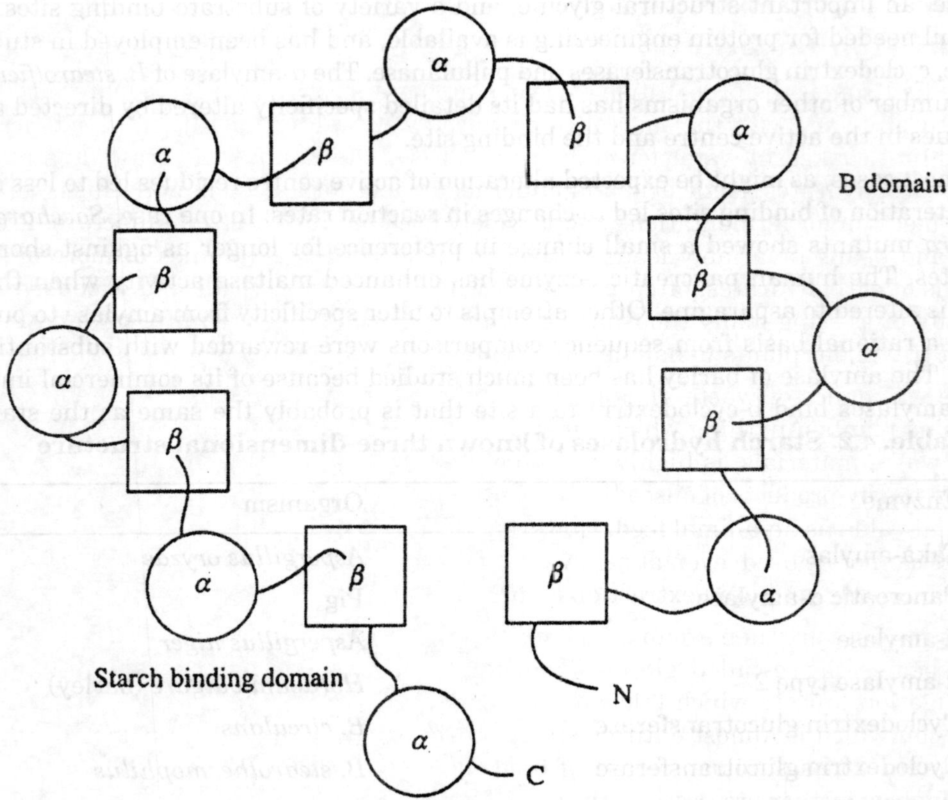

Fig. 3.3. Schematic structure of the α-amylases and related enzymes. They are based on the α_8/β_8 arrangement, i.e. have eight α helices and eight β sheets alternating around a drum shape and have a starch binding domain near the C terminus and another characteristic domain between the third α and β structural elements.

Glucose Isomerase

This enzyme is used in the final stage in the conversion of starch to fructose, and while it is not strictly speaking involved with *poly*saccharides it cannot be ignored. When the idea of converting glucose to fructose, as an alternative sweetener to compete with sucrose, first arose the best-known glucose isomerase was the mammalian enzyme. This involves a phosphorylation reaction and ATP. As already pointed out reactions involving nucleotide cofactors are just not feasible, and it was not until some bacterial enzymes capable of carrying out the isomerization were found that an enzyme based process could be developed.

The enzyme in question is probably a xylose isomerase, and a number of sources have been found and approved for food use. It is relatively expensive to produce which makes it worthwhile developing a column based immobilized enzyme system, so that it can be re-used as much as possible. In practice the enzyme is used in the form of the disrupted producing organism. The enzyme appears to adhere to the cell fragments in an active form.

Process Scale Starch Degradation

Starch Grain Isolation

The first step is the preparation of clean starch grains from the source material. The process will naturally vary with the source. In a typical 'wet milling' operation, based on maize, the grain is first soaked in water for 48 hours at 50 °C. This softens the grain, and also extracts soluble carbohydrates, lactic acid and minerals to yield corn steep liquor. This is an important growth medium for fermentations and is widely used for antibiotic manufacture. Processing of grains and beans invariably leads to the production of large volumes of more or less dilute solutions of carbohydrates, often pentosans which are very difficult and expensive to dispose of. Plants processing many thousands of tonnes of grain a year require, nowadays, dedicated effluent treatment plants to handle what remains a growing problem.

The economics of waste disposal are now a very significant part of the overall costs of processing plants. Anything which can reduce the cost of disposal, preferably by finding a use for the waste materials, is highly desirable. A substantial cattle food industry has grown up based on the by-products of oilseed processing, where the 'meal' remaining after lipid extraction has found value as an animal feed. However, much remains to be done since some of the waste products cannot be used, even at zero value, as cattle food. The next step is mechanical removal of the hypocotyls, and lipid extraction by pressing or solvent extraction.

The residues are then ground, and are effectively a mixture of cellulose cell wall fragments, protein bodies--the so-called 'gluten' - and starch grains. They are separated by a flotation and centrifugation process which takes advantage of the high density of starch grains, to yield a suspension which is almost entirely starch. It contains, as does the starch grain, about 1 per cent protein and a similar amount of lipid. It is drum dried for storage and transport, but increasingly is further processed in the same factory without drying. The details of the processing are very specific to a particular crop. It is said for example that wet milling is not suitable for European grown varieties of maize.

There are possibilities of using enzymes during wet milling, though this mostly involves cellulases as an aid to cell wall disruption, and will be discussed under cellulose. There are also possibilities of modifying the cell wall by suitable mutations with a similar objective. Several hundred thousand tonnes of maize are processed in this way, but even this is only a small proportion of the total maize production and it may be difficult to justify growing a special variety of maize just for this application. Since the process is already highly efficient the gains would need to be substantial.

Enzyme digestion *:* Starch is mixed with water to give a 30-35 per cent suspension (at high starcrh contents the mixture is thixotropic, and this has been exploited as the supporting medium for a preparative-scale electrophoresis method). There is little possibility at this stage of using column based continuous methods because the mixture is too viscous and all the processes are based on batch methods with no attempt made to recover the enzymes. The tank containing usually at least 500 litres is then steam heated, the pH adjusted to between 6 and 7 by addition of lime, and an amylase preparation added.

The most widely used one is from *B. licheniformis* and has the unusual property of long-term stability at 100°C despite the fact that the source organism is not a thermophile. It is activated by calcium ions, and if the pH is not adjusted with calcium hydroxide, then these must be added. Steam heating causes gelatinization of the starch but the simultaneous action of the enzyme causes a thinning out which reaches completion in about three hours. The amount of enzyme added, expressed in terms of some convenient activity measure, is adjusted empirically so as to produce this result with rninimal enzyme content.

Loss of activity is believed to be mainly due to hydrolysis of amide groups from the enzyme asparagine and glutamine residues. The now popular SHP (starch hydrolysis products) are made in the same way but without such extensive gelatinization and are less extensively hydrolyzed. Some processes stop at this point and the *maltodextrins* are sold as such and are characterized by their 'DE' value (this means 'dextrose equivalent' and *not* the 'degree of esterification' used in pectin terminology): the reducing power as a proportion of that which would be shown by pure glucose. It is a rough estimate of the number of reducing ends released by the enzyme attack. The higher the number the lower the average molecular weight.

Obviously the precise mixture of dextrans will depend on the nature of the starch used, particularly the degree of branching, but it is said that the precise α-amylase used can also affect the outcome. An alternative treatment makes use of fungal α-amylases, often from *Aspergillus oryzae,* which release maltose and produce syrups with a relatively high maltose and lower oligomer content. β-amylases and pullulanases may also be used on the starch slurry, but in all cases the temperature must be reduced to 50°C since they do not have the temperature resistance of bacterial α-amylases.

The solution is clarified by filtration and usually dried to a solids content of 65–75 per cent and stored in drums. Thus a variety of mixtures of maltose and higher oligomers is available and used mostly on a semi-empirical basis. Nearly complete conversion to maltose can be achieved with these enzymes. Some typical compositions produced by the various methods of hydrolysis are shown in Table 4.3 and an outline of the way in which processes have developed since 1950 is shown in Fig. 4.4.

Table 4.3 Starch hydrolysis methods

	Products of starch hydrolysis (per cent)			
Method	*Glucose*	*Maltose*	*Triose*	*Oligos*
Acid	91.5	2.5	6.0	
Acid enzyme	93.0	2.5	3.5	
Enzyme	96.0	1 to 2		0.1
Maltose/dextrose	18.0	43.0	?	?
High maltose	5.0	52.0	15.0	?

More recently there has been a demand for more extensively hydrolyzed starch as a substrate for conversion to fructose.

Glucose and fructose manufacture

The liquefied starch is cooled to 50–60 °C and the pH adjusted to 4 – 4.5 with hydrochloric acid. Then a suitable glucoamylase is added and the mixture maintained for two to three days, depending on the amount of enzyme added. Liquefied starch which has been treated with β-amylase or pullulanase can also be used as a feed for this stage. The *Aspergillus* enzyme is most often used and leads to a syrup which is 96 per cent glucose. It is usual to refine the syrup by charcoal adsorption to remove small amounts of colouring material, and it is then sold as a concentrate or crystalline material.

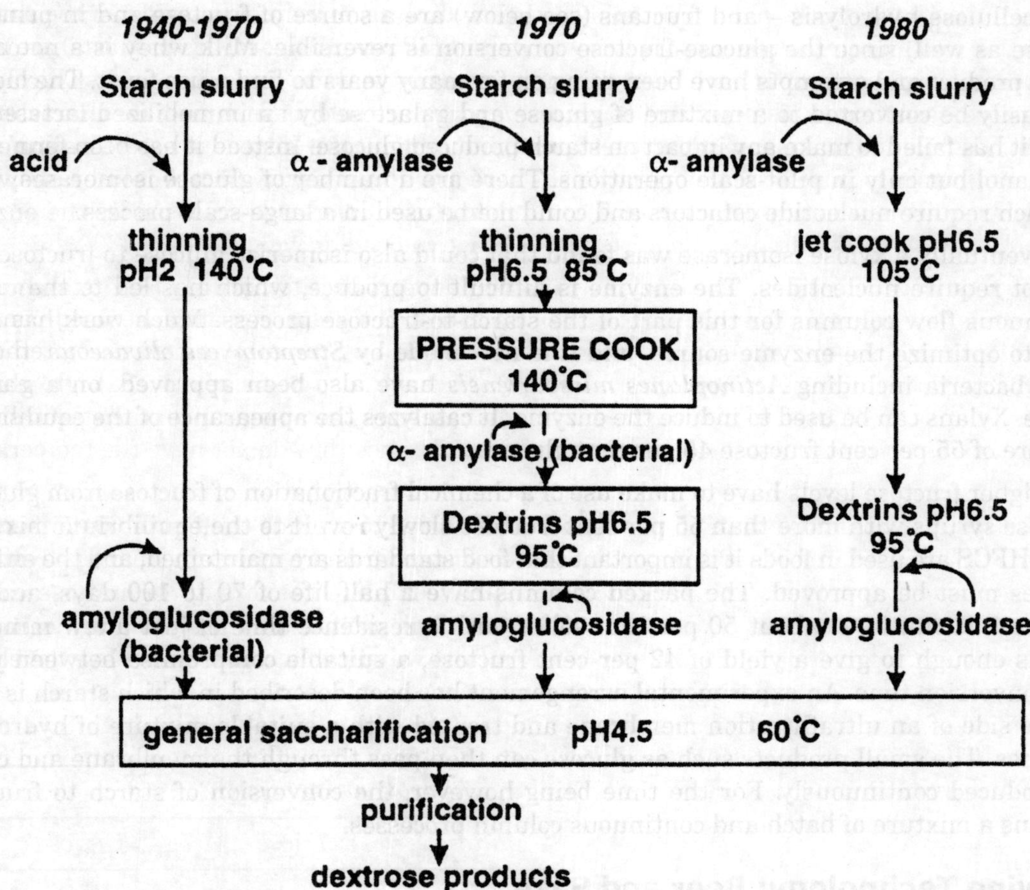

Fig. 4.4. The development of starch degradation processes over the last fifty years. The objective is to produce glucose or a range of polyglucose syrups.

The major use for glucose syrup is now for conversion to high-fructose corn syrup, 'HFCS'. In 1985 about half of the world production was converted to fructose : most of the remainder went into confectionery. The rise in demand for fructose syrups is a reflection of the world-wide changes in sweetener consumption. Sucrose which in terms of tonnages is pre-eminent has been partly displaced by fructose for a variety of reasons. Fructose is sweeter than glucose, has a higher solubility which makes it more convenient to use and is able to compete with sucrose both in terms of price and applications. This has been particularly so in the USA where at least a billion dollars worth of imported sucrose has been displaced. In other parts of the world fructose has been held back by deliberate support for sucrose for political reasons, and it is still unclear whether fructose can compete in a completely free market. Since it is based on what is frequently a by-product, that is starch, from some other process, it might be able to.

At present the starch raw material makes up about 50 per cent of the total costs of making HFCS. It should also be noted that there are other potential sources of glucose – for example from cellulose hydrolysis – and fructans (see below) are a source of fructose and in principle glucose as well, since the glucose-fructose conversion is reversible. Milk whey is a notorious waste product and attempts have been going on for many years to find a use for it. The lactose can easily be converted to a mixture of glucose and galactose by an immobilized lactase, but so far it has failed to make any impact on starch produced glucose. Instead it has been fermented to ethanol but only in pilot-scale operations. There are a number of glucose isomerases, most of which require nucleotide cofactors and could not be used in a large-scale process.

Eventually a xylose isomerase was found that could also isomerize glucose to fructose and did not require nucleotides. The enzyme is difficult to produce, which has led to the use of continuous flow columns for this part of the starch-to-fructose process. Much work has been done to optimize the enzyme source, and it is now made by *Streptomyces olivaceous,* though other bacteria including *Actinoplanes missouriensis* have also been approved, on a glucose source. Xylans can be used to induce the enzyme. It catalyzes the appearance of the equilibrium mixture of 55 per cent fructose 45 per cent glucose.

Higher fructose levels have to make use of a chemical fractionation of fructose from glucose; fructose syrups with more than 55 per cent fructose slowly revert to the equilibrium mixture. Since HFCS are used in foods it is important that food standards are maintained, and the enzyme sources must be approved. The packed columns have a half life of 70 to 100 days, and the substrate is run in at about 50 per cent solids with a residence time of just a few minutes. This is enough to give a yield of 42 per cent fructose, a suitable compromise between yield and conversion time. An experimental arrangement has been described in which starch is kept on one side of an ultrafiltration membrane and treated with a suitable mixture of hydrolytic enzymes. The small products such as glucose can then pass through the membrane and could be produced continuously. For the time being however, the conversion of starch to fructose remains a mixture of batch and continuous column processes.

Brewing Technology: Beer and Sake

Brewing is usually thought of as the production of beer from barley by the use of yeast, but more properly includes a variety of sources of starch, and whilst yeast is always involved, other micro-organisms may also play a part. Thus the source of starch can be virtually any plant with significant levels, and in the absence of sufficient endogenous amylases to break it down to sugars accessible to yeast, fungal amylases may be brought into action. Rice based drinks such as *sake* are included as well as traditional African beers based on sorghum, but not the small number of alcoholic drinks based on the fermentation of lactose, which is not used by yeast.

Barley

The first point of application of biotechnology will clearly be the barley itself. Varieties have been selected over hundreds of years on the basis of their agronomic performance, that is such factors as yield and return to the farmer, as well as their malting quality. There have been some attempts in the last few years to transform barley by injection of exogenous DNA, though as we discussed elsewhere as a monocotyledon it is in a category in which less progress has been made. Barley protoplasts have been successfully regenerated, containing a gene for β-glucanase derived from the fungus *Trichoderma reesei,* and progress in this area is likely to be rapid.

Some targets for modified barley have been identified and are indicated in Table 4.4. In many cases no reported progress has been made towards these targets, but since they have been identified by industrial research organizations as of some priority it is to be expected that they will be achieved in the near future. This does not of course mean that the modified barleys will then at once go into large-scale use, which will depend on many other factors and may take years.

Malting

Malting is the process in which barley, and possibly other grains, are made to germinate under controlled conditions of humidity and temperature, usually in piles on special malting floors. After a suitable time interval (typically days) the germination process is halted by heating in a kiln and the malted grain dried. At this stage the starch degrading enzymes have been produced within the seeds, but have not completed their action on the starch. Drying halts the activity but does not damage the enzymes, which resume activity when the malt is ground and extracted with hot water to produce the wort.

After a suitable interval extract of hops is added and the wort boiled which stops all enzyme activity, followed by cooling and the addition of yeast for fermentation. The wort solids contain 90–92 per cent saccharides of which maltose represents ~70 per cent of the total, while glucose, fructose, sucrose, maltotriose, maltotetraose and higher dextrins make up the rest. In addition β-linked glucans originally present in the barley persist to this stage. Many other components are also present including amino acids and proteins which promote the growth of yeast used in the fermentation. Gibberillic acid is known to be a regulator of the production of mRNA

Tables 4.4. Targets for modified barley

Target	*Status*
1. Better disease resistance	Genes for resistance now being transferred between varieties, e.g. mildew resistance from *Hordeum bulboswn* to *H. vulgaris*
2. Reducing the amount of non-starch polysaccharides	Low β-glucan barley varieties. Integration of heat-stable β-glucanase
3. Optimized amylose/amylopectin ratios	Not yet
4. High foaming	Not yet
5. Reduced haze levels	Low antocyanogen varieties
6. Low protein/high starch	Low hordein varieties
7. Enhanced flavour components	Not yet
8. Undesirable flavour components minimized	Not yet

for both α-amylase (of which there are several isozymes) and a β-glucanase in barley during germination.

A gibberillic acid sensitive region of an α-amylase promoter has been identified in barley, and is potentially useable as a general promoter for the expression of heterologous genes during germination. It might thus find a role in the modification of barley to make it more suitable for malt production. The way in which amylases attack intact starch grains has already been discussed above, and is relevant to the malting process since, unlike the processing of isolated starch grains, this takes place during normal germination. A high level of hordein, the barley storage protein, is regarded as undesirable in malting barley not just because it is accompanied by a relatively lower level of starch but because it appears to form a gel coat over the starch grains and makes them inaccessible to the amylases. This results in poorer yields and slower malting. On the other hand the hordeins contribute to the surface tension properties of beer and such important consumer attributes as foam stability and glass cling, so some caution is needed in modifying them. Barley varies in the amount of endogenous amylase and it is customary to add extra hydrolytic enzymes to the wort in order to optimize the extent of hydrolysis, so as to make the right balance between the smaller fermentable maltose and polymers up to maltotetraose and the higher dextrins on a consistent basis. Bacterial and fungal β-glucanases improve the separation of residues from the wort, and improve the efficiency of filtration, as well as reducing the haze level during and after fermentation. This is particularly useful when sorghum is the source of the wort.

Fungal pentosanases are also useful in improving extraction and reducing haze, especially from wheat and sorghum based worts. In addition to these, all the starch degrading enzymes listed above have been used to improve the fermentability of the wort. This is largely a matter of adjusting the ratio of the various maltose polymers. Once the relatively small amounts of glucose and fructose have been used up, maltose is swiftly taken up by the yeast, followed by maltotriose.

Brewing yeasts do not significantly use the higher dextrins. The extent to which these remain is important to the consumer attributes of the beer. It would be possible by appropriate additions to maximize the maltose level and eliminate most of the dextrins. This would result in an improved ethanol yield but significantly alter the acceptability of the beer. It is common practice to add additional fermentable carbohydrate such as isolated starch to the wort as a partial replacement for malt. The motives are economic, in reducing ingredient cost. In principle it would be possible to replace all the endogenous barley enzymes by added amylases, eliminating the malting step all together, and some maltsters have experimented along these lines. All these attempts run the risk of altering the consumer acceptability of the product, but there is a whole range of possibilities between what might be described as the 'traditional' method where nothing but barley yeast and hops are used to the other end of the scale where isolated barley starch is broken down with the aid of added bacterial and fungal enzymes, though yeast would still be used for the fermentation, with the addition of the necessary yeast growth promoters. Consumers are apparently prepared to accept beer in which some added enzymes and some added carbohydrate are used. Although wort production and fermentation is on a batch basis, not least because in some countries the taxation system is based on the carbohydrate content of each batch, continuous flow systems have been tried.

They are inflexible and difficult to adapt to varied production requirements – most brewers make several different beers – and also suffer seriously when something goes wrong such as microbial contamination. Instead greater efficiency has been achieved by using larger batches, typically 10^6 litres. Continuous flow has been more exploited in processes for ethanol manufacture, for use as fuel or feedstock.

Sake Manufacture

Sake brewing from rice makes an interesting contrast with beer from barley, since the overall process is much the same – an alcoholic drink from starch – but the details are quite different. Brown rice is first polished to remove the bran and steeped in water, followed by steaming for about an hour. After cooling, part of the rice – about 20 per cent – is then innoculated with spores of *Aspergillus oryzae.* The variety is chosen partly on the basis of its ability to penetrate the starch grains. The starch grains must be partly gelatinized by this treatment, but it must also largely eliminate any endogenous amylases present in the rice. Growth continues until, just before sporulation, it is stopped by cooling to 5 °C. Another portion of the steamed rice is used to prepare a yeast culture, and after about two weeks is ready for use as an innoculum.

Finally the two cultures are mixed with the remainder of the steamed rice at about 35 per cent water content for the main fermentation, with a complex system of back mixing and fresh additions, all at 10 – 15°C. Both fermentations take place together, and the combination of α amylase and glucoamylase from the *Aspergillus* and the *Saccharomyces cereivisiae* leads to a very high ethanol level of around 20 per cent.

After about thirty days the brew is filtered, pasteurized and bottled hot. Extensive work has now been done on transformed rice, but since the *sake* process appears to treat the rice simply as a source of starch and even eliminates the endogenous amylases, transformed varieties have not so far found any applications in this field. The α-amylase from *Aspergillus oryzae* is a monomeric glycoprotein of 478 residues of known crystallographic structure. It has a very high degree of homology (98 per cent) with the *niger* enzyme and has been expressed in yeast. A glucoamylase from *A. usamii* that is good at attacking raw starch has been moved into *oryzae* and improved its ability to attack starch grains. There is little doubt that there are many opportunities to optimize the properties of the *Aspergillus* stage of the fermentation.

Yeasts

The genetic transformation of yeasts is now a well-established technology, and yeasts are used as the production organism for one biotechnology process for the synthesis of human albumin. Originally established by a brewing company, the idea was to exploit the large amounts of yeast produced by all brewers as a source of secondary products of high value. Thus yeast transformed so as to make human albumin was to be used to brew beer, and then harvested and the albumin extracted. In the event the yeast is now designed to secrete the albumin into the growth medium – extraction from intact yeast would be very difficult – and it was felt that the general manufacturing standards of a brewery while perfectly adequate for beer manufacture, would not be appropriate for pharmaceutical products to be injected as a blood extender.

The process exists in its own right but no longer has any connections with brewing. This is an interesting example of where the original concept finished up as a rather different process in the face of practical difficulties. Another one is the growth of bacteria on petroleum residues that was originally intended to be a way of removing waxes as part of the refining process with a by-product of edible bacteria, but finished up as a way of making cattle feed by using all the oil for growth in very large reactors and as such was widely exploited. Yeast is fairly limited in the carbohydrates that it can use, and most of the targets for genetic engineering involve widening the capabilities of the yeast.

Yeast cannot use dextrins, which are about 25 per cent of the sugars present in wort, and has no amylolytic activity of its own. Interestingly enough yeast itself, if allowed to grow aerobically accumulates glycogen as its own storage carbohydrate, and this can in some circumstances reduce the yield in brewing operations, which become at least in part a process for converting starch into glycogen. There have been a number of attempts to insert amylolytic enzymes into yeast, and this is on the face of it a desirable target.

Rather than adding extraneous amylases, either as crude isolates as in current beer practice, or secreted by fungi as in the *sake* kind of process, let the yeast do the whole job – or at the least serve as the source for the extra amylase activity needed from time to time. Mammalian, wheat, rice and barley α-amylases have all been expressed in *Saccharomyces cerevisiae* but none of them was able to generate significant amounts of ethanol when given starch as a substrate. The reason for this is not clear, but probably depends on the levels of expression and secretion and the state of the starch substates tried.

A more successful attempt by A. G. Kumagai *et al.* (1993) used a promoter sequence associated with alcohol oxidase sequences in *Pichia pastoris,* and achieved a fairly high level of secreted rice α-amylase with its use. They constructed a shuttle vector containing cDNA for the rice α-amylase as well as the appropriate promoter regions. Yeast transformed with this vector produced a precursor of a amylase which has a signal peptide leading to the secretion of the functional enzyme that was isolated and identified in the culture medium. This yeast was able to produce ethanol when grown on 'soluble', i.e. partly hydrolyzed, potato starch and the level could probably be optimized to get near to the 6–10 per cent usually made in yeast fermentations. Whether this strain of yeast would be useful for brewing is a quite different question, since so many of the essential flavour components are due to the traditional yeasts and the traditional malting process.

A great deal of development work will be required before yeasts producing amylases are likely to be used in drinks manufacture. The first application may be to use them as part of a mixed culture with the standard yeasts, if this proves feasible. They might also prove to be a convenient source of isolated amylases. Their real usefulness will be in a process for the general degradation of polysaccharides to produce ethanol. Obviously a case could be made for the expression of some of the other dextrin degrading enzymes in yeast, such as glucoamylases. cDNA from *A. Oryzae* for both α-amylase and β-glucoamylase has been prepared and expressed in yeast but not yet put to use in making *sake*.

Baking Technology

Baking is a process in which flour, usually obtained from wheat, is mixed with water and then fermented with yeast, followed by heating to make bread. Many variants exist, some of them based on other cereals, and some of them not using yeast. In all cases however the endogenous enzymes of the cereal are involved, and in our case the α-amylase, and the β-glucanase are most important. The properties of the dough produced by mixing flour with water tend to be dominated by the protein components, the gliadin and glutenin. The status of the starch grains became important in a quite unforeseen way when the United Kingdom was negotiating to enter the European Economic Community. It was anticipated that this would lead to a decreased use of hard North American wheat, and an increase in the use of soft Continental wheats. The former were believed to lead to a longer shelf life, while the latter notoriously produce bread that stales very quickly. Thus the nation's daily bread was at risk, and much was made of this by politicians. North American wheat is characterized by a glutenin-gliadin ratio different from Continental varieties, and a number of research projects were started to see if adjustment of this factor could be used to overcome the perceived difficulties.

In fact it turned out that the harder wheat led to different grinding requirements, which in their turn led to greater fragmentation of the starch grains, and this is the factor that determines the staling rate. It is common practice to modify the state of the starch grains by adding additional amylases. Malt was widely used, but since it also contains many other enzymes greater control is desirable. This has been obtained by using bacterial and fungal a-amylase from *A. niger* and *B. subtilis,* both of which are permitted food additives. They produce an improved colour as well as improved crumb structure. Addition of amylases does retard staling, but the structure is due to better gas production by the yeast during the fermentation stage.

As with brewing, the key function of the enzyme is the way in which it attacks intact starch grains and modifies their properties, as well as the level of fermentable maltose it produces. Glucoamylases can also be used to boost yeast growth and loaf volume, though this enzyme is not yet widely used. It should be noted that addition of xanthans and guar gums also retards staling, and there are undoubtedly physical effects to do with the movement of water involved in what is a complicated process.

There are substantial unsolved problems in microwaving bread based products, which are also dependant on the status of the starch grains. Pentosanases (or hemicellulases) are also widely used in baking practice. They derive from *A. niger* and *Trichoderma viride,* and are present as minor activities in standard α-amylase preparations (indeed it is thought that some of the effects of amylases are really due to the contaminating pentosanases). Commercial materials are anything but homogeneous, and contain a number of different though related activities.

Wheat contains fairly high levels of pentosans, which are insoluble in water, but bind it. Hydrolyzing them leads to a reduction in bound water and increase in the free water which in its turn leads to softer doughs. While there is a clear improvement in loaf volume and general properties, it is not possible to go much further than that as an explanation of the effect, which remains obscure. There is some evidence for complex formation between glutens and pentosans, which would certainly affect the dough structure. Pentosanases are especially important in controlling the water binding in high fibre breads, where water binding is more difficult to control.

GLYCOGEN

Glycogen is the characteristic storage carbohydrate of animals, and in higher organisms is found mainly in the liver and musculature. Glycogen-like materials are however also found in plants. It is an $\alpha(1 \rightarrow 4)$ glycan, as is amylopectin, but differs in being much more highly branched. Like starch it is synthesized and degraded *in vivo* by way of a phosphorylase, to give glucose-1-phosphate. There is little or no biotechnological interest in glycogen as such – and it is highly unlikely that animal sources would replace plant in the foreseeable future. The main interest is in a series of rare human hereditary variants, in which branching enzyme deficiencies lead to the synthesis of glycogens with different structures. They have close parallels to those found in maize mutants and were useful in disentangling the various enzyme activities.

The best-known condition is *Gauchers* disease in which linear unbranched glycogen is formed and in *Pompes* disease where an $\alpha(1 \rightarrow 4)$ glucosidase activity deficiency leads to deposition of glycogen in the lysosomes. In the latter case attempts to alleviate the condition by administration of the enzyme in liposomes has had only limited success. While this may be a candidate for gene replacement therapy it is not likely to be available in the near future.

MINOR STORAGE POLYSACCHARIDES

Starch is the most important storage polysaccharide, and has the all-important property of insolubility in water which is a characteristic of all storage materials. Nevertheless many other polysaccharides are found at lower levels in the storage organs, and some of them are water soluble. They should probably be considered as storage materials too, if only because they have no other obvious function. Some of them are mainly located in the cell wall and might contribute something to its integrity, but again the general view is that they also have a storage function. They are mostly neutral sugars, and are polymers of galactose, arabinose, rhamnose, mannose and glucose, together with small amounts of other sugars, and are very widespread. These, like glycogen, have no special interest for biotechnology and are best known rather for the problems that they can cause. This is because they tend to turn up in the effluents from seed processing and are mainly responsible for the extremely high BOD (biological oxygen demand) values which are so expensive to deal with.

Pentosans

Pentosans, mostly arabinosans – though xylose is also found – occur at significant levels in wheat flour (about 2 per cent) and in soy meal (about 7 per cent) and thus derive from the two most important agricultural commodities in international trade. They have been regarded as important to the baking process in flour, and a variety of pentosanases have been investigated as possible process moderators.

Lentinans

Lentinans are $\beta(1 \rightarrow 3)$ and $(1 \rightarrow 6)$ branched polysaccharides found in fungi, particularly basidiomycetes. They have attracted interest since they are said to have anti-tumour activity via stimulation of lymphocytes and macrophages to produce interleukins.

FRUCTANS: INULIN AND LEVAN

Fructans are important for two reasons. The first is that they are a potential source of fructose, for which there is growing demand as a sweetener. The second is that their biosynthetic mechanisms are different from most of the polysaccharides and may offer the best opportunity for manipulation. In addition they occur very widely in commercial plants such as cereals and must be considered to be components of raw materials which may affect useful properties. They always contain at least one glucose residue but are otherwise made up of fructofuranose five-membered rings which leads to rather different polymer behaviour as compared with the pyranose ring-based glucans. They are the only polysaccharides based on the furanose structure.

Occurrence and Uses

Fructans (which used to be called fructosans, and are described as 'inulin' in commercial practice) are found in large amounts in only a small number of plants, but are widely distributed. Both monocotyledons and dicotyledons have them. The reason for this variation in content is unknown, and while their primary function is clearly as a reserve carbohydrate, their advantage in comparison with glucans may be connected with chill resistance. They tend to occur instead of glucans such as starch, but many species contain both as well as mannans.

Unlike starch they are mostly localized in vacuoles, and probably in solution, with the larger species possibly present as colloidal sols. This means that they exert a far greater osmotic effect than starch and to some extent break the 'rule' that storage materials should be insoluble. Fructans were first isolated from *Inula helenium,* hence the name inulin and many other *Compositae* have since been found to make them. They occur in all parts of the plant, notably, from the biotechnology point of view, in the seeds of the *Gramineae* though only to the extent of 1 or 2 per cent. The best-known cultivated dicotyledon is the Jerusalem artichoke *(Helianthus tuber-osus),* the tubers of which, together with cereals, are the major source of inulin used as a special food for diabetics.

In practice the fructose syrups isolated from such sources by hydrolysis isomerizes and contains up to 20 per cent glucose. With monocotyledons, in the grasses there is a suggestion that fructans and starches are mutually exclusive, with fructans found in temperate species and starches in the tropical varieties. They are widely distributed, and in commercial crops occur in wheat, barley and asparagus, as well as being the major storage carbohydrate of garlic (*Allium sativum*). In wheat, fructans appear to be synthesized before starch during seed maturation. Apart from use in diabetic diets some herbal extracts which are probably high in inulin have been advocated as diuretics, while inulin itself is used in kidney clearance tests since it remains intact in the blood after injection for long periods. Although there is now a substantial market for fructose as a sweetener (6 million tonnes in 1994 compared with 3.4 million in 1985) it is almost entirely made from maize (corn) starch by hydrolysis and isomerization of the resulting glucose.

This is a matter entirely of economics and since corn starch is a very cheap and plentiful raw material it is unlikely to be displaced from its major markets. Nevertheless there are smaller markets in various parts of the world where fructans may be the preferred source of fructose. There are clearly possibilities of manipulating potential crop plants to improve the efficiency of production of fructans. Fructans have been found in the marine algae *Cladophora* and *Rhizoclonium,* but are not present in other species. Starch is also present. Fructans have not as yet been described in fungi. There are also microbial sucrose-dependent fructosyl transferases which may mean that fructans are present in microbes.

Biosynthesis

Fructans are synthesized by the transfer of a fructose residue from sucrose onto another sucrose receptor molecule. Sucrose is a 'high energy' compound – the ΔG of hydrolysis is –

27.6 kJ compared with UDP (31.8 kJ) and phosphoryl derivative values of 20.8 kJ – and it can thus act widely as a donor in polysaccharide biosynthesis. In this instance, although UDP-fructose has been found it is not involved in fructan synthesis. Enzyme systems capable of doing this in three different ways are known to exist leading to the main three types of fructans:

(i) The isokestose series – linear chains with $\beta(1 \rightarrow 2)$ links of the general form

$$\text{Glc} - (1 \rightarrow 2) - \text{Fru} - 1 \rightarrow (2 - \text{Fru} - 1)_n \rightarrow 2 - \text{Fru}$$

with a maximum degree of polymerization (DP) of about 35. These are sometimes referred to as *inulins,* but commercial 'inulins' are likely to be a mixture of all types of fructans depending on their source.

(ii) The neokestose series – in which the glucose residue is linked directly, through the 2 and 6 positions, to two fructose residues and thus the glucose residue is not terminal:

$$\text{Fru} - 2 \rightarrow (1 - \text{Fru} - 2)_m \rightarrow 1 - \text{Fru} - (2 \rightarrow 6) - \text{Glc} - (1 \rightarrow 2) - \text{Fru}-1$$

$$\rightarrow (2 - \text{Fru} - 1)_n \rightarrow 2 - \text{Fru}$$

(iii) The kestose series - where the linkages are $\beta(2 \rightarrow 6)$ bonds with a general formula

$$\text{Glc} - (1 \rightarrow 2) - \text{Fru} - 6 \rightarrow (2 - \text{Fru} - 6)_n \rightarrow 2 - \text{Fru}$$

Fructans of this type are sometimes called levans, or phleins since they are found in the *Pooideae* grasses. There have also been reports of *branched fructans* though they have not been fully characterized.

The initial synthetic step in all cases is a condensation:

$$\text{Glc} - (1 \rightarrow 2) - \text{Fru} + \text{Glc} - (1 \rightarrow 2) - \text{Fru} = \text{Glc} - (1 \rightarrow 2) - \text{Fru} - (1 \rightarrow 2) - \text{Fru} + \text{Glc}$$

Sucrose

Isokestose

by the enzyme sucrose sucrosefructosyl transferase (SST) which is active even where the fructans are neokestose based or are levans. The enzyme has been purified to homogeneity and is capable of using some receptors other than sucrose, though it never uses isokestose. It is clearly one point at which site directed mutagenesis might be used to vary specificity and increase the range of fructans synthesized. Two further enzymes are required to form the fructans. A fructan fructan-fructosyl transferase can transfer fructose residues from one fructan to another, including isokestose as either donor or acceptor.

The first of these leads to the formation of $(1 \rightarrow 2)$ links, and thus the inulin or isokestose series, while the second can form the Fru–$(2 \rightarrow 6)$–Glc links found in the neokestose series. These two enzymes have been purified from *Asparagus* and thus here too the first steps in genetic engineering have already been taken. The enzymes responsible for the levan series have not as yet been isolated, though their activity has been demonstrated in cell-free extracts with sucrose as substrate.

A fructosyl transferase from *B. subtilis* has been successfully transferred to tobacco plants where it causes accumulation of fructans up to 8 per cent of the dry weight. This is clearly a development of considerable potential significance for fructose production, and while tobacco is an experimental plant in this context, there would seem to be no barrier to its expression in other plants such as potatoes. These enzymes are attractive as subjects for biotechnology simply because sucrose is readily available in large amounts.

Sucrose-dependent transferases which can transfer the glucosyl residue of sucrose are also known and a combination of these with the fructosyl transferases could clearly lead to novel polysaccharides. The use of sucrose as a donor is all the more attractive since the other possibilities involving nucleotide derivatives would certainly be much more difficult. Cyclic fructans have not been described but would be of considerable interest as a contrast with cyclodextrins.

Degradation

Fructan hydrolases have been described and appear to function by removing terminal fructose residues until sucrose remains. There is no evidence of internal group hydrolysis. It is likely that a separate enzyme is involved in splitting the (2 → 6) links in neokestose and one specific for Fru-(2 → 6)–Fru links in the levan series has been described. Since acid hydrolysis is easy it may be the preferred method rather than any use of enzymes in process applications. It is also worth noting that fructan hydrolases are absent from the human gut, though they may be present in the microbial flora.

Characterization

Most structural determinations on fructans have been done by methylation. Some solution properties have been measured, and most analytical methods now depend on gel filtration/ sizes exclusion chromatography. The molecular weights tend to be ~50000 and are often less, and there is always some polydispersity based on a single fructose residue difference. This is what would be expected from the synthetic pathway. Table 4.5 summarizes some solution properties (in dimethyl sulphoxide) of Jerusalem artichoke inulin.

Table 4.5. Fundamental properties of Jerusalem artichoke inulin

Property	
Solubility	Soluble in dimethyl sulphoxide (DMSO)
Molecular weight	3400 Da
Intrinsic viscosity [η]	9 m lg^{-1}
Sedimentation coefficient, $s^{\circ}_{20,w}$	~0.4 S

CHAPTER

5

Nucleosides

All cells in living organisms contain the large nucleic acid molecules of ribonucleic acid (RNA) and deoxyribonucleic acid (DNA). Both these molecules are polymers of nucleotides. DNA is found in chromosomes, and genes are unique sequences of DNA nucleotides. The genes contain the inheritable information which together with RNA directs the synthesis of all the cell's proteins. The realization that DNA, and thus the inheritable information, can be transferred from one organism to another has led to the important field of recombinant DNA technology. The ability to open up the DNA molecule and insert another gene is fundamental to these manipulations and requires the use of two special classes of enzymes. Restriction endonucleases cut the DNA molecule between specific nucleotides while ligases rejoin pieces of cut DNA.

Plasmids, the non-chromosomal DNA found in microorganisms, are the usual acceptors for these transferred pieces of DNA and the new molecule that is formed is known as a recombinant DNA molecule. In addition, another group of enzymes, the DNA polymerases, which synthesize a new strand of DNA using an existing DNA template, are vitally important in the analysis of gene structure and expression. Recombinant DNA technology is a major growth industry and is being exploited commercially to produce a range of products, including vaccines, drugs and enzymes.

NUCLEIC ACID COMPOSITION AND STRUCTURE

The nucleotides of RNA and DNA consist of three components: a carbohydrate, a phosphate group and an organic nitrogenous base. There are two types of carbohydrate molecule in nucleic acids, both of which are D-pentoses, i.e. contain five carbon atoms. The carbohydrate in RNA is ribose, while DNA contains deoxyribose, which has a hydrogen atom instead of a hydroxyl group attached to the carbon in the 2 position (Fig. 5.1).

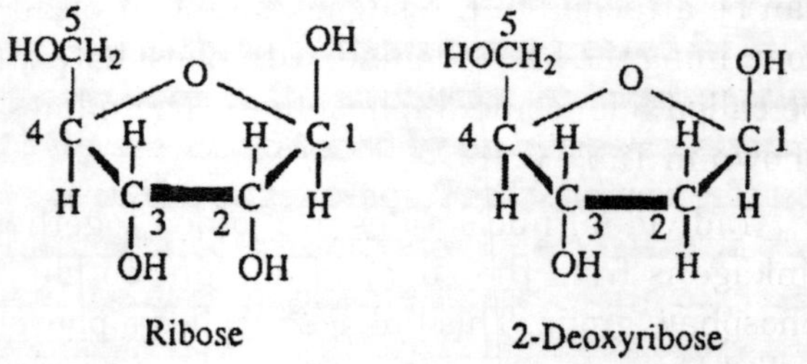

Fig. 5.1. The structures of ribose and 2-deoxyribose.

There are five common bases found in nucleic acids. Adenine (A), guanine (G) and cytosine (C) are found in both DNA and RNA. Uracil (U) is found only in RNA and thymine (T) only in DNA. The structures of these bases are shown in Fig. 5.2. Adenine and guanine are purine bases while uracil, thymine and cytosine are the pyrimidine bases.

Alkaline hydrolysis splits the nucleotide into its phosphate and sugar-base residues. The sugar-base is known as a nucleoside. The nucleosides are named according to the type of base present. If a purine base is present it will end -osine, e.g. adenosine, while if a pyrimidine is present the name will end -idine, e.g. uridine.

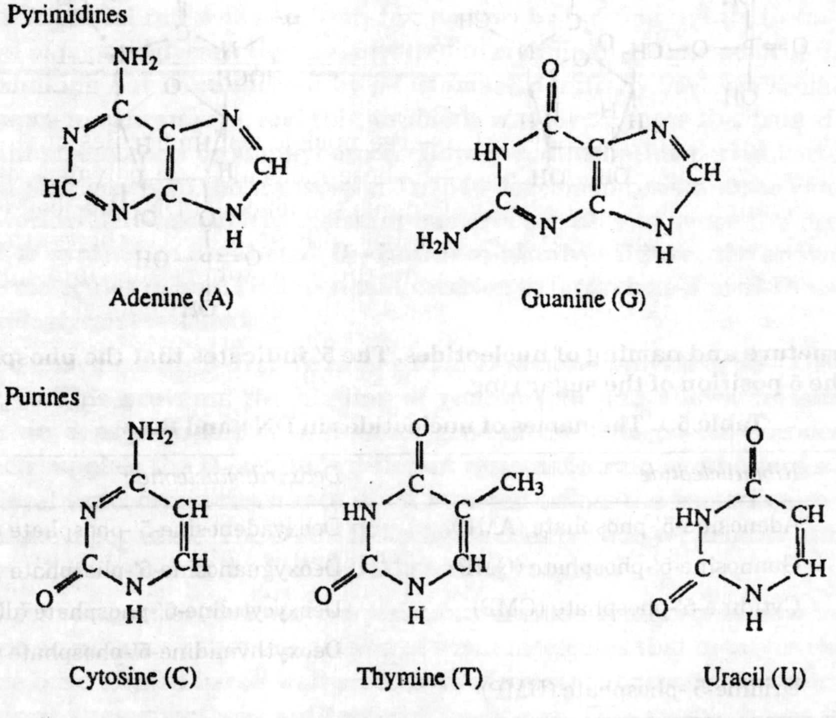

Pyrimidines

Adenine (A)

Guanine (G)

Purines

Cytosine (C)

Thymine (T)

Uracil (U)

Fig. 5.2. The structures of the pyrimidine and purine bases of nucleic acids.

The nucleotides are named according to the nucleoside and the position on the sugar molecule to which the phosphate is attached. To differentiate between the sugar ring system and the base ring system the sugar positions are designated with a superscript. The phosphate can be attached to the sugar residue at either the 3′ or the 5′ position (Fig. 5.3). A nucleotide containing the base adenine, with phosphate attached to the 5′ position of ribose, would therefore be called adenosine-5′-phosphate. The names of the nucleotides found in DNA and RNA are shown in Table 5.1.

Individual nucleotides are joined together to form oligonucleotides or nucleic acids. The linkage is from the sugar of one nucleotide to the sugar of the next nucleotide through the phosphate group. This linkage is called a phosphodiester link. The phosphate between the sugars is linked to the 5′ position of one sugar molecule and the 3′ position of the second molecule. This repeating sugar-phosphate sequence forms the backbone of the nucleic acid. The five different bases, attached to the sugars at the 1′ position, stick out away from the sugar-phosphate backbone. By convention, the end of the nucleic acid containing the free 5′ hydroxyl or phosphate group is written to the left and is called the head, while the end with the free 3′ phosphate is the tail and is written to the right.

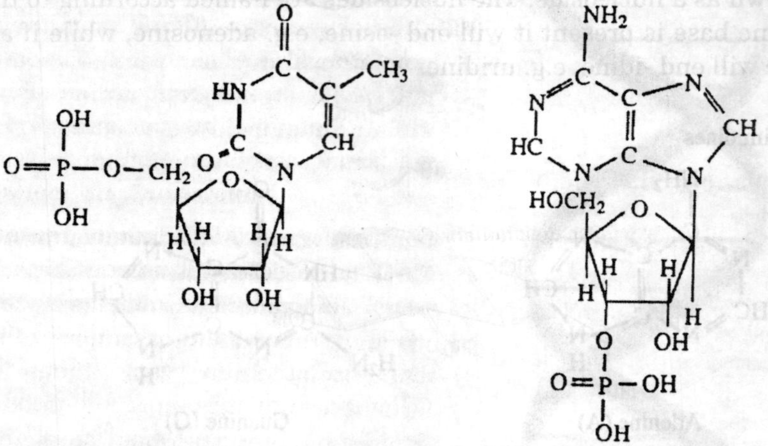

Fig. 5.3. The structure and naming of nucleotides. The 5′ indicates that the phosphate group is attached to the 5 position of the sugar ring.

Table 5.1. The names of nucleotides in DNA and RNA

Base	*Ribonucleotide*	*Deoxyribonucleotide*
Adenine	Adenosine-5′-phosphate (AMP)	Deoxyadenosine-5′-phosphate (dAMP)
Guanine	Guanosine-5′-phosphate (GMP)	Deoxyguanosine-5′-phosphate (dGMP)
Cytosine	Cytidine-5′-phosphate (CMP)	Deoxycytidine-5′-phosphate (dCMP)
Thymine		Deoxythymidine-5′-phosphate (dTMP)
Uracil	Uridine-5′-phosphate (UMP)	

It is the sequence of different bases in the oligonucleotide that is of functional importance and therefore the nucleic acid chain is usually expressed simply as its base sequence, e.g.

The structure of DNA

Watson and Crick showed that the normal structure of DNA consists of a double helix made from two single oligonucleotide strands. The two sugar-phosphate strands of the helix run in opposite (anti-parallel) directions and the bases point inwards and pair very precisely with bases on the opposite strand. Base pairing occurs between a purine and pyrimidine base; guanine always pairs with cytosine and adenine with thymine (Fig. 5.5).

This helical structure is held together by hydrogen bonding between the base pairs. The hydrogen bond is formed between a pair of electrons on the keto group or ring nitrogen of one base and a hydrogen atom on a ring nitrogen or amino group of another base (Fig. 5.6). Three such bonds are formed between a cytosine/guanine base pair and two bonds between an adenine/thymine pair. Individual hydrogen bonds are very weak but the presence of a large number of bonds in the double helix produces an extremely stable configuration. The structure results in the base pairs being stacked parallel to each other, thus allowing hydrophobic interactions between the stacked base pairs, which also helps to stabilize the structure.

AGCTAAGGTCCATG...

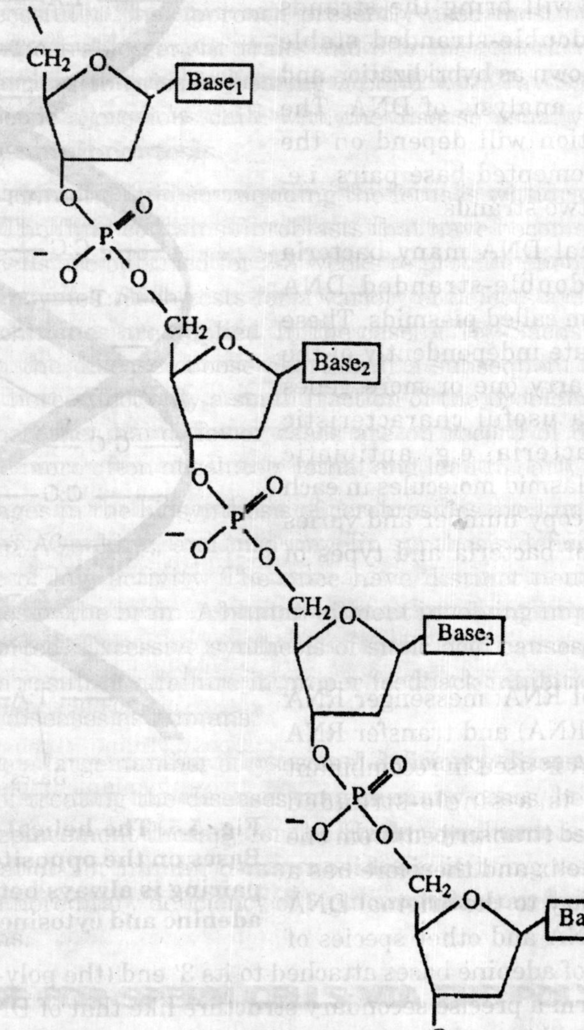

CH_2 O Base$_1$ O P O O O CH_2 O Base$_2$ O P O O O CH_2 O Base$_3$ O P O O O CH_2 O Base$_4$ O

Fig. 5.4. Oligonucleotide structure. The nucleotides are joined via the phosphodiester bridges between sugar residues. The bases stick out away from the sugar-phosphate backbone.

In most DNA molecules the two ends of the DNA double helix are joined to form a closed circle. When this occurs the DNA then forms a super-coil by winding back on itself. Supercoiled DNA depends on the integrity of the closed circle. Any breaks in the circle will cause unwinding of the super-coil.

The two anti-parallel DNA strands are complementary to each other. Thus where there is an adenine residue in one strand there is a thymine in the other, while guanine and cytosine complement each other. When two complementary strands of nucleic acids (DNA or RNA) are mixed in the test-tube the hydrogen bonding between the base pairs of opposite strands will bring the strands together and produce a double-stranded stable structure. This process is known as hybridization and is used extensively in the analysis of DNA. The strength of the hybridization will depend on the number of correctly complemented base pairs, i.e. the homology between the two strands.

As well as chromosomal DNA many bacteria contain closed circular double-stranded DNA structures in their cytoplasm called plasmids. These plasmids are able to replicate independently of the chromosome and usually carry one or more genes that are responsible for a useful characteristic displayed by the host bacteria, e.g. antibiotic resistance. The number of plasmid molecules in each bacterium is known as the copy number and varies between different species of bacteria and types of plasmid.

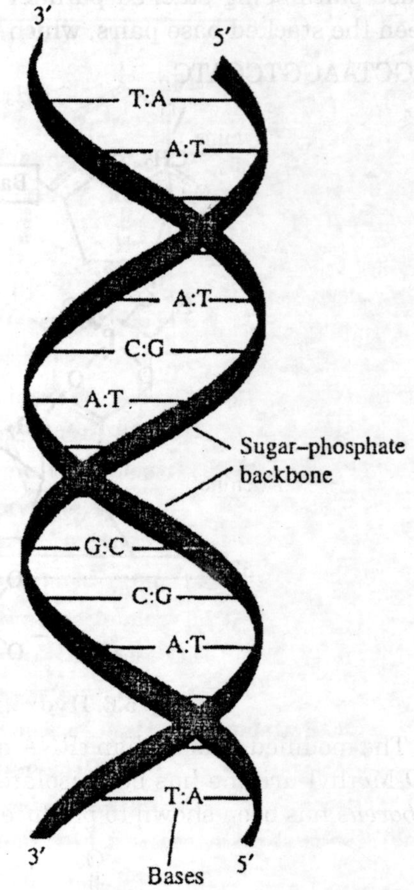

Fig. 5.5. The helical structure of DNA. Bases on the opposite strands pair. The pairing is always between thymine and adenine and cytosine and guanine.

The structure of RNA

There are three types of RNA: messenger RNA (mRNA), ribosomal RNA (rRNA) and transfer RNA (tRNA). Of these only mRNA is used in recombinant DNA technology. mRNA is a single-stranded oligonucleotide which is copied (transcribed) from one of the strands of the DNA helix and therefore has a base sequence complementary to the original DNA sequence. It differs from DNA and other species of RNA because of a sequence of adenine bases attached to its 3′ end (the poly-A tail). The single-stranded mRNA does not form a precise secondary structure like that of DNA. However, some base pairing between nucleotides on the same strand probably occurs to produce short helical areas which are separated by extensive non-helical regions.

NUCLEOSIDES

Reviews have appeared on the general chemistry of nucleosides and on the conformation and chromatography of nucleosides and nucleotides. In addition, symposium papers on the chemistry and biology of nucleosides and nucleotides, including reviews on the synthesis of nucleosides, the application of ^{13}C n.m.r., and C-glycosyl nucleosides, have been published. Nucleoside X (1) has been isolated from the variable loop of *E. coli* tRNA; the analogues (2) and (3) were prepared from uridine.

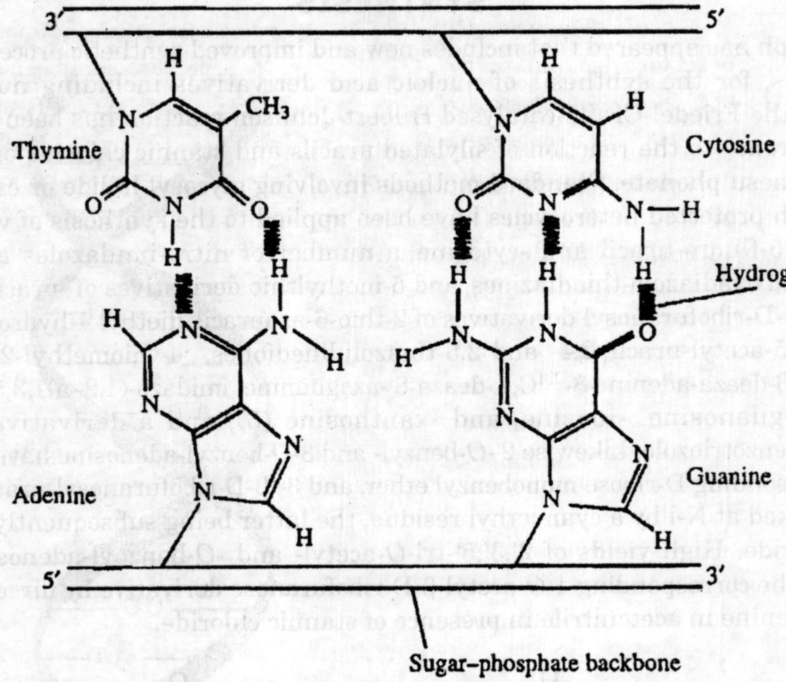

Fig. 5.6. Hydrogen bonding between the base pairs.

The modified mammalian tRNA nucleoside (4) has been characterized by mass spectrometry. 2′-*O*-Methyl-uridine has been isolated from human urine; and a new species of *Streptomyces herbaceus* has been shown to produce Ara-A(9-β-D-arabinosyladenine).

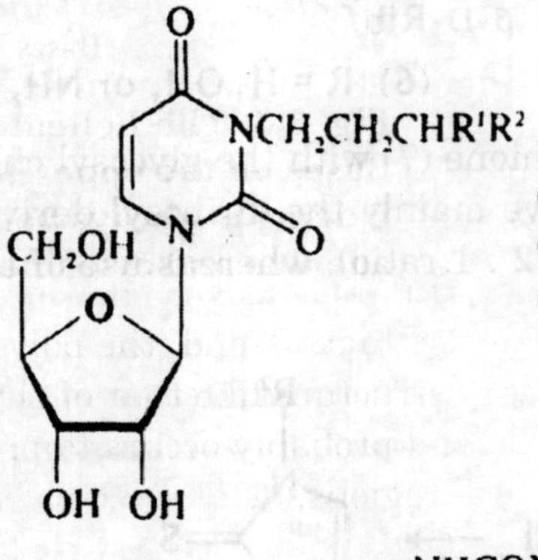

$NCH_2CH_2CHR^1R^2$

(1) $R^1 = NH_2$, $R^2 = CO_2H$
(2) $R^1 = NH_2$, $R^2 = H$
(3) $R^1 = H$, $R^2 = CO_2H$

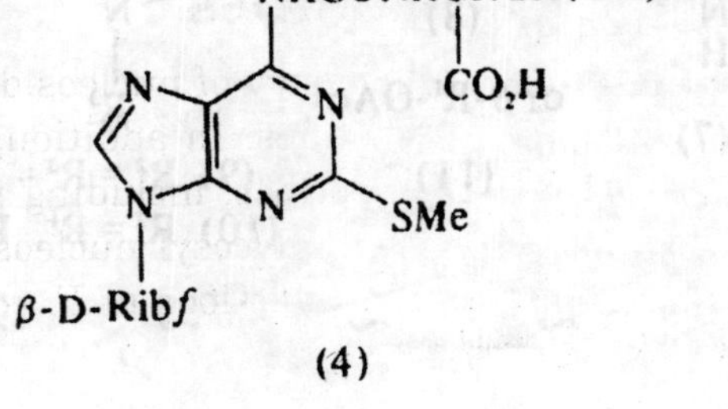

NHCONHCHCH(OH)Me

CO_2H

SMe

β-D-Rib*f*

(4)

SYNTHESIS

A monograph has appeared that includes new and improved synthetic procedures, methods and techniques, for the synthesis of nucleic acid derivatives including nucleosides. The mechanism of the Friedel-Crafts-catalysed Hilbert-Johnson reaction has been reviewed, with particular reference to the reaction of silylated uracils and stannic chloride or trimethylsilyl trifluoromethanesulphonate. Standard methods involving glycosyl halide or ester derivatives in reaction with protected heterocycles have been applied to the synthesis of various glycosyl derivatives of 5-fluoro-uracil and -cytosine, a number of nitro-imidazoles and -pyrazoles, thiadiazines and oxadiazolo-thiadiazines, and 5-methylthio derivatives of -uracil, -4-thiouracil, and -cytosine, β-D-ribofuranosyl derivatives of 2-thio-6-azauracil, diethyl 4-hydroxypyrazole-3,5-dicarboxylate, 5-acetyl-uracil, 2,4- and 2,5-thiazolidinediones, 4-thiomethyl-2-azapurine, 2 - hydroxypurine 3-deaza-adenine-8- ^{14}C, 1-deaza-8 -azaguanine, imidazo-(1,2–a)1,3,5-triazenes, *e.g.*, (5), *lin*-benzo-guanosine, -inosine, and -xanthosine (6), and a derivative of *N*-2-(β-D-ribopyranosyl)benzotriazole. Likewise 2′-*O*-benzyl- and 3-*O*-benzyl-adenosine have been prepared from the corresponding D-ribose monobenzyl ether, and 3-(β-D-ribofuranosyl)uracil from a uracil derivative blocked at N-l by a cyanoethyl residue, the latter being subsequently removed with sodium methoxide. High yields of 2′,3′,5′-tri-*O*-acetyl- and -*O*-benzoyl-adenosine have been obtained from the corresponding 1-*O*-acetyl-β-D-ribofuranose derivative by direct reaction with unprotected adenine in acetonitrile in presence of stannic chloride.

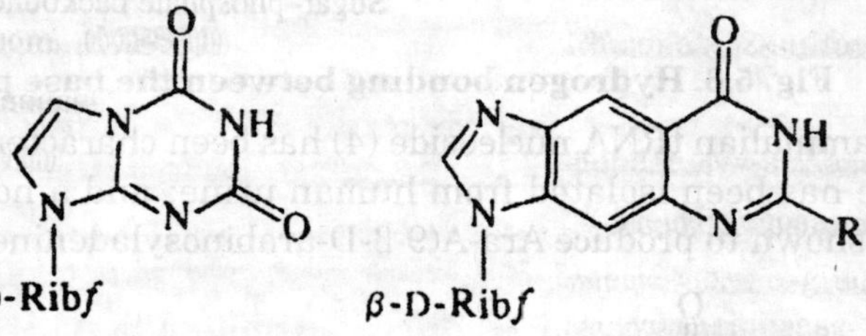

β-D-Rib*f*

(5)

β-D-Rib*f*

(6) R = H, OH, or NH_2

Ribosylation of the silylated imidazolethione (7) with the glycosyl chloride (8) in presence of stannic chloride or silver perchlorate gave mainly the diribosyl derivative (9) rather than the expected mono-ribosyl nucleoside (10) (2 : 1 ratio), whereas use of the tetra-acetyl sugar (11) gave mainly (10).

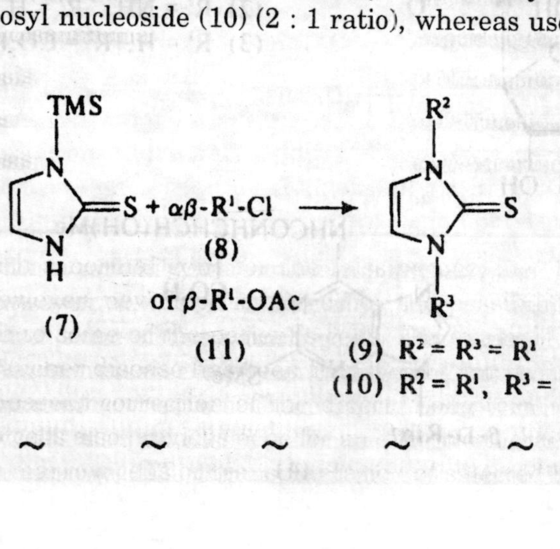

TMS

S + αβ-R^1-Cl (8) or β-R^1-OAc (11) → S

(7)

(9) $R^2 = R^3 = R^1$

(10) $R^2 = R^1$, R^3 = H

~ ~ ~ ~ ~ ~ ~

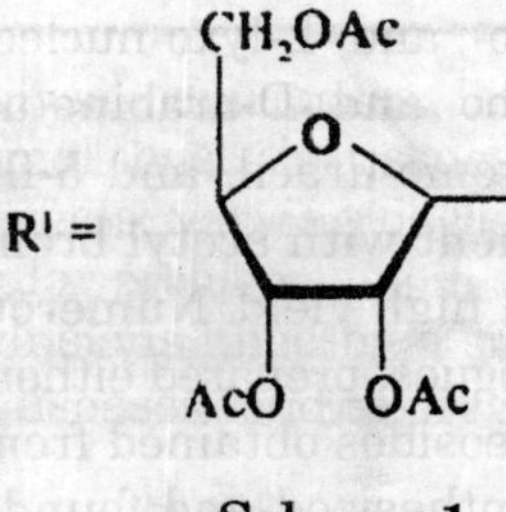

Scheme 1

Nucleosides prepared conventionally from deoxy-sugars have included the antiviral nucleoside dihydro-5-azathymidine (12), the tritiated analogue (13), and the 6-substituted derivatives (14) (Scheme 2), [^{14}C-2] and [^{14}C-4]-2′-deoxy-uridine, and 2′-deoxy-ribonucleosides from 5-azapyrimidine derivatives, 5-substituted uracils including 5-ethynyl- and (E)-5-(2-bromoethenyl)

Reagents: (i) $SnCl_4$-MeCN; (ii) $NaBH_4$ or Grignard Reagent

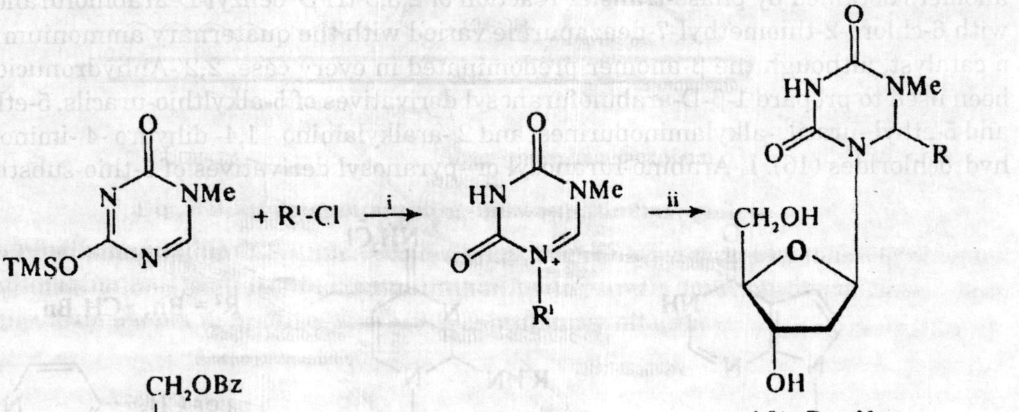

(12) R = H
(13) R = ^{3}H
(14) R = Me or Ph

Reagents : i, $SnCl_4$-MeCN; ii, $NaBH_4$ or Grignard Reagent

Scheme 2

-uracil, and a number of 5-polyfluoroalkyl uracils, including some 4-thio- and 4-amino-analogues of these 2′-deoxynucleosides. The 2′-deoxyribonucleotide of benzimidazole has also been prepared.

Nucleosides have also been obtained from standard bases with 6-deoxy-D-allose to give 5′-C-methyl analogues of uridine, cytidine, and adenosined from 2-, 3-, and 5-deoxyallose derivatives giving the corresponding deoxy-*ribo*-hexofuranosyl cytosine, and from 5-chloro-tetra-hydrofuran-2-carboxylic acid esters to give corresponding 2′,3′-dideoxy-α-nucleoside analogues of uridine. The synthesis of 3′- and 4′-hydroxy derivatives of (RS)-1-(tetrahydro-2-furanyl)5-fluorouracil (FTORAFUR) has also been reported. The deoxygenation of intact nucleosides has been used to prepare 2′- and 3′ -deoxyguanosine from guanosine *via* chloro-deoxy sugar intermediates 5′-deoxy-D-ribo-, -D-arabino-, -D-lyxo-, and -L-lyxo-nucleosides of 5-fluoro-uracil from 5-fluoro-uridine, and similarly 5′-deoxy-D-ribo- and -D-arabino-nucleosides of 5-fluorocytidine, and 2′-deoxyribonucleosides of uracil, 5-bromo-uracil, and 5-fluoro-uracil from uridine or 5-fluoro-uridine, which on sequential treatment with acetyl bromide and tributyltin hydride gave the corresponding 2′-deoxynucleoside in high yield. Numerous syntheses have also been reported for arabinofuranosyl nucleoside analogues, prepared either conventionally from arabinofuranosyl derivatives or *via* 2,2′-anhydro-nucleosides obtained from appropriate ribonucleosides. 5-Aza-cytosine-D-arabinoside has been synthesized and found to show similar antiviral activity to Ara-C(arabinosyl-cytosine). 7-α- 7-β-, 9-α-, and 9-β-arabinofuranosyl derivatives of 3-deazaguanine have also been prepared, but none showed any anti-tumour activity. 9-(α-D-Arabinofuranosyl)-8-aza[2-^{14}C]-adenine, 7-(β-D-arabinofuranosyl)-pyrrolo[2,3-*d*]pyrimidine-4(3*H*)-one (15), 1-(α-D -arabinofuranosyl)- and 1-(β-D-xylofuranosyl)-4-nitropyrazole, and α-arabino-nucleosides of 5-fluoro-cytosine and -uracil derivatives have also been prepared. An improved synthesis of 9-(β-D-arabinofuranosyl)-2-fluoro-adenine has been reported. The ratio of a to β anomers obtained by phase-transfer reaction of 2,3,5-tri-*O*-benzyl-D-arabinofuranosyl bromide with 6-chloro-2-thiomethyl-7-deazapurine varied with the quaternary ammonium salt used as a catalyst, although the β-anomer predominated in every case. 2,2′-Anhydronucleosides have been used to prepare 1-β-D-arabinofuranosyl derivatives of 5-alkylthio-uracils, 5-ethyl-cytosine, and 5-ethyl -uracil, -alkylaminopurines, and 2 -aralkylamino - 1,4- dihydro -4 -imino -pyrimidine hydrochlorides (16). L-Arabino-furanosyl or -pyranosyl derivatives of 5-thio-substituted indole

and 5- and 6-nitro-indole have been prepared from L-arabinose by reaction with indoline derivatives followed by dehydrogenation of the indolinyl *N*-glycoside ; the nitro-derivatives were also reduced to corresponding amino-compounds. The dimeric nucleoside analogue (17) has been obtained from *NN*^{1}-bis(6-aza-5-uracilyl)ethylene diamine by reaction with 2,3,4,6-tetra-*O*-

acetyl-α-D-gluco-pyranosyl bromide followed by deacetylation. The course and scope of the sugar-exchange reaction of purine nucleosides has been investigated; the exchange, promoted by a glycosylhalide in presence of mercuric cyanide, appears to involve an *N*-7, *N*-9-bis-glycosylated purine intermediate, from which the ribose moiety is eliminated as a 1,5-anhydro-sugar, with the resulting *N*-7-glycoside undergoing an N-7 → N-9 transglycosylation, yielding a mixture of *N*-7- and *N*-9-nucleosides. The cyclization of glycosulamin derivatives continues to offer a useful route to nucleoside analogues: Condensation of the *N*-imidazolyl ribofuranosylamine (18) with aryloxy-carbonyl isocyanates gave the corresponding ribofuranosyl-imidazo[1,2-a] 1,3,5-triazenes (19) (Scheme 3).

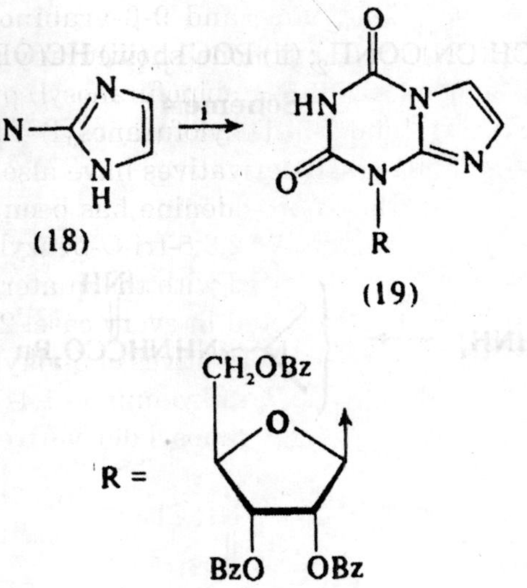

(18) (19)

R = (CH_2OBz, BzO, OBz)

Reagents: i, ArOCO(NCO)

Scheme 3

Likewise 3-hypoxanthosine (20) had been obtained by reaction of 2,3-O-isopropylidene-D-ribofuranose with 5-amino-4-carbamoyl-imidazole followed by condensation of the glycosylamine product with ethyl formate and deprotection; the 4-thio analogue was similarly prepared. The α-and β-D-mannofuranosyI-adenines have been prepared from αβ-D-mannofuranosyl-amine as outlined in Scheme 4, and guanine analogues were also synthesized. Scheme 5 summarizes a

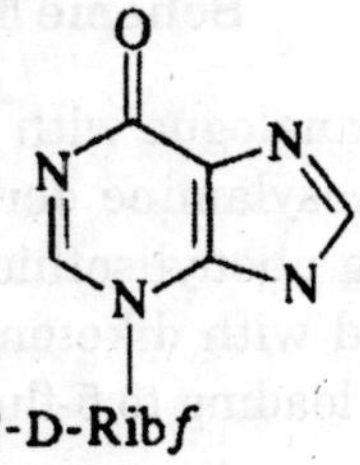

β-D-Rib*f*

(20)

synthesis of virazole (21) from 2,3,5-tri-*O*-benzyl-D-ribofuranosylhydrazine using a reagent derived from benzyl cyanoformate; alternatively the heterocycle could be formed first, and then condensed with peracetylated ribofuranose conventionally. Treatment of 2,3,4-tri-*O*-acetyl-D-

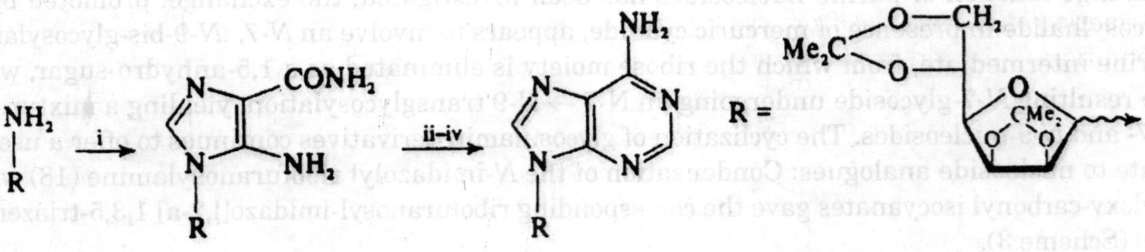

Reagents: (i) $EtOCH{=}NCH(CN)CONH_2$; (ii) $POCl_3$; (iii) $HC(OEt)_3$; (iv) NH_3

Scheme 4

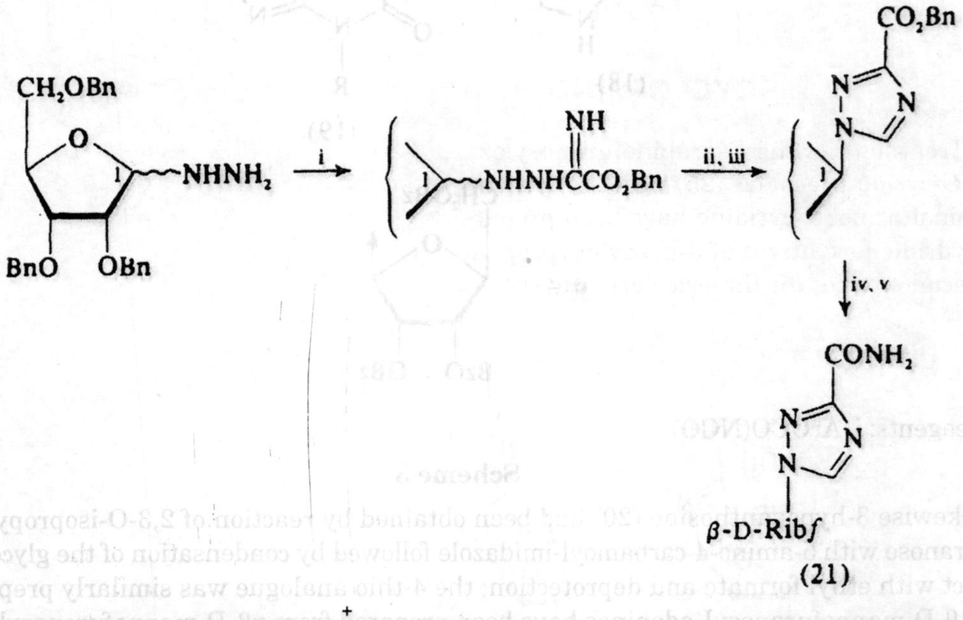

Reagents: (i) $BnOCOC(SEt){=}\overset{+}{N}H_2BF_4^-$; (ii) $HC(OEt)_3$; (iii) TsOH; (iv) NH_3; (v) H_2-Pd

Scheme 5

ribopyranosyl chloride or its furanosyl analogue with anthranilonitrile using a phase-transfer procedure gave the corresponding glycosylamine derivative, which could be cyclized to give quinazoline nucleoside analogues using phenylisothiocyanate. 2-Amino-oxazoline derivatives of pentofuranoses have been condensed with diketene to give nucleosides of 6-methyl-uracil (22), and with methyl [2-^{13}C]propynoate leading to 5-fluoro-2′-deoxy-[4-^{13}C]uridine.

Glycosyl isothiocyanates have been converted to the iv-glycosides of a series of heterocycles by similar condensation-cyclization reactions. 2,3,4,6-Tetra-*O*-acetyl-β-D-glucopyranosyl azide undergoes cycloaddition with dihalo-propynes or dihalo-butynes to give the corresponding triazole derivatives (23). Nucleoside Q has been synthesized by a procedure that attaches the cyclo-pentene ring to a 7-deaza-9-β-D-ribofuranosyl-purine derivative, and the natural product was thus shown to have the *3S,4R,5S* configuration in the cyclo-pentenyl substituent. The formation of nucleosides from cyanamide and potassium nitrite under possible primitive earth conditions has been investigated.

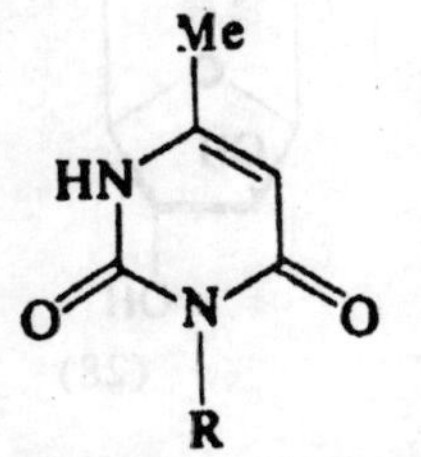

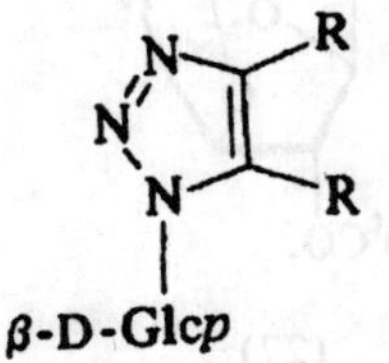

(22) R = β-D-Araf
α-D-Ribf

(23) R = CH_2Cl, CH_2Br, H

CYCLO- AND ANHYDRO-NUCLEOSIDES

Treatment of the D-arabinofuranosyloxazoline (24) with 1-chloro-l-cyano-ethene gave the 2,2′-*O*-cyclonucleosides (25) and (26) (Scheme 6). 3,5-Di-*O*-acyl derivatives of 2,2′ -*O*-cyclo-β-D-arabinofuranosyl-cytidine have been prepared by reaction of cytidine with the acid chloride or anhydride derivatives of 3-heptyloxypropanoic, 3-dodecyloxypropanoic, and 11-butyrylamino-undecanoic acid; the dimeric derivative (27) was also prepared.

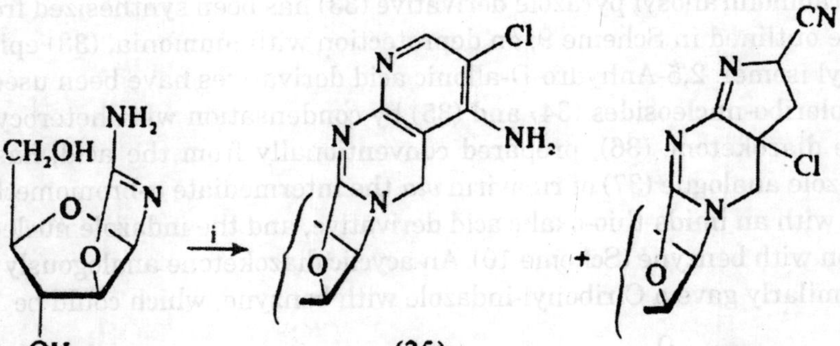

(24) (25) (26)

Reagent : (i) CH_2=CClCN

Scheme 6

Base-catalysed equilibration of 2,2′-*O*-cyclo-β-D-arabinofuranosyl-uracil with the 2,3′ -*O*-cyclo-β-D-xylofuranosyl isomer strongly favours the former compound; 2′,3′-anhydrouridine is an intermediate in the reaction, and with sodium methoxide in boiling methanol the *xylo* and *arabino* isomers of uridine could also be isolated, together with the 2-*O*-methyl-nucleoside (28). 8,2′- and 8,3′-(5′)-Cyclonucleosides have been prepared from 8-bromo-5′-deoxy-adenosine by sequential tosylation and treatment with sodium hydro-sulphide; desulphurization of these anhydro compounds gave the corresponding 2′,5′- and 3′,5′-dideoxyadenosines.

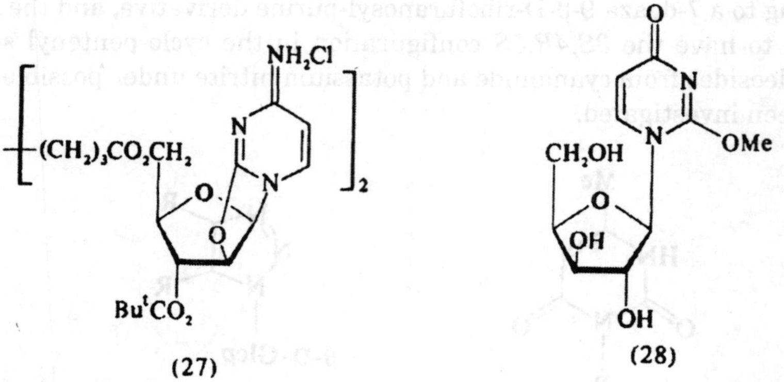

(27) (28)

A new class of 2,4′-*O*-cyclonucleoside (29) has been obtained as outlined in Scheme 7; 4′-substituted nucleosides prepared from this compound included the methoxy derivative (30), obtained using methanolic silver nitrate, and hence the uridine analogues (31) by reduction. The 2,5′-*O*-cyclonucleoside (32) has been prepared conventionally from the corresponding ribosyl-imidazole derivative. Anhydro-adenosine has been isomerized to 2′,3′-anhydro-β-D-lyxo-furanosyl-adenine in 42% overall yield by the sequence outlined in Scheme 8. Deamination of 1-β-amino-3-deoxy-β-D-xylofuranosyl)-uracil with nitrous acid did not yield any 2′,3′-anhydro-uridine or products derived from this anhydro-sugar derivative, uracil being the major product.

C-NUCLEOSIDES

The 3-β-D-arabinofuranosyl pyrazole derivative (33) has been synthesized from D-mannose by the procedure outlined in Scheme 9; on deprotection with ammonia, (33) epimerized to the β-D-ribofuranosyl isomer. 2,5-Anhydro-D-allonic acid derivatives have been used to synthesize the C-*sym*-triazoloribo-nucleosides (34) and (35) by condensation with heterocyclic hydrazine derivatives. The diazoketone (36), prepared conventionally from the acid, has been used to prepare the thiazole analogue (37) of ribavirin *via* the intermediate α-bromomethyl ketone (38) by condensation with an imido-thio-oxalic acid derivative, and the indazole nucleoside analogue (39) by cyclization with benzyne (Scheme 10). An acyclic diazoketone analogously prepared from D-ribonic acid similarly gave a C-ribonyl-indazole with benzyne, which could be

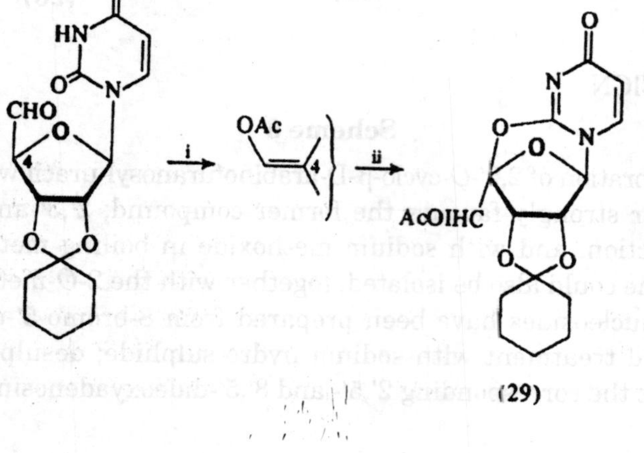

(29)

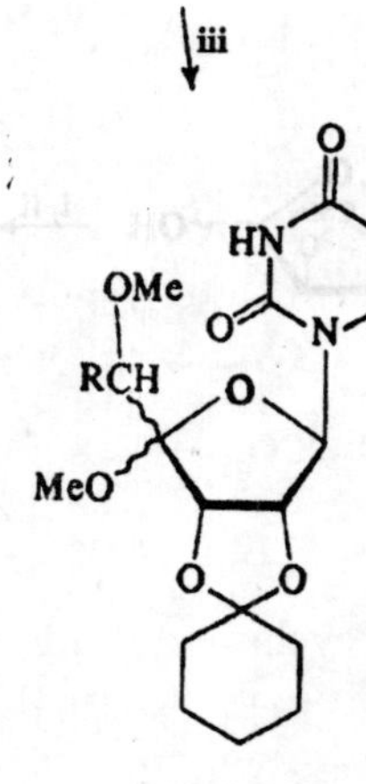

(30) R = OH
(31) R = H

Reagents: (i) Ac_2O-Et_3N-p-$Et_2NC_5H_4N$; (ii) NIS-NaOAc; (iii) MeOH-$AgNO_3$

Scheme 7

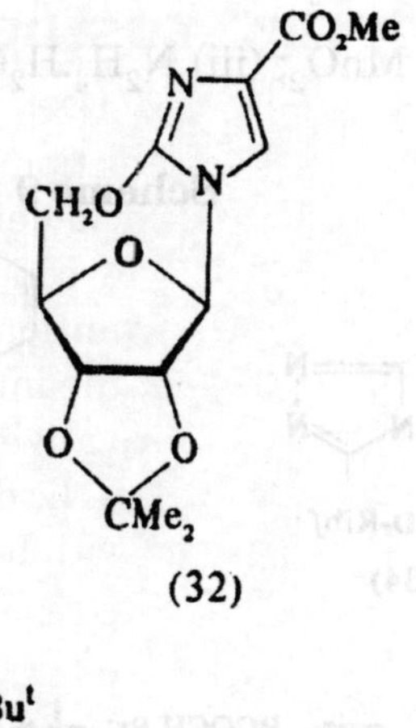

(32)

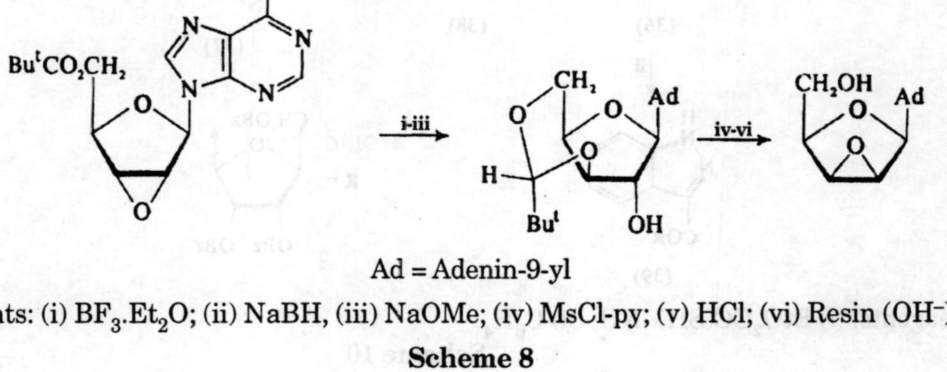

Ad = Adenin-9-yl

Reagents: (i) $BF_3.Et_2O$; (ii) NaBH, (iii) NaOMe; (iv) MsCl-py; (v) HCl; (vi) Resin (OH^-)

Scheme 8

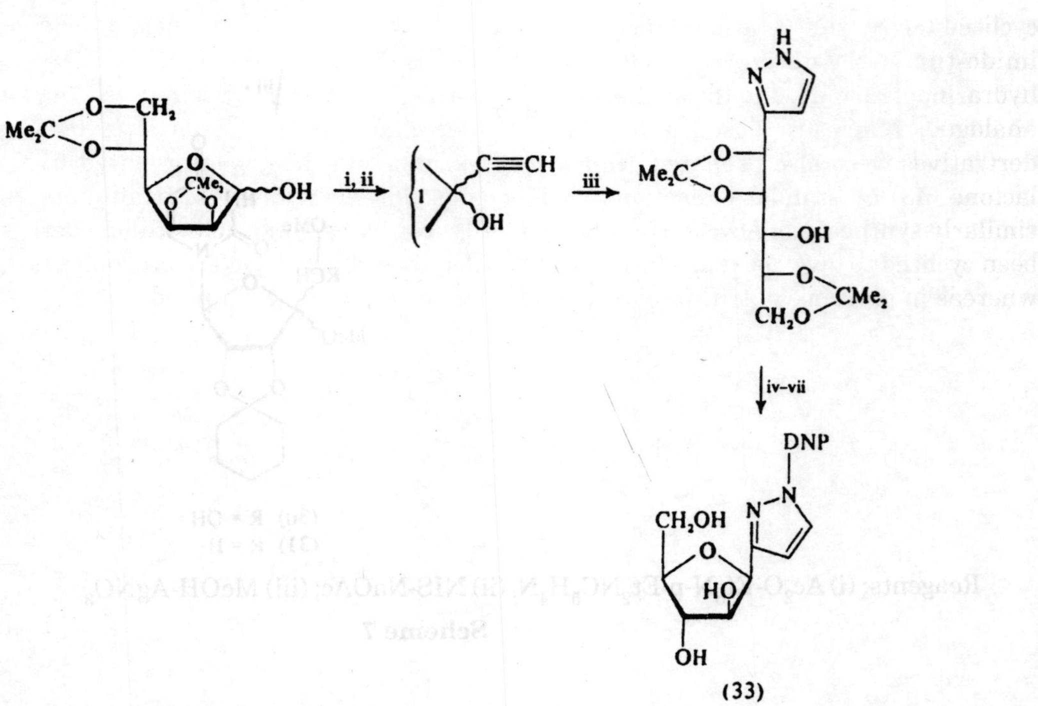

(33)

Reagents: (i) HC≡CMgBr; (ii) MnO_2; (iii) $N_2H_4.H_2O$; (iv) Me_2CO-H^+; (v) DNPF-Et_3N; (vi) MsCl-py; (vii) $BF_3.Et_2O$

Scheme 9

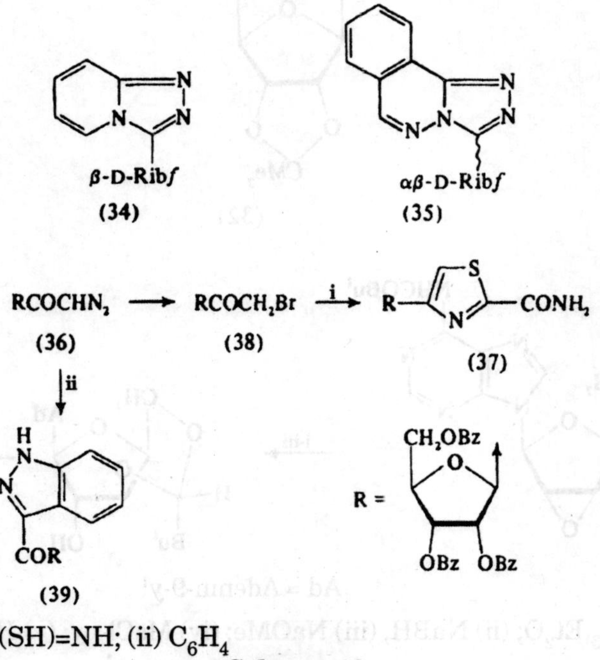

β-D-Rib*f*
(34)

αβ-D-Rib*f*
(35)

RCOCHN₂ (36) → RCOCH₂Br (38) →(i) (37)

(39)

R =

Reagents: (i) $EtO_2CC(SH)$=NH; (ii) C_6H_4

Scheme 10

cyclized to give the corresponding C-nucleoside (40) (with some α-anomer). Condensation of an imido-thiolester derivative of 2,5-anhydro-D-allonic or D-altronic acid with 2-chloro-3-hydrazinopyrazine led to the synthesis of the triazolo-pyrazine-C-nucleoside (41) and its 2′-deoxy analogue, which are closely related to adenosine and formycin, and an extensive series of derivatives were also prepared. Homopseudo-uridine (42) has been prepared from the chiral lactone (43) by standard reactions, and homopseudo-isocytidine (44) (tautomer shown) was similarly synthesized. Acyclic D-arabino-tetrahydroxybutyl-imidazolinethione derivatives have been cyclized to give D-erythrofuranosyl C-nucleosides (45); β-anomers predominate on heating, whereas in presence of trifluoroacetic acid, α-anomers could be obtained.

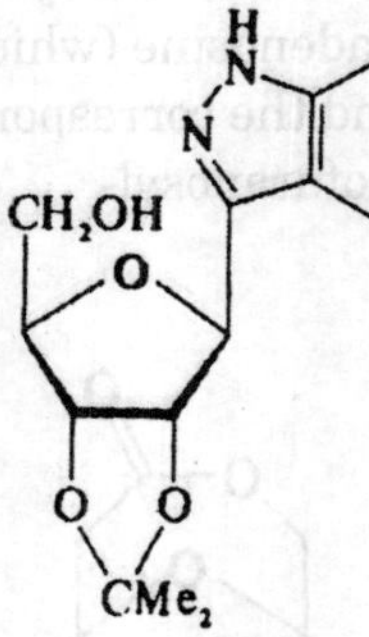

(40)

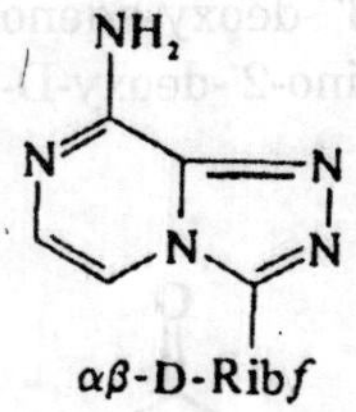

αβ-D-Ribf

(41)

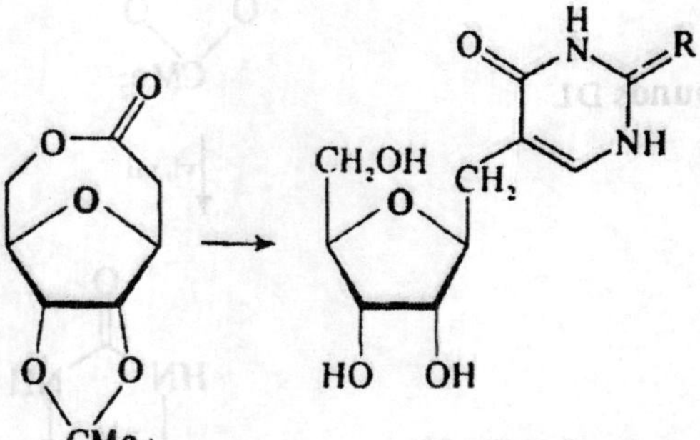

(43)

(42) R = O

(44) R = NH

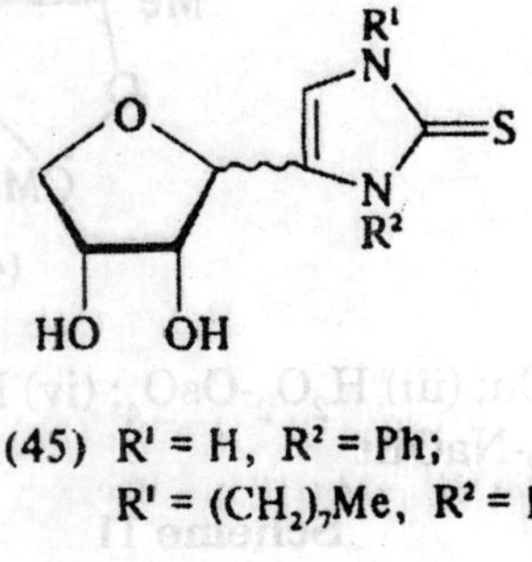

(45) $R^1 = H$, $R^2 = Ph$;

$R^1 = (CH_2)_7Me$, $R^2 = H$

The reaction of 1,1,3,3-tetrabromopropanone with furan derivatives has been used to prepare C-methyl analogues (46) and (47) of pseudo-uridine (Scheme 11), and the related C-hydroxymethyl analogues (48) were similarly prepared.

AMINO-SUGAR NUCLEOSIDES AND RELATED COMPOUNDS

Syntheses of nucleosides containing amino-sugars either utilize preformed amino- or azido-sugars in reaction with heterocyclic bases, or modify preformed nucleosides. Syntheses involving standard condensation procedures for heterocyclic bases with amino-sugar derivatives have included the preparation of 2′-azido-, 3′-azido-2′,5′-diazido-, and 3′,5′ -diazido-derivatives of *arabino*-uridine 3′-amino-3′-deoxy-adenosine, -uridine, and cytidine 5′-phosphates, 2′,3′-bis(2-chloroethyl)-aminophosphoryl-3′-amino-3′-deoxyadenosine (which has anti-tumour activity), 3′-*N*-methyl-*N*-nitrosoureido-3′ -deoxy-adenosine and the corresponding 5′-substituted isomer and 2′-azido-2′-deoxy- and 2′-amino-2′-deoxy-D-arabinofuranosyl-

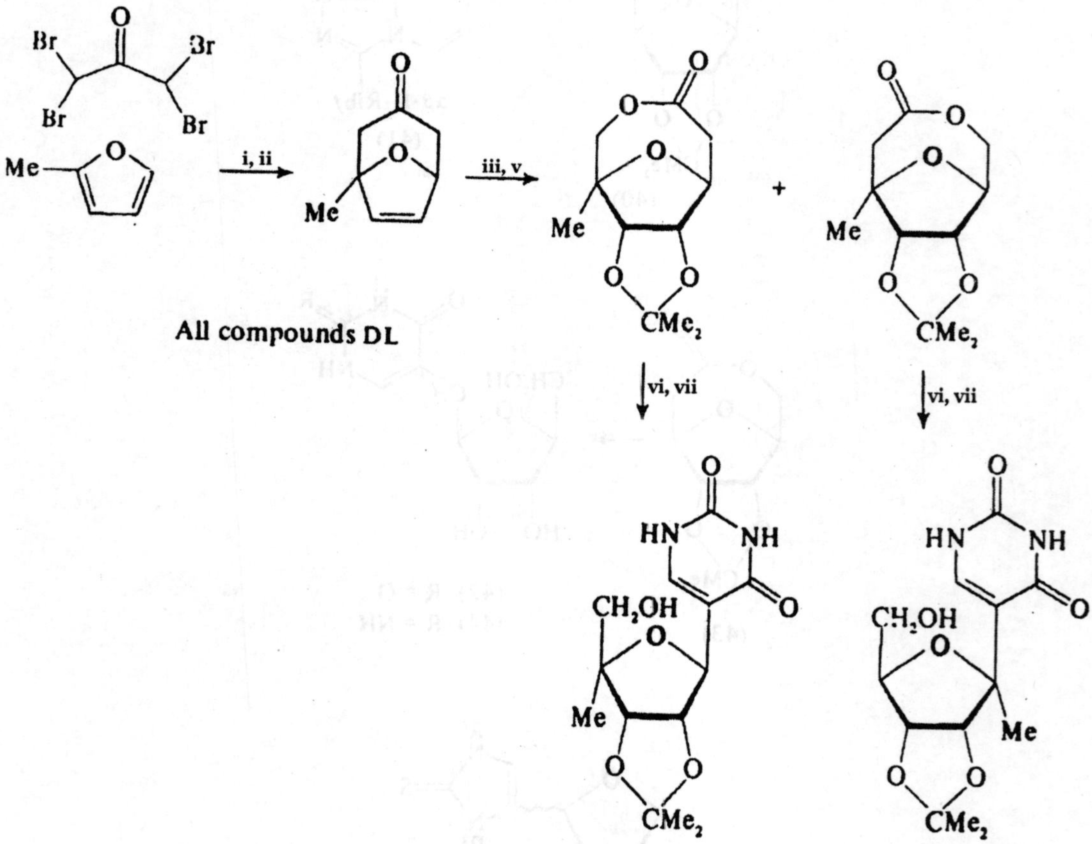

Br Br Br Br O Me O i, ii O O Me iii, v O O O Me O O CMe₂ + O O O Me O O CMe₂ All compounds DL vi, vii vi, vii

(46) (47)

Reagents: (i) $Fe_2(CO)_9$; (ii) Zn-Cu; (iii) H_2O_2-OsO_4; (iv) Me_2CO-$CuSO_4$-TsOH; (v) CF_3CO_3H; (vi) $Bu^tOCH(NMe_2)_2$; (vii) $CO(NH_2)_2$-NaOEt

Scheme 11

adenine and -guanine. Alternatively nucleosides have been modified by azide-displacement procedures to give 2-amino-2-deoxy-β-D-arabinofuranosyl nucleosodes and l-(5-amino-5-deoxy-β-D-arabinofuranosyl-)-5-substituted-uracils.

A convenient conversion of 2′-azido-2′-deoxy-uridine to 2′-amino-2′-deoxy-ribofuranosyl purines has been described. The conversion of the 3-azido-3-deoxyarabinofuranosyl uracil derivative (49) to 3′-amino-3′-deoxy-uridine involves the cyclonucleoside (50), which is considered to arise by oxygen attack on an aziridine intermediate formed during the triphenylphosphine reduction step (Scheme 12). Catalytic reduction of the 3′-azido-3′-deoxy-2′-*O*-mesyl derivative prepared from 2′,3′-anhydro-adenosine yielded an 2′,3′-aziridine of inverted *(lyxo)* configuration (Scheme 13), and the anhydro-lyxose isomer likewise gave the inverted ribo-aziridine; these aziridines showed unusual c.d. and ^{1}H n.m.r. spectra.

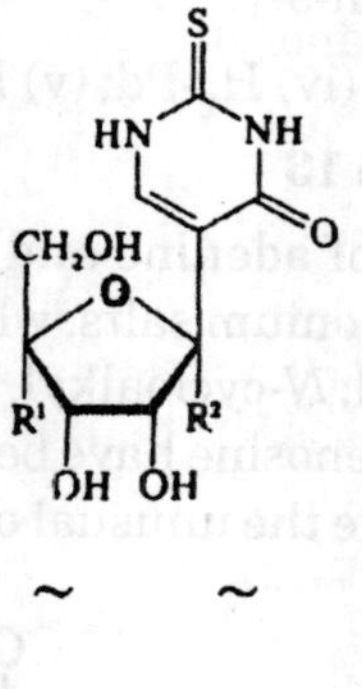

(48) $R^1 = CH_2OH$, $R^2 = H$; $R^1 = H$, $R^2 = CH_2OH$

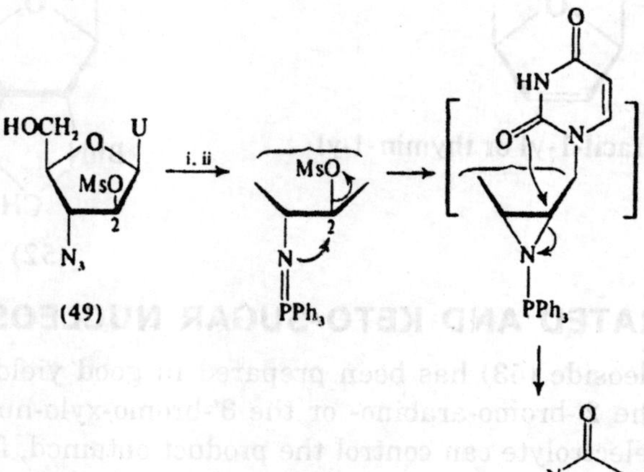

~ ~ ~ ~ ~ ~ ~ ~

(49) (50)

U = Uracil-1-yl

Reagents: (i) PPh_3-py; (ii) NH_4OH

Scheme 12

Treatment of 2′-deoxy-3′,5′-di-*O*-mesyl-uridine or -thymidine with aqueous alkali yielded the corresponding 3′,5′-oxetan derivative, which with alkali in hexamethylphosphortriamide gave the 2′,3′-unsaturated nucleoside, and this was then converted to the corresponding 5′-amino derivative (51) by a standard sequence.

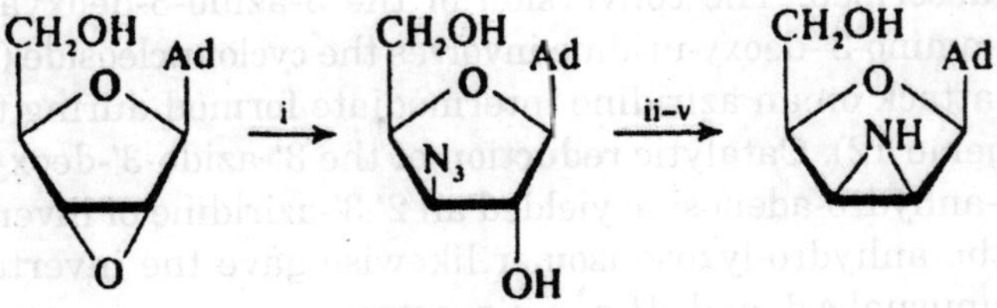

CH_2OH Ad O → (i) CH_2OH Ad O N_3 OH → (ii–v) CH_2OH Ad O NH

Ad = Adenin-9-yl

Reagents: (i) LiN_3; (ii) Bi[t]COCl-py; (iii) MsCl-py; (iv) H_2-Pd; (v) NaOMe

Scheme 13

Treatment of 5′-deoxy-5′-halogeno nucleosides of adenine and uracil with trimethylamine yielded the corresponding 5′-deoxy-5′-trimethylammonium salts, whose interaction with anionic polynucleotides was then examined. Various *N*-alkyl, *N*-cycloalkyl, and *N*-alkaryl derivatives of 3′-amino-3′-deoxy-and 3′,5′-diamino-3′,5′-dideoxy-adenosine have been prepared from the amino-sugar nucleosides; the 3′-benzylamino derivative gave the unusual oxazolidine derivative (52).

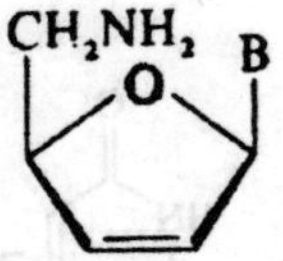

CH_2NH_2 O B

(51) B = Uracil-1-yl or thymin-1-yl

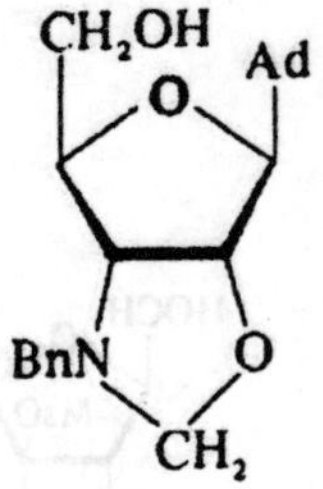

CH_2OH O Ad BnN O CH_2

(52)

UNSATURATED AND KETO-SUGAR NUCLEOSIDES

The unsaturated nucleoside (53) has been prepared in good yield by electro-chemical elimination from either the 2′-bromo-arabino- or the 3′-bromo-xylo-nucleoside indicated in Scheme 14; however, the electrolyte can control the product obtained, for whereas 2′-bromo-2′-deoxy-3′,5′-di-*O*-propanoyl-uracil gave the 2′,3′-unsaturated nucleoside in presence of tetraethylammonium toluene-*p*-sulphonate in methanol, with sodium acetate a mixture of products was obtained; 3′-deoxy-3′-iodo-adenosine yielded 3′-deoxy-adenosine (cordycepin) besides the 2′,3′-unsaturated nucleoside. 1-Acetyl-glycenose derivatives have been condensed with purine and pyrimidine derivatives in presence of antimony pentachloride to give αβ-mixtures of nucleosides, *e.g.*, (54). The 4′,5′-unsaturated nucleosides (55) and (56) have been prepared by treatment of corresponding 5′-deoxy-5′-iodo-nucleosides with DBU. The Wittig reaction has been used to prepare unsaturated nucleosides, *e.g.*, (57), from the dialdose derivative obtained by Pfitzner-Moffatt oxidation of 2′,3′-*O*-isopropylidene-uridine; the *N*-phenyl succinimido derivative (58) showed significant antiviral activity.

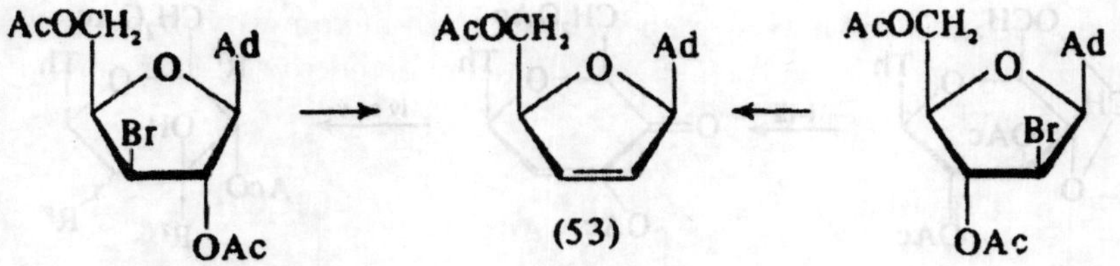

AcOCH$_2$ Ad Br OAc → (53) ← AcOCH$_2$ Ad Br OAc

Scheme 14

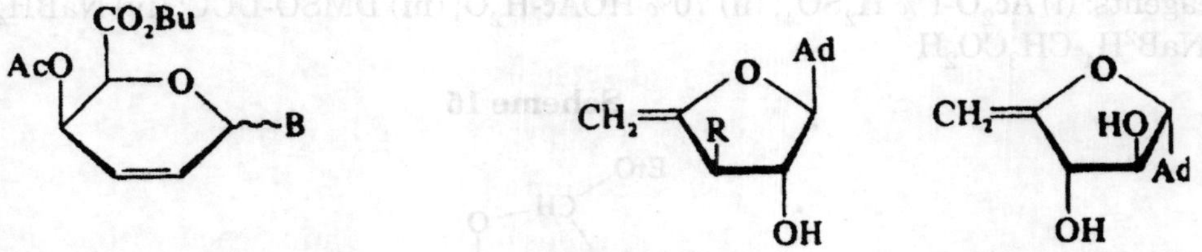

CO_2Bu AcO B

CH_2 R Ad OH (55) R = H, OH

CH_2 HO Ad OH (56)

Ad = Adenin-9-yl

X=CH U O O CMe$_2$

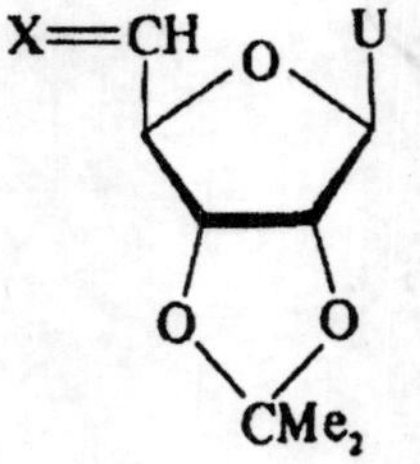

(57) X = CHCOMe or $CHCO_2Me$

(58) X = [N-phenylsuccinimide-3-ylidene] NPh

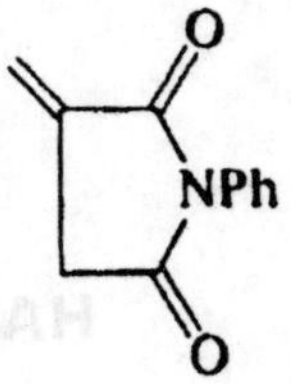

U = Uracil-1-yl

On acetolysis, the 4,6-*O*-ethylidene nucleoside analogue (59) gave the corresponding 2,3,6-tri-*O*-acetyl derivative, which, on Pfitzner-Moffatt oxidation, yielded the unsaturated keto nucleoside (60). The reduction of this intermediate to 2′-deoxy-D-*arabino*-hexo-nucleosides using sodium borohydride in acetic acid was also investigated using deuterium-labelled reagents; sodium borohydride in perdeuteriated acetic acid gave the 2′-deuterio compound (61), whereas sodium borodeuteride in acetic acid gave the 3′,4′-dideuterio-labelled derivative (62) (Scheme 15).

A useful method for preparing enol-acetates from uracil 4′-keto-nucleosides uses dimethyl formamide diethyl acetal followed by acetic anhydride; 2′-keto isomers, or N-3-substituted analogues, did not react. Further studies supported a mechanism involving O-4 esterification of the nucleoside followed by enolization, which is intramolecularly catalysed by the dimethylamino group, as illustrated in (63).

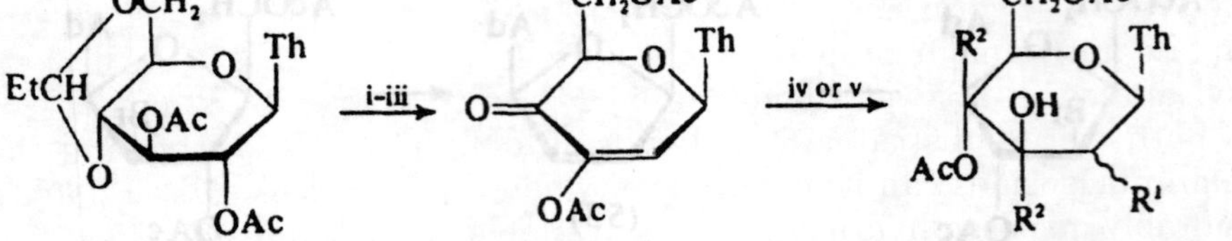

(59) (60) (61) $R^1 = {}^2H$, $R^2 = H$
(62) $R^1 = H$, $R^2 = {}^2H$
Th = Thymin-1-yl

Reagents: (i) Ac_2O-1 % H_2SO_4; (ii) 70% HOAc-H_2O; (iii) DMSO-DCC; (iv) $NaBH_4$-$C^2H_3CO_2{}^2H$; (v) NaB^2H_4-CH_3CO_2H

Scheme 15

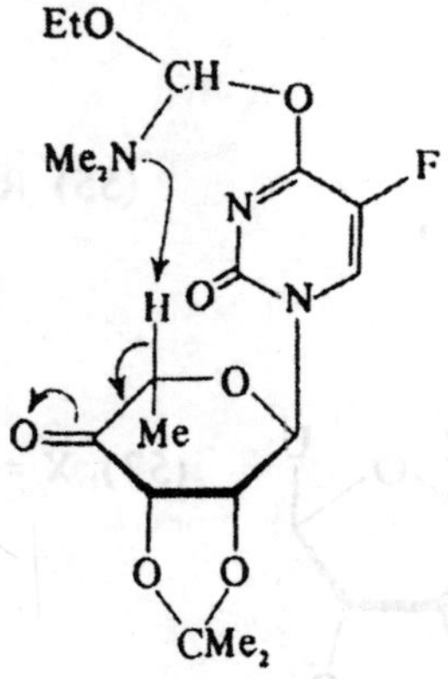

(63)

HALO-NUCLEOSIDES

4′-Deoxy-4′-fluoro-D-glucopyranosyl thymine and 2′-deoxy-2′-fluoro-D-arabino-furanosyl derivatives of 5-substituted-cytosine and -uracil have been prepared by standard methods from the corresponding fluoro-sugar; the uracil derivatives stopped replication of Herpes simplex virus (HSV-1) *in vitro.* 5′-Deoxy-5′-iodo-adenosine derivatives have also been prepared from iodo sugars.

Treatment of nucleosides with thionyl chloride in HMPT gave 5′-chloro-5′-deoxy-2′,3′-*O*-sulphinyl derivatives, which could be converted to 5′-chloro-5′-deoxy-nucleosides on treatment with sodium methoxide in methanol, or to 2,2′-anhydro-nucleosides with imidazole in DMF.

Likewise 2′-deoxy-adenosine yielded 5′-chloro-2′,5′-dideoxy-adenosine on treatment with a limited amount of the thionyl chloride reagent followed by methanolic ammonia, with thionyl chloride in pyridine, 2′,5′- and 3′,5′-di-*O*-acyl or ether derivatives of adenosine yielded 3′- or 2′-chloro-deoxy analogues with inversion of configuration; the acyl derivatives of these chlorinated sugars gave 2′,3′-anhydro-adenosine with base, and the corresponding deoxy-adenosines with tributyltin hydride; β-D-xylofuranosyladenine with thionyl chloride in HMPT gave the corresponding 5′-chloro-5′-deoxy-nucleoside, not the 2′-isomer previously suggested.

BRANCHED-CHAIN NUCLEOSIDES

9-(2,3-Dideoxy-3-*C*-(hydroxymethyl)-β-D-*erythro*-pentofuranosyl)-4-thio-guanine (64) has been prepared, and its anti-tumour properties investigated. Likewise, the 3-C-hydroxymethyl-β-D-xylofuranosyl analogues (65) of adenosine have been prepared by standard condensation procedures. Both were substrates for adeno-sine aminohydrolase, showing that the C-3′ branched-chain nucleosides can be deaminated when the branch is *trans* related to the base provided a suitably disposed hydroxy-group is available on the *endo-side* of the furanose ring. The isomeric 4′-*C*-hydroxymethyl-adenosine (66) and analogous derivatives of other purine and

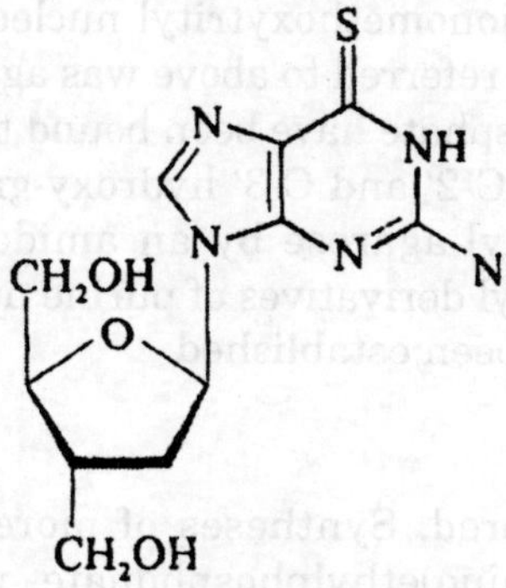

(64) 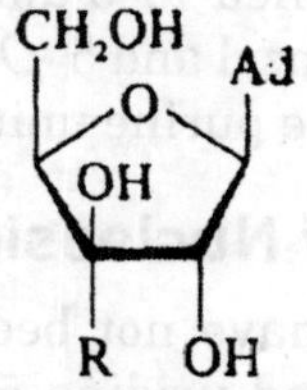 (65) R = Me or CH_2OH 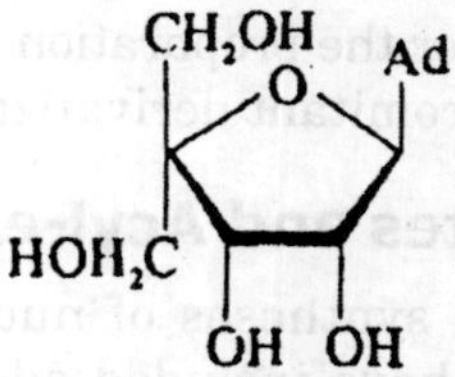 (66)

pyrimidine nucleosides have also been synthesized either conventionally from the peracetyl-branched-chain sugar or by base-catalysed reaction of the nucleoside 5′-aldehydes with formaldehyde; in the latter case, epimerization occurred at C-3′ if the vicinal hydroxy -groups at C-2′ and C-3′ were unprotected. l-(β-D-Apiofuranosyl)-5-fluoro-uracil has been synthesized as part of a study of the ^{13}C n.m.r. spectra of apiosyl pyrimidine nucleosides, and the mass spectra of peracetylated apiosyl-adenine and uracil nucleosides have been compared with ribosyl analogues.

ETHER AND ACETAL DERIVATIVES

Bis-chlorodi-isopropylsiloxane has been recommended as a reagent for forming 3′,5′-disilyl derivatives of nucleosides, *e.g.*, (67); 5′-monomethoxytritylated nucleosides gave the corresponding 2′,3′-cyclic ether, from which the trityl ether group could be selectively removed using toluene-*p*-sulphonic acid in dioxan. These reactions provide useful routes to 2′- or 5′-mono-substituted nucleossides.

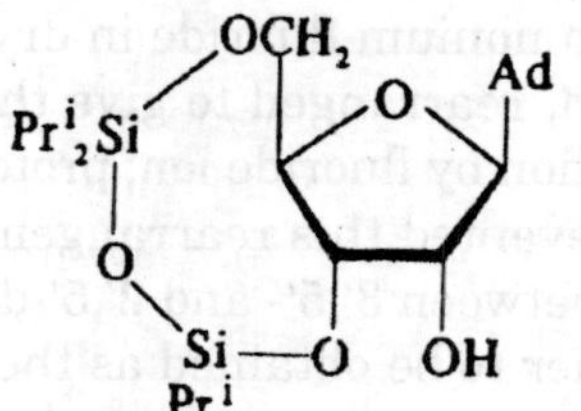

(67)

t-Butyldimethylsilyl ethers continue to provide valuable blocking groups in nucleoside chemistry. Various mono- and di-silyl ethers of cytidine and guanosine have been prepared, and mono-ethers from 5′-*O*-monomethoxytrityl nucleosides provide useful routes to 2′- or 3′-substituted nucleoside derivatives, including 2′,5′-nucleotides, since they are readily separated from each other, and can also be equilibrated in methanol. The 2-*O*-t-butyldimethyl-silyl derivative of l,3,5-tri-*O*-benzoyl-α-D-ribofuranose has also been used to prepare nucleosides, which may be debenzoylated to give derivatives suitable for the synthesis of nucleotides; the silyl ether group is readily removed using potassium fluoride with 18-crown-6 ether in THF.

Treatment of 4-*N*-benzoyl-cytidine with varying ratios of t-butyldimethylchlorosilane in presence of imidazole allowed the synthesis of all the possible ethers having the 5′-position substituted; the remaining ethers were made from the 5′-monomethoxytrityl nucleoside, and the base-catalysed equilibration of the 2′- and 3′-mono ethers referred to above was again noted. Tubericidin and nebularin and also 6-azauridine-5′ -monophosphate have been bound to polymer supports through cyclic ketal linkages formed between the C-2′ and C-3′ hydroxy-groups and laevulinic acid, which in turn is attached to 6-aminohexyl-agarose by an amide linkage. Conditions for the preparation of 5′-*O*-benzyl and 5′-*O*-glycosyl derivatives of purine nucleosides without concomitant derivatization of the purine unit have been established.

Phosphates and Acyl-esters of Nucleosides

Routine syntheses of nucleotides have not been covered. Syntheses of more unusual derivatives have included adenosine and uridine 5′-(1-amino)ethylphosphonate, which are unaffected by alkaline phosphatase, and 5-fluoro-2′-deoxy-uridine 5′-*p*-azidophenylphosphate, used as a photo-affinity label for thymidylate synthetase. New phosphate ester derivatives of nucleosides useful for oligonucleotide synthesis have included 3′-(2-cyanoethyl, 5-chloro-8-quinolinyl)phosphates and 2,2,2-tribromoethyl, 2,2,2-trichloroethyl phosphates, the latter being selectively dealkylated by electrochemical reduction to give the corresponding 2,2,2-trichloroethyl phosphates.

A one-flask method for the synthesis of 5′-phosphordiamidates of nucleosides has been described, and a new direct synthesis of phosphate-blocked nucleotide derivatives of arabinosylcytosine reported. Thymidine has been selectively phosphorylated at the 5′-position using SS-diaryl phosphorodithioates in good yield, the protecting thiol ester groups being readily removed by silver acetate in aqueous pyridine.

3′- and 5′-Mononucleotides can be readily distinguished by their e.i. mass spectra; in all cases the m/e 501 ion was far more abundant in the 3′-isomer. 2′-3′- and 3′ ,5′-Di-*O*-acylderivatives of 9-(β-D-arabinofuranosyl)adenine have been prepared and evaluated as potential antiviral agents; in presence of tetrabutylammonium fluoride in dry THF, the 2′, 3′-di-*O*-acyl, 5′-*O*-silyl ether derivatives initially prepared, rearranged to give the 3′,5′-di-*O*-acyl nucleoside by 2′ → 5′-acyl migration following desilylation by fluoride ion; protonation of the first formed O-5-anion by a solvent such as acetic acid prevented this rearrangement. The regioselective deacylation of ribonucleosides is equilibration between 3′,5′- and 2′,5′-di-*O*-acyl derivatives can be achieved on silica gel allowing the 3′,5′-isomer to be obtained as the major product on silica-gel column chromatography.

MISCELLANEOUS NUCLEOSIDES

A further report on the reactions of 5′-azido-5′-deoxy uridine describes its cyclization in presence of DDQ to give the bridged compounds (68) or (69).

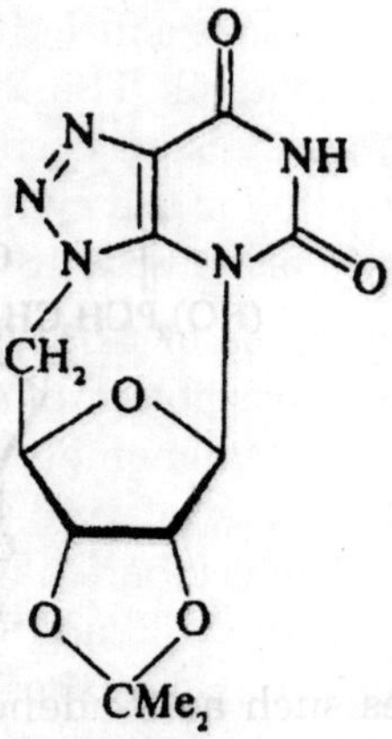

(68) 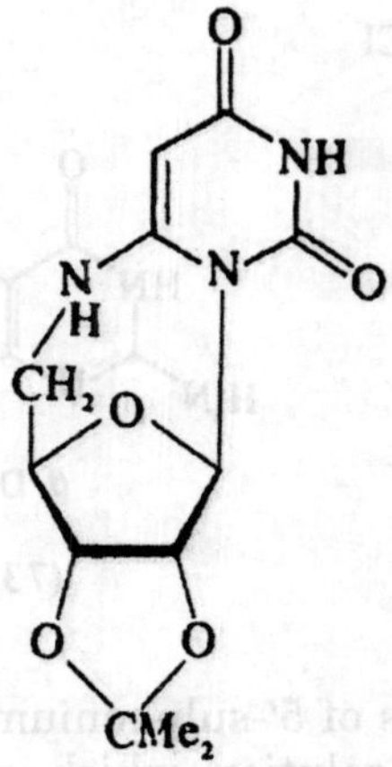 (69)

Treatment of the acetylenic nucleoside (70) with dipolar reagents gave the double-headed nucleoside analogues (71) and (72) (Scheme-16). The double nucleoside 8,8′-biguanosine (73) has been prepared by irradiation of guanosine with its 8-bromo-derivative, and its crystal structure was also determined.

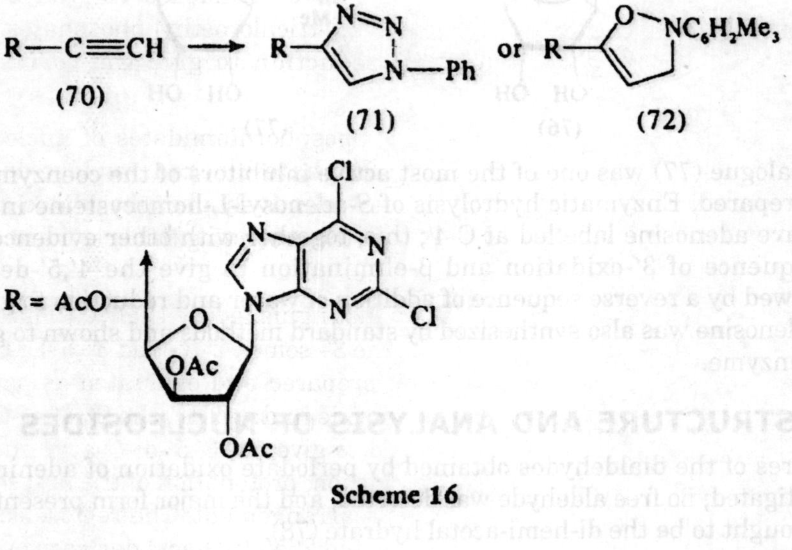

(70) (71) (72)

Scheme 16

2-Seleno-cytidine gives the novel 2′-cyclo-2-selenocytidine (74) on treatment with 2-acetoxyisobutyryl chloride.

A Wittig condensation has been used to prepare the phosphonate analogue (75) of 2′-deoxy-5-fluoro-uridylic acid from the corresponding 5′-aldehyde nucleoside; the compound is a potent inhibitor of thymidylate synthetase.

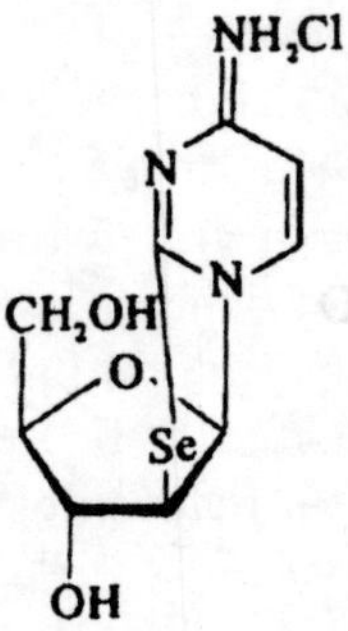

(74) 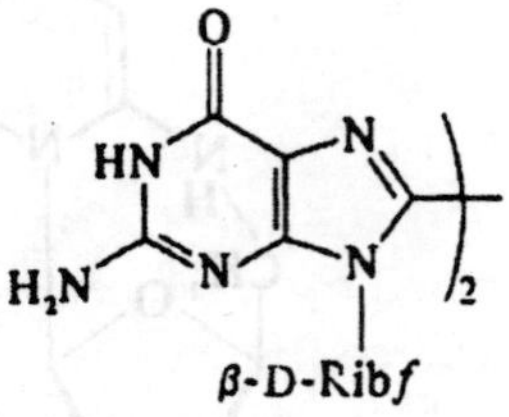 (73) 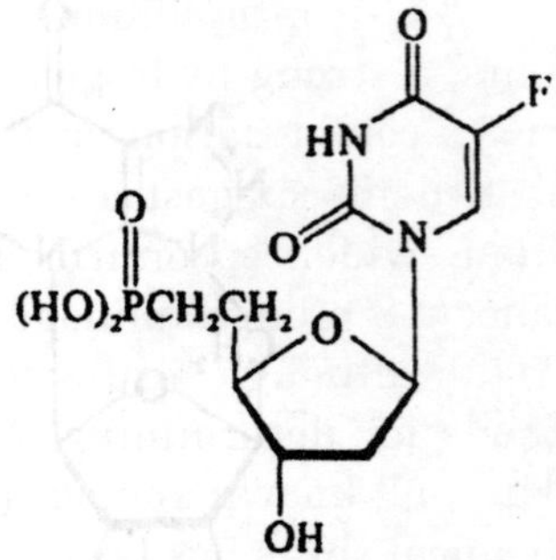 (75)

Glycoside hydrolysis of 5′-sulphonium nucleosides such as *S*-adenosyl-methionine occurs very readily in alkaline solution, which is rationalized by a mechanism involving H-5′ proton abstraction and ring opening illustrated in structure (76). A range of analogues of S-adenosyl-homocysteine have been prepared, with varying substituents at S-5′, O-2′, O-3′, and N-6; the

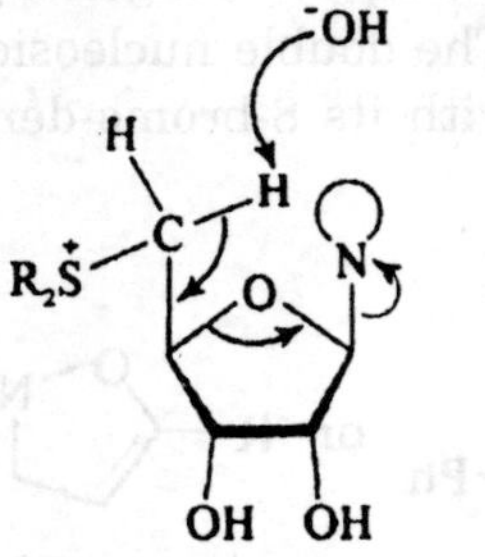

(76) 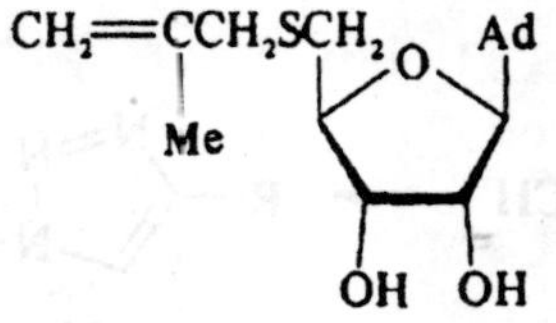 (77)

S-isobutenyl analogue (77) was one of the most active inhibitors of the coenzyme *S*-adenosyl-L-methionine prepared. Enzymatic hydrolysis of *S*-adenosyl-L-homocysteine in deuterium or tritium oxide gave adenosine labelled at C-4′; this, together with other evidence, suggested a mechanistic sequence of 3′-oxidation and β-elimination to give the 4′,5′-dehydro-3′-keto nucleoside, followed by a reverse sequence of addition of water and reduction to give adenosine. 4′,5′-Dehydro-adenosine was also synthesized by standard methods and shown to give adenosine with the same enzyme.

STRUCTURE AND ANALYSIS OF NUCLEOSIDES

The structures of the dialdehydes obtained by periodate oxidation of adenine nucleosides have been investigated; no free aldehyde was detected, and the major form present in crystalline products was thought to be the di-hemi-acetal hydrate (78).

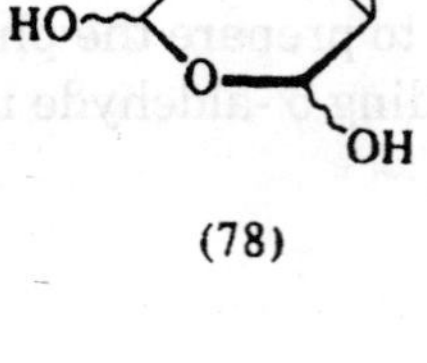

(78)

The conformation of nucleosides and nucleotides has been reviewed. The conformation of 2-alkylthio-2′,3-*O*-isopropylidene-adenosine derivatives has been studied by i.r. and c.d. measurements; a strong hydrogen bond exists between *N*-3 and *O*-5′ in carbon tetrachloride, indicating a *syn* conformation for the base; c.d. curves, which are markedly dependent on the sulphur substituent, suggest a *syn* conformation in chloroform and an *anti* arrangement in water. H N.m.r. evidence, confirmed by an *X-ray* crystal structure study, indicates that 9-β-D-arabinofuranosyl-8-n-butylaminoadenine adopts the *anti*-conformation in dimethyl sulphoxide and the crystal, although 8-substituted adenosines are usually expected to be *syn* owing to steric. Methods for determining the anomeric configurations of ribonucleosides have been reviewed. The well known anomeric-proton rule for determining anomeric configuration from ^{1}H n.m.r. chemical shifts has been supplemented by the rule, which states that the heterocyclic proton on the carbon next to the glycosidically linked nitrogen is downfield in the isomer in which the base and the C-2′-hydroxy-group are *trans* to each other relative to the same proton in the *cis* isomer. The rule holds for unprotected and acetalated nucleosides, but not for acyl or benzyl derivatives.

The ^{13}C n.m.r. spectra of inosine, 1-methylinosine, and 6-methoxypurine riboside have been completely assigned. Quantum mechanical calculations on some imidazole and benzimidazole nucleosides do not predict the conformations observed in the crystal, the difference being attributed to stacking forces and intermolecular hydrogen bonds in the latter case. An *ab initio* molecular orbital study of adenosine has been reported.

CHAPTER 6

Antibiotics

New antibiotics reported this year have included several groups of closely related pseudo disaccharides; fortimicin C, a carbamoyl derivative of fortimicin A (1), fortimicin D, the 6′-demethylated analogue of fortimicin A, and fortimicin KE, the 6′-demethylated derivative of fortimicin B (2), have been isolated from the culture broth of Mcromcwospora o/ivoasferospora, which also yields fortimicin A and B and sporaricin A (3, $R^2 = NH_2CH_2CO$-) and sporaricin B (3, $R^2 = H$) have been obtained from the culture broth of *Saccharopolyspora hirsuta;* these are all 1,4-diamino-cyclitol α-glycosides of 6-*epi*-purpurosamine B.

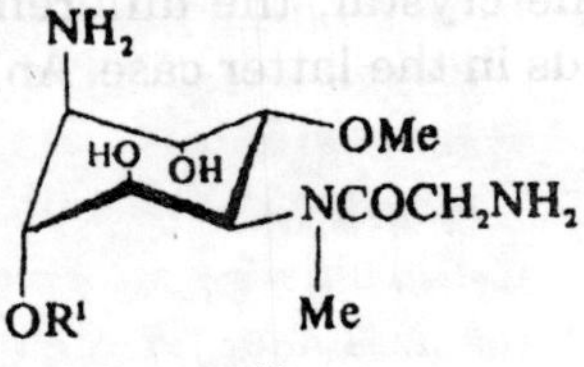

(1) 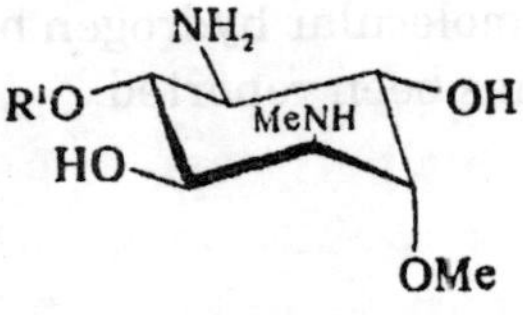 (2) 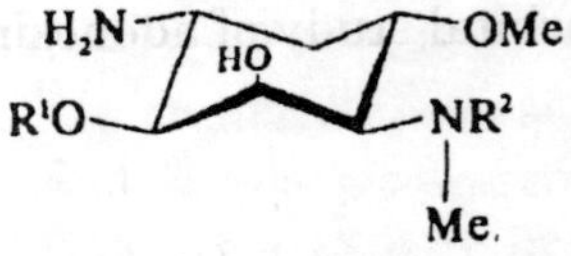 (3)

R^1 =

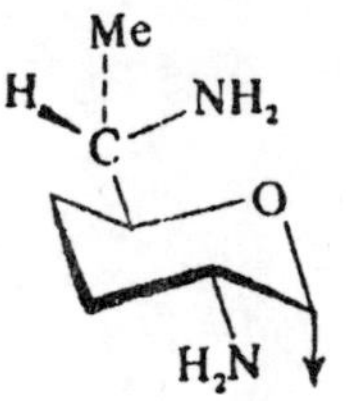

Istamycin A (4) and istamycin B (5) have been isolated from the culture broth of *Streptomyces tenjimariensis,* and sannamycin A (identical with istamycin A) and sannamycin B (6) from the culture broth of *Streptomyces sannanensis*; these are likewise 1,4-diamino-cyclitol α-glycosides of 6-*N*-methyl-purpurosamine C. Derivatives of the amino-sugar and cyclitol constituents of fortimicin B have been prepared by benzyloxycarbonylation and alcoholysis of the antibiotic.

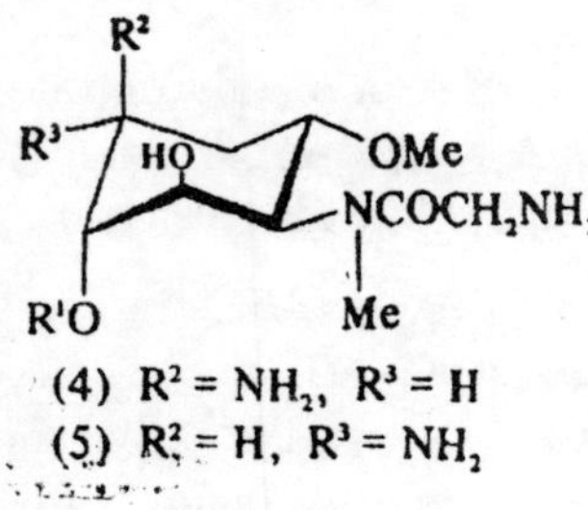

(4) $R^2 = NH_2$, $R^3 = H$

(5) $R^2 = H$, $R^3 = NH_2$

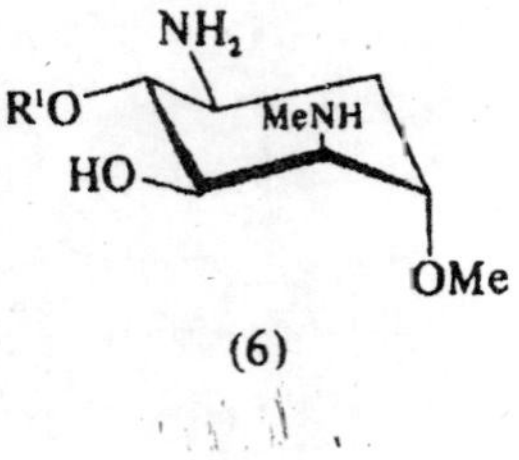

(6)

R^1 =

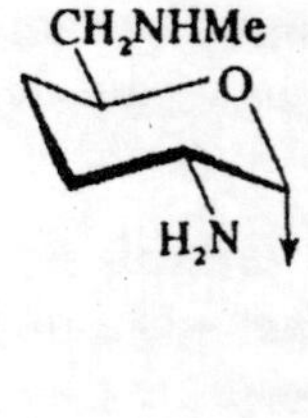

The structure (7) has been proposed for the heteropentasaccharide, virido-pentaose B, isolated from the antibiotic sporoviridin, which also contains two other pentasaccharides. A tetrasaccharide and a trisaccharide (8) composed of quinovose and viosamine were obtained by partial methanolysis of the pentaose (7) by removal of the 2-deoxy sugar units, the latter being characterized as the 3-amino-sugar, D-acosamine.

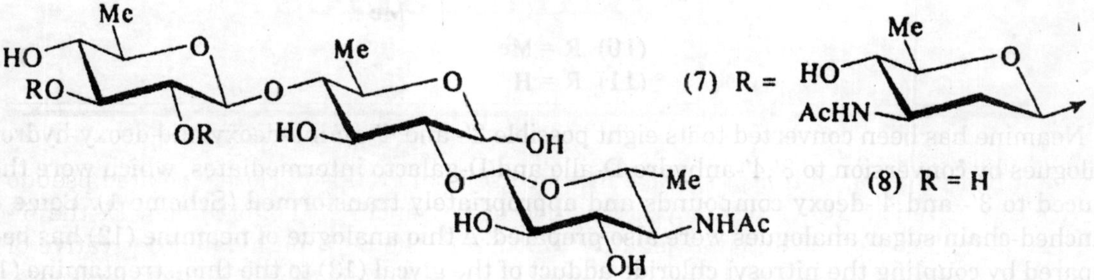

(7) R =

(8) R = H

Recent advances in the synthesis of aminoglycoside antibiotics have been reviewed. Derivatized dihydrostreptomycin has been hydrolysed to give a chiral streptidine derivative (9), which could be condensed with a dihydrostrepto-biosaminyl chloride derivative to give α- and β-glycosides, the α-form yielding dihydrostreptomycin again on deprotection.

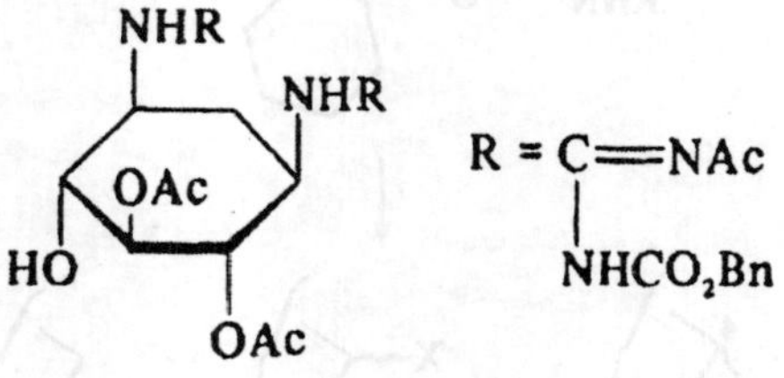

R = C(=NAc)NHCO$_2$Bn

(9)

3′,4′-Dideoxykanamycin B can be prepared on a large scale from kanamycin B by hydrogenation of the corresponding 3′-ene derivative obtained by established procedures *via* the 3′,4′-epoxide. 5-Epikanamycin B has been prepared using the methanesulphonate displacement procedure to invert the C-5 configuration in the parent antibiotic, and 3′-deoxyseldomycin 5 together with the corresponding 2′-deoxy regioisomer have been obtained by reduction of a 2′,3′-epimino derivative prepared from seldomycin 5.

4′-Deoxy-6′-*N*-methyl-amikacin has been prepared from 1,3-ureido-kanamycin A by Raney nickel treatment of its 4′-thioacetate followed by 6′-*N*-methylation; the activity of this derivative suggests that the 4′-hydroxy- and 6′-amino-groups in amikacins play little part in their defence against inactivating enzymes. Other derivatives of pseudotrisaccharides reported include *NN*-dialkyl derivatives of kanamycins A and B, 1-*N*-(1,3-dihydroxy-2-propyl)kanamycin B, which showed good antibacterial activity 6′-*N*-[^{14}C] acetyl-tobramycin, which was used to study the binding of the antibiotic to *E. coli* ribosomes and 2′-*N*-propyl-butirosin A, which proved to be as good an antimicrobial as butirosin A itself. Lividamine has been converted to 2-deoxy-4-*N*-glycyl-6-*O*-α-nebrosaminyl-fortamine (10) and its 3-de-*O*-methyl analogue (11), which are weaker anti-bacterials than the related sporaricin A mentioned above. 4-*N*-Acyl- and 4-*N*-alkyl derivatives of fortimicin B have also been prepared, and their antibacterial behaviour investigated.

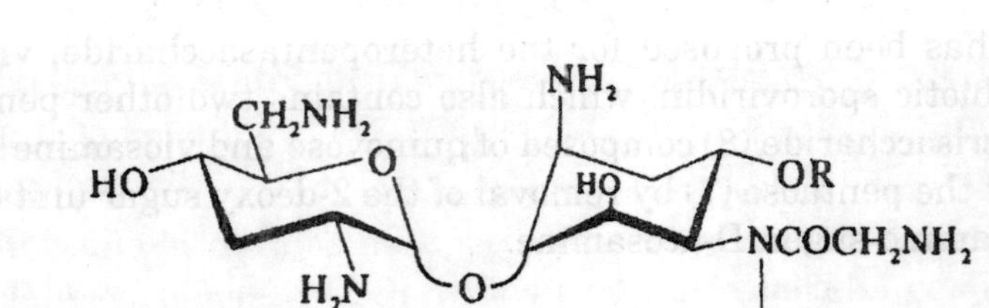

(10) R = Me

(11) R = H

Neamine has been converted to its eight possible 3′- and 4′-amino-deoxy and deoxy-hydroxy analogues by conversion to 3′,4′-anhydro-D-allo and D-galacto intermediates, which were then reduced to 3′- and 4′-deoxy compounds and appropriately transformed (Scheme 1). Some 4′-branched-chain sugar analogues were also prepared. A thio analogue of neamine (12) has been prepared by coupling the nitrosyl chloride adduct of the glycal (13) to the thio-streptamine (14) followed by reduction and deprotection.

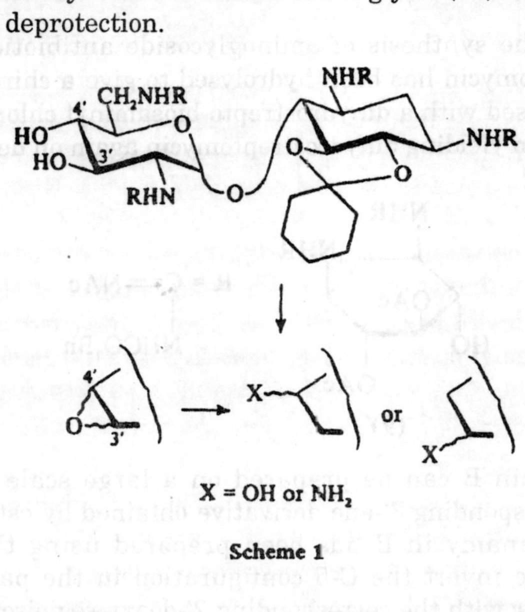

X = OH or NH_2

Scheme 1

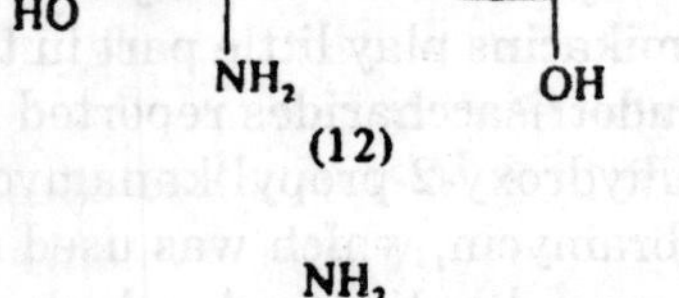

(12) 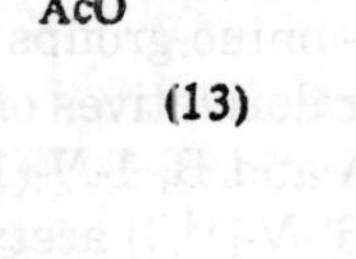 (13) 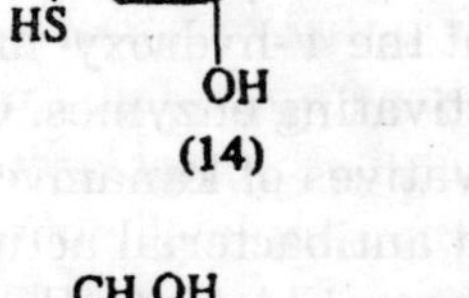 (14)

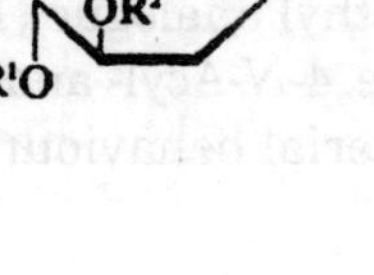

(15) R^1 = R, R^2 = H; R^1 = H, R^2 = R

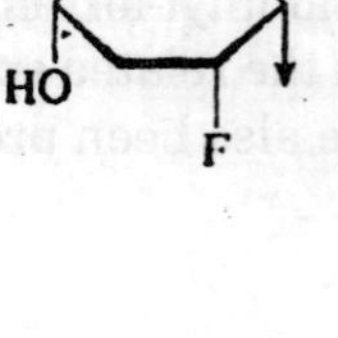

2′,3′-Dideoxy-2′-fluoro-pseudodisaccharides (15) have been synthesized using the glycal addition route to glycosides with 3,4,6-tri-*O*-acetyi-2-fluoro-D-glucaal. 4′-N-propyl sorbistin A_1 has been prepared, and found to be as effective as sorbistin A1 against Gram positive bacteria. The mycinose moiety of tylosin can be specifically removed under mild conditions by oxidation of the free hydroxy-group of the sugar in the otherwise protected antibiotic to the corresponding glycos-4-ulose, which with a primary amipne hydrochloride is degraded to a pyrrole derivative, liberating the aglycone.

Selective *N*-acylation of kanamycin A has been studied, and conditions found for the preferential formation of the mono-6″-*N*-acyl or the 3-*N*-acyl derivative. Base treatment of 3′-chloro-3′-deoxy-pseudotrisaccharides derived from butirosin have given new cyclic derivatives containing a 3′,5″-anhydro bridge. An improved procedure for the preparation of triacetyloleandomycin has been reported.

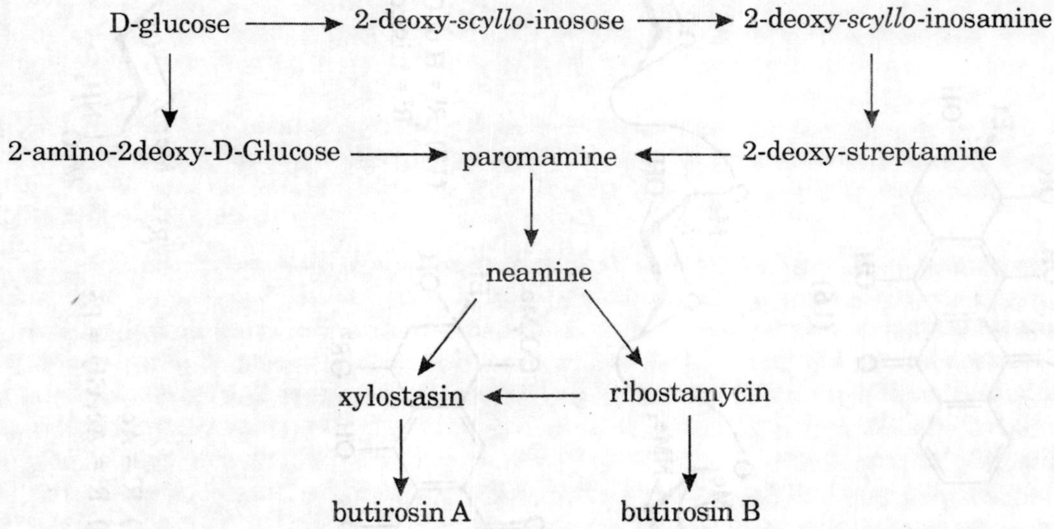

MACROLIDE ANTIBIOTICS

The synthesis of stereoisomers of the amino-sugar component of megalomycins, together with full details on the revised structure of these antibiotics reported. The antifungal polyene antibiotics levorin A_2 and candicidin D, which contain mycosamine, are identical. The unusual deoxysugar L-digitoxose (2,6-dideoxy-L-ribohexose) has been isolated from the acid hydrolysis products of the macrolide antibiotics nystatin A_3, candidinin, and polyfungin B. Maridomyin III, in which both ester groups are propionyl derivatives, is the principal component of the maridomycin complex elaborated by a valine resistant mutant of *Streptomyces hydroscopicus* (the six components differ depending on whether acetyl, propionyl, or isovaleryl ester groups are present). The deltamycins A_t-A_4 are closely related compounds, being 9-oxo analogues of the maridomycins, with the mycarose residue substituted at *O*-4 by acetyl, propionyl, butyryl, or isovaleryl residues, respectively, and deltamycin A_4 was shown to be identical to carbomycin A.

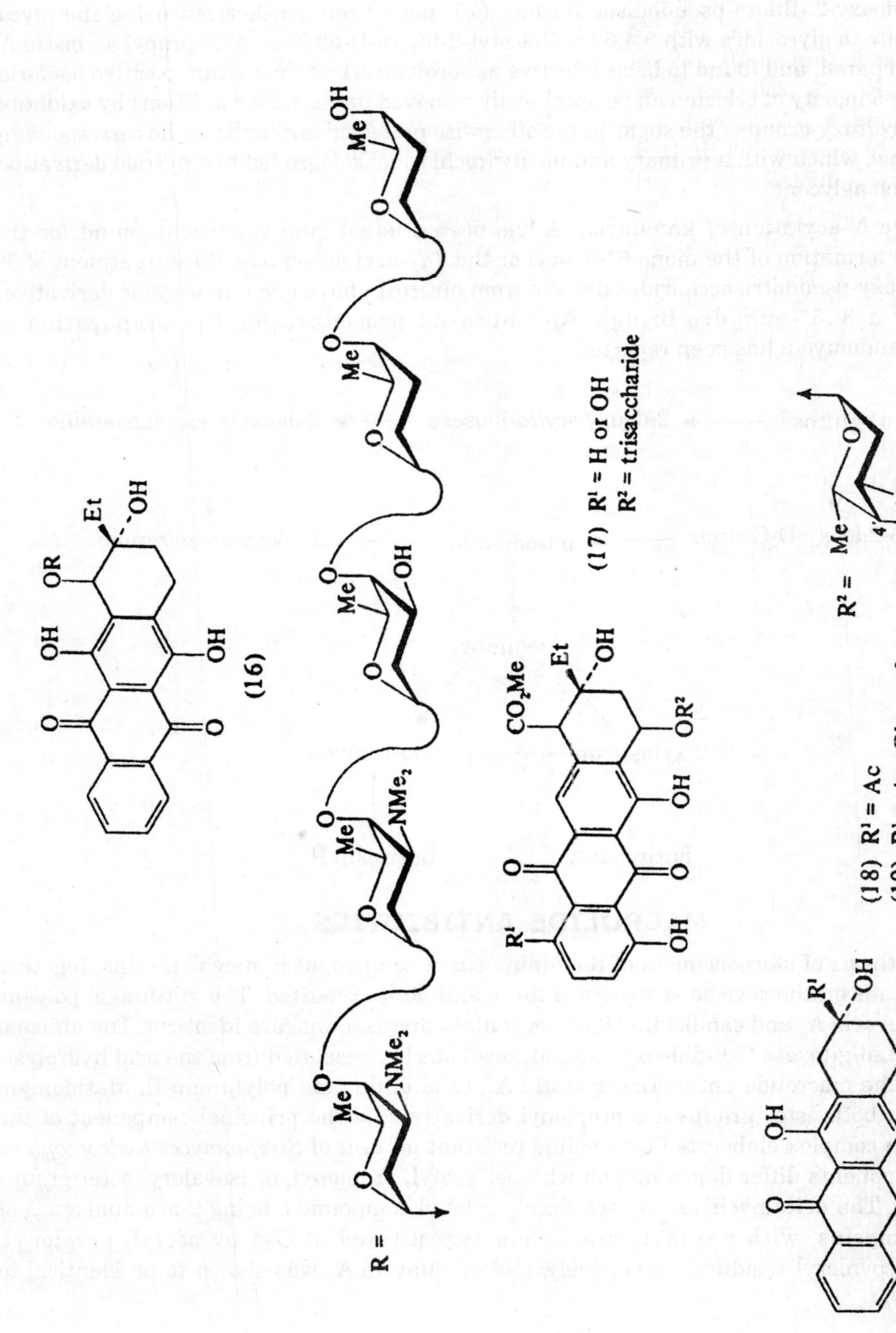

(16)

(17) R^1 = H or OH
R^2 = trisaccharide

(18) R^1 = Ac
(19) R^1 = Ac, Gly = R^2

(20) R^1 = CH_2OH, Gly = R^2

ANTHRACYCLINE ANTIBIOTICS

Two new compounds, roseorubicin A and B, have been isolated from the culture broth of *Actinomyces roseoviolaceus.* Roseorubicin A (16) contains a penta-saccharide chain, L-rhodinose → L-rhodinose → 2-deoxy-L-fucose → L-rhososamine → L-rhodosamine, which on mild acid hydrolysis gave roseorubicin B, the corresponding tetracycline disaccharide, which in turn could be selectively hydrolysed to γ-rhodomycin 1, the simple rhodosaminyl glycoside. Structures have been reported for the anti-tumour compound aclacinomycin A and its analogues, which are elaborated by *Streptomyces galilaeni;* they contain trisaccharide chains attached to the aglycone (17), in most cases having a deoxysugar (L-amicetose, L-rhodinose) or a ketosugar (D- or L-cinerulose A, L-aculose) attached through 2-deoxy-L-fucose to L-rhodosamine by α-1 → 4 links. Horton's group has reported syntheses of daunorubicin analogues (18) by appropriate condensation of modified glycosyl halides or glycals with dauno-mycinone-, these include derivatives of 3-amino-2,3,6-trideoxy-α-D-*ribo*-hexo-pyranoside, 3-amino-2,3,6-trideoxy-α- and -β-D-arabinopyranoside and 3,6-diamino-2,3,6-trideoxy-α-D-*ribo*-hexopyranoside, and 3,6-dideoxy-α- and -β-D-*ribo*- and -L-*lyxo*-hexopyranoside. 4′-*O*-Methyl-daunorubicin (19) and 4′-*O*-methyl-adriamycin (20) together with their 4′-*epi*-isomers have likewise been prepared from appropriately modified sugars by Koenigs-Knorr condensation with the tetracycline.

Daunorubicin and adriamycin have been converted to *N*-alkyl and *NN*-dialkyl analogues by cyanoborohydride reduction of derivatives formed from aldehydes or ketones; the *N*-alkyl derivatives showed enhanced anti-tumour activity with reduced cardiotoxicity. 4′-*O*-Tetrahydro-pyranyl derivatives of daunorubicin, daunomycin, and adriamycin have also been prepared.

Nogalamycin has been converted to a series of aglycone analogues, and the stereochemistry of the anthracycline ring investigated.

NUCLEOSIDE ANTIBIOTICS

A new antibiotic neosidomycin, isolated from *Streptomyces hygroscopicus,* has been shown to be the indoyl-uronic acid derivative (21) (the sugar may be D or L). A new active polyoxin constituent, polyoxin N, has been isolated from the culture broth of *Streptomyces piomogenus* and assigned structure (22). The antibiotic nikkomycin X has been shown to have the structure (23), an analogue of nikkomycin Z, which is the corresponding uracil-1-yl derivative of the same sugar unit. Sinefungin, an antifungal antibiotic that is a potent inhibitor of tRNA and protein methyl transferases *in vitro,* has the structure (24). Tunic-aminyl uracil, a degradation product of tunicamycin, has been shown to have the structure (25).

2′-Deoxycoformycin and cordycepin have been isolated from a new strain of *Aspergillus nidulans* Y 176-2 cultured on wheat-bran, the first example of a potent adenosine deaminase inhibitor (2′-deoxycoformycin) occurring in a fungus.

Tubercidin (7-deaza-adenosine) has been synthesized by conventional procedures from 4-chloro-5,7-diazaindoline and tetra-acetyl ribofuranose and the α-anomer was also prepared. A series of glycosyl naphtha [2,3-c]pyrazole-4,9-dione derivatives have been prepared from glycals and the parent hetero-cycle (26), giving either 1-*N*- or 2-*N*-substituted compounds; the former show cytostatic activity against HeLa cells, irrespective of the particular pyranosyl substituent, whereas the 2-*N*-substituted compounds were inactive.

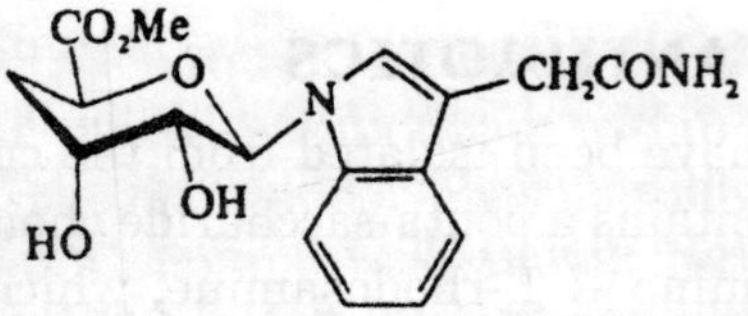

(21)

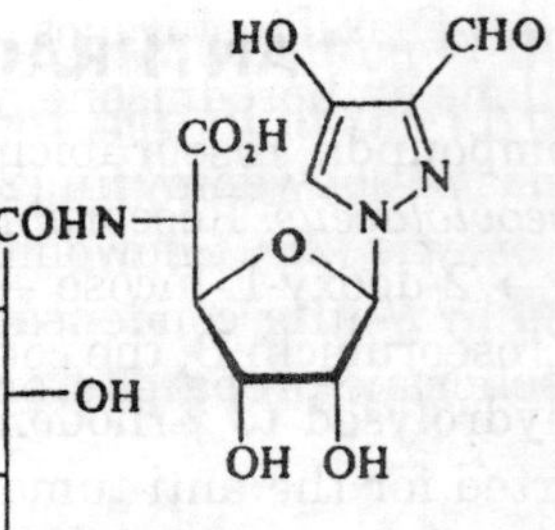

(22)

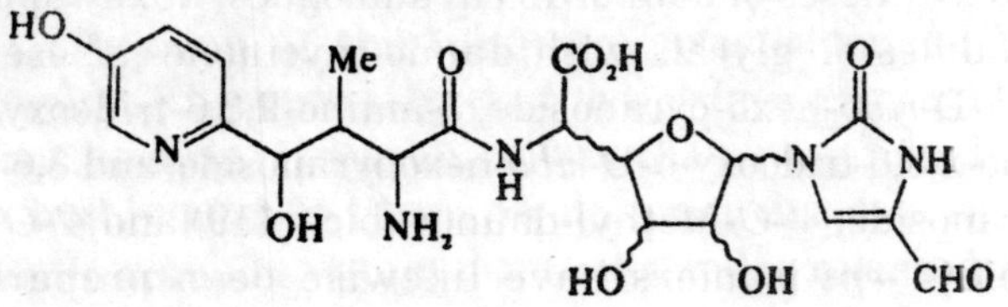

(23)

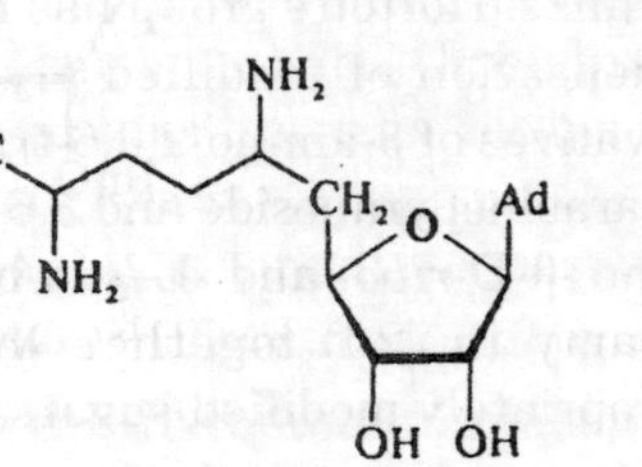

(24)

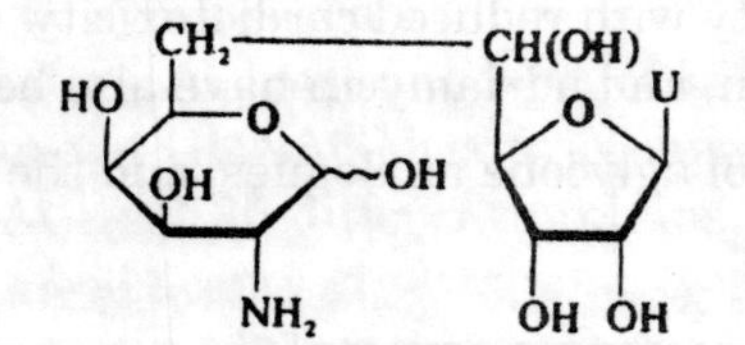

(25)

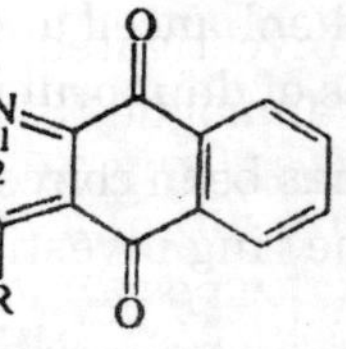

(26) R = H or CO_2Et

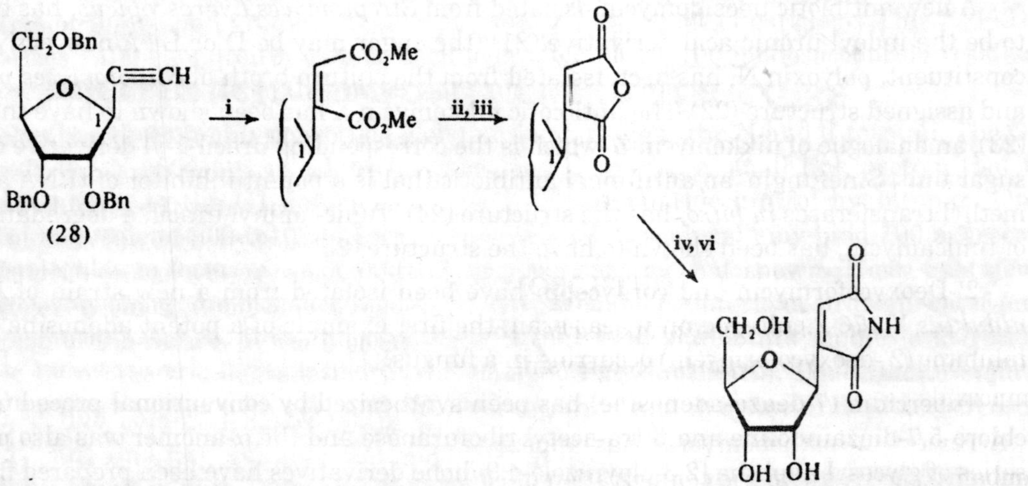

(28)

Reagents: (i) MeOH-CO-$PdCl_2$; (ii) KOH; (iii) $(CF_3CO)_2O$; (iv) NH_3; (v) AcCl-DMF; (vi) H_2-Pd

Scheme 3

Conjugates of 1-β-D-arabinofuranosyl-cytosine have been prepared with cortisol and cortisone and with prednisolone and prednisone, which show good anti-tumour activity.

A new synthesis of showdomycin (27), outlined in Scheme 3, utilizes the carboxylation of the ribofuranosyl ethyne (28). Showdomycin, labelled with carbon-14 at the β-carbonyl group, has been prepared by Wittig condensation of the 3,6-anhydrohept-2-ulosonic acid derivative (29) with the phosphorane prepared from [l-^{14}C]chloroacetamide (Scheme 4).

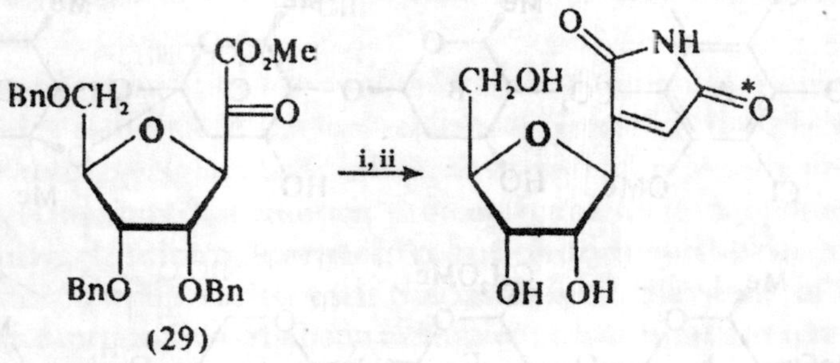

(29)

Reagents: (i) $Ph_3P{=}CH^{14}CONH_2$; (ii) BCl_3

Scheme 4

A number of isoformycins (30) have been prepared from appropriate glyco-furanosyl thioformidate derivatives by condensation with 3-cyano-4-amino-pyrazole, and their conformations were investigated. glycosyl thioimidate has similarly been used to prepare the triazole C-glycoside analogue (31) of ribavirin by condensation with the hydrazone of ethyl oxamate, and the thiazole *C*-glycoside analogue was also synthesized.

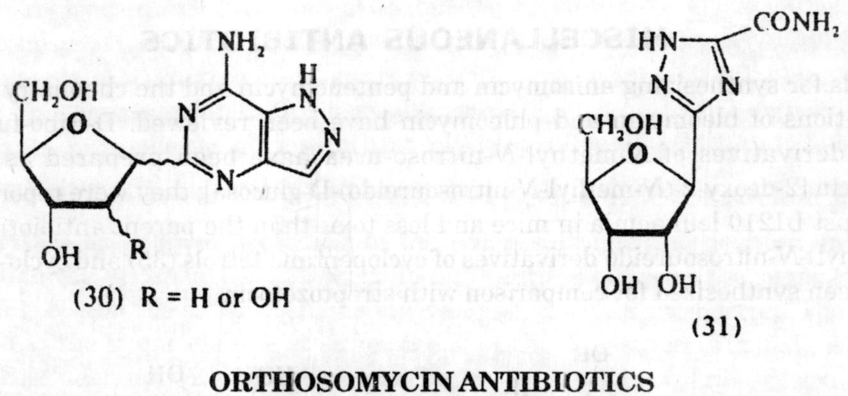

(30) R = H or OH

(31)

ORTHOSOMYCIN ANTIBIOTICS

The above title has been proposed for the oligosaccharide antibiotics, which include the everninomicins, flambamycins, destomysins, avilamycin, curamycin, hygromycin, and related compounds.

A review on orthosomycins has appeared. The structure (32) has been proposed for flambamycin, the anomeric configuration of rings, D, F, and G being established by high field ^{13}C n.m.r.; for the latter studies, labilose, (6-deoxy-2,4-di-*O*-methyl-D-galactose), the glycoside component of labilomycin, was also synthesized for comparative purposes. Other papers have recorded ^{13}C n.m.r. and mass spectral data on flambamycin and its derivatives. Avilamycin A

(33) has the same structure as flambamycin except that the sugar evermicose (2,6-dideoxy-3-C-methyl-D-*arabino*-hexose) (unit D) in the latter is replaced by evalose (6-deoxy-3-C-methyl-D-mannose). Avilamycin C differs from avilamycin A in having a 1-hydroxy ethyl side-chain instead of acetyl in unit H. A structure has been proposed for curamycin A (34) based on chemical and spectroscopic studies and comparison with flambamycin and avilamycin.

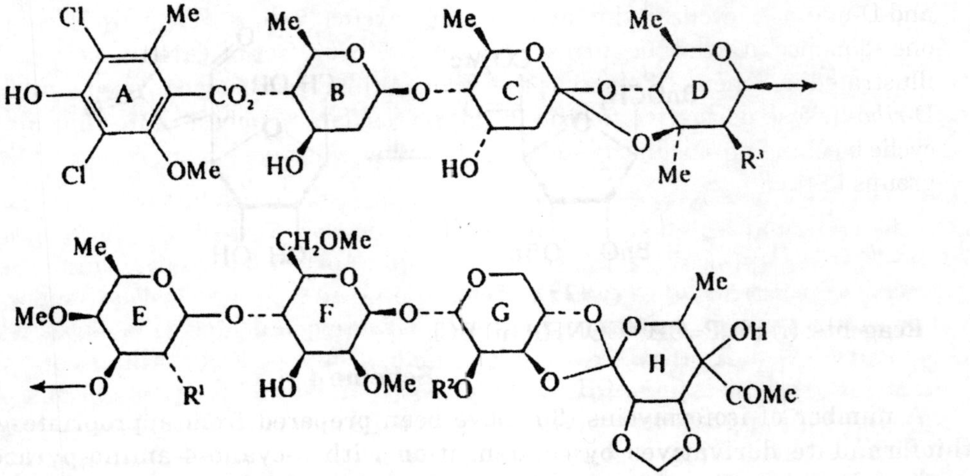

(32) $R^1 = OH$, $R^2 = Me_2CHCO^-$, $R^3 = OH$
(33) $R^1 = H$, $R^2 = Me_2CHCO$-, $R_3 = OH$
(34) $R^1 = OH$, $R^2 = Ac$, $R^3=H$

MISCELLANEOUS ANTIBIOTICS

Methods for synthesizing anisomycin and pentenomycin and the chemistry and biological transformations of bleomycin and phleomycin have been reviewed. D-Ribo-furanosyl and -pyranosyl derivatives of *N*-methyl-*N*-nitroso-urea have been prepared as analogues of streptozotocin [2-deoxy-2-(*N*-methyl-*N*-nitrosoureido)-D-glucose]; they were reported to be more active against L1210 leukaemia in mice and less toxic than the parent antibiotic. Likewise *N*-(2-chloroethyl)-*N*-nitrosoureido derivatives of cyclopentane tetrols (35) and cyclo-hexane tetrols (36) have been synthesized for comparison with streptozotocin.

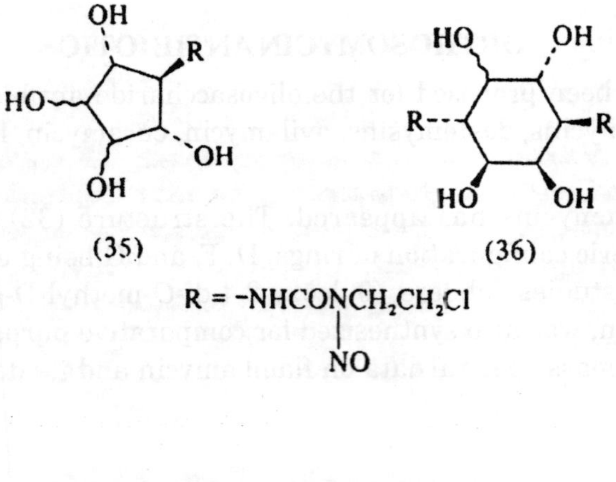

(35) (36)

$R = -NHCON(NO)CH_2CH_2Cl$

Biosynthetic studies have established that D-glycosamine is incorporated into streptozotocin, and L-citrulline and methionine furnish the amide side-chain and N-methyl group, respectively, but the source of the nitroso-group was not established. 2-Deoxy-2-ureido-D-glucose, which shows weak antimicrobial activity against some Gram negative bacteria and fungi, has been isolated from *Streptomyces halstedii;* in aquecus solution it slowly cyclized to give the corresponding 1,2-diazolidone derivative. Isomeric ureido derivatives of D-galactose and D-mannose cyclized similarly. Spectinomycin (37) has been synthesized in two laboratories; one sequence utilized the nitrosyl chloride adduct of glucal with the actinamine derivative (38) illustrated in Scheme 5, whereas the other, outlined in Scheme 6, used a preformed 4,6-dideoxy-D-*ribo*-hexose derivative for the initial glycosidation step; a route that makes elegant use of cyclic blocking groups in the sugar ring to allow selective oxidation of the C-4 and C-3 hydroxy-groups in turn.

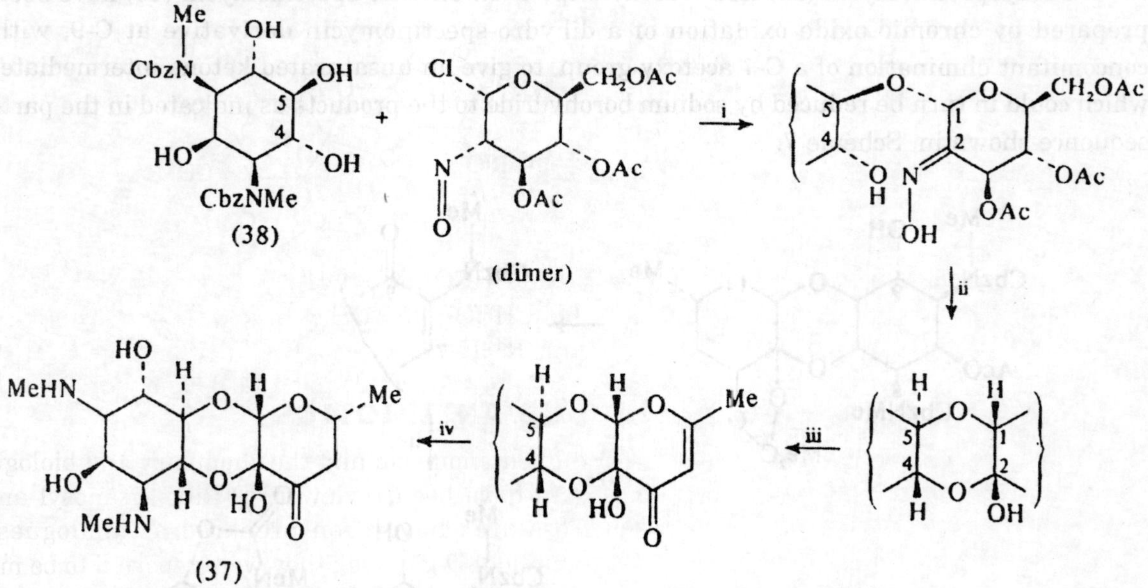

(38) (dimer) (37)

Reagents: (i) DMF; (ii) MeCHO-MeCN-HCl; (iii) $KHCO_3$-MeCN; (iv) H_2-Pd

Scheme 5

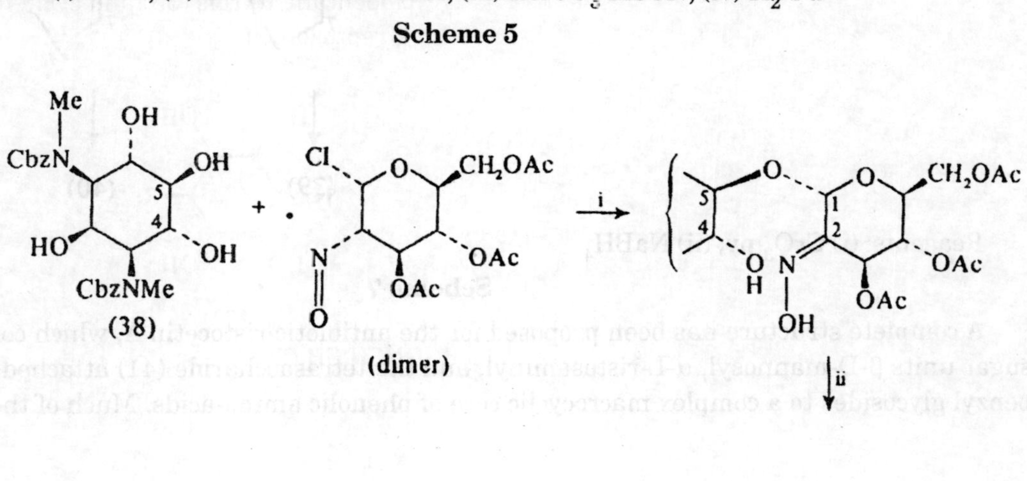

(38) (dimer)

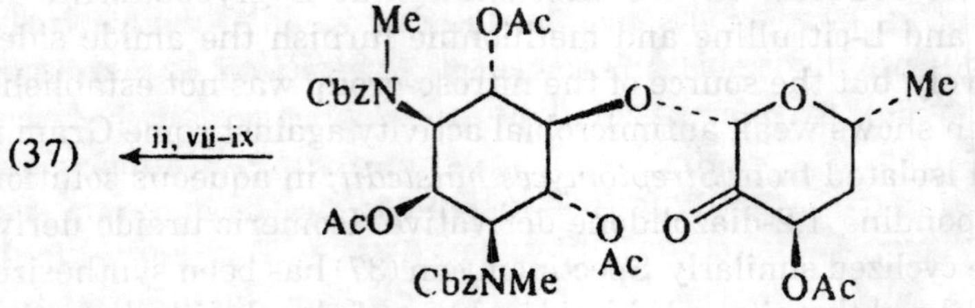

Reagents: (i) CF_3SO_3Ag; (ii) NaOMe; (iii) $MeC(OMe)_3$–H^+; (iv) Ac_2O-py-DMAP; (v) HOAc-H_2O; (vi) CrO_3-py-HCl; (vii) Bu_2SnO; (viii) Br_2-Bu_3SnOMe; (ix) H_2-Pd

Scheme 6

7-Deoxyspectinomycin (39) and 7-deoxy-8-*epi*-4-(R)-dihydro-spectinomycin (40) have been prepared by chromic oxide oxidation of a dihydro-spectinomycin derivative at C-9, with concomitant elimination of a C-7 acetoxy-group, to give an unsaturated ketone intermediate, which could in turn be reduced by sodium borohydride to the products as indicated in the part-sequence shown in Scheme 7.

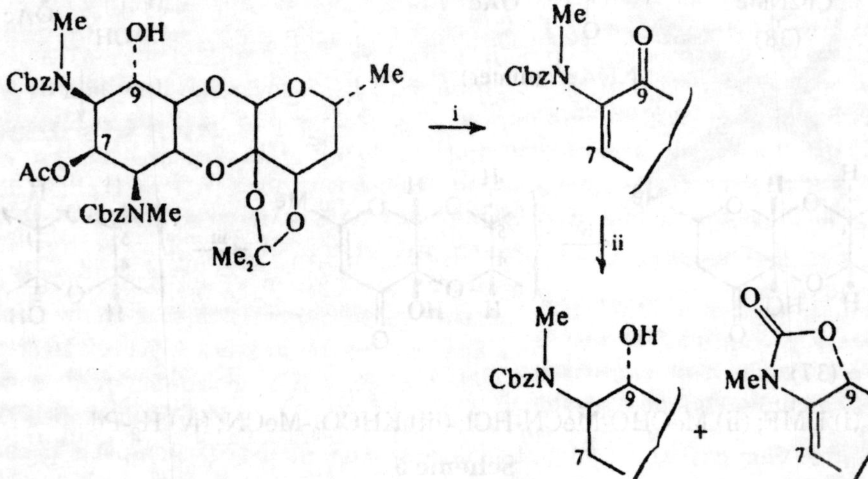

Reagents: (i) CrO_3-py; (ii) $NaBH_4$

Scheme 7

A complete structure has been proposed for the antibiotic ristocetin A, which contains the sugar units β-D-mannosyl, α-L-ristosaminyl, and the tetrasaccharide (41) attached as aryl or benzyl glycosides to a complex macrocyclic core of phenolic amino-acids. Much of the evidence came from a 270 MHz ^{1}H n.m.r. study on the ristosaminyl-core obtained after removal of the other sugar residues by mild acid hydrolysis.

β-D-Ara*p*-1 ⟶ 2-α-D-Man*p*-1 ⟶ 2-β-D-Glc*p*
6
↑
β-L-Rhamp-1

(41)

Two further degradation products have been obtained from moenomycin A, suggesting that when the part structures (42) and (43), are linked together they form the complete antibiotic. Another report describes the glucose-glucosamine disaccharide present in fragment (42), which was obtained from moenomycin A by acid hydrolysis. Earlier work on the structure and properties of moenomycin A has been reviewed.

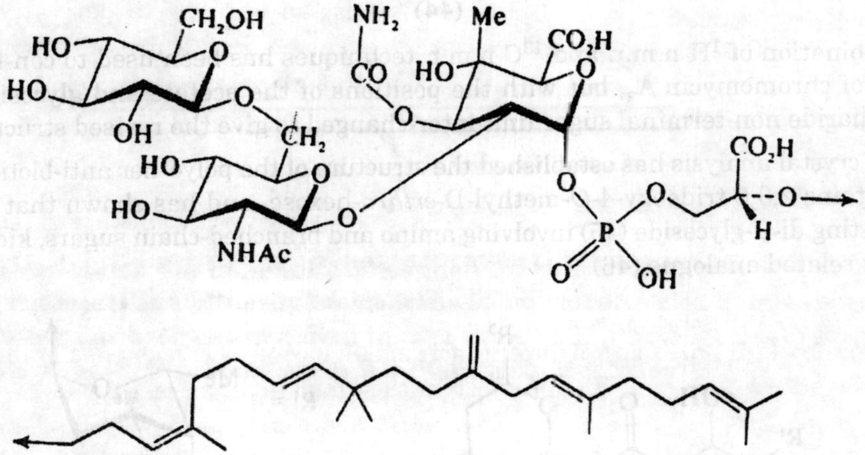

(42)

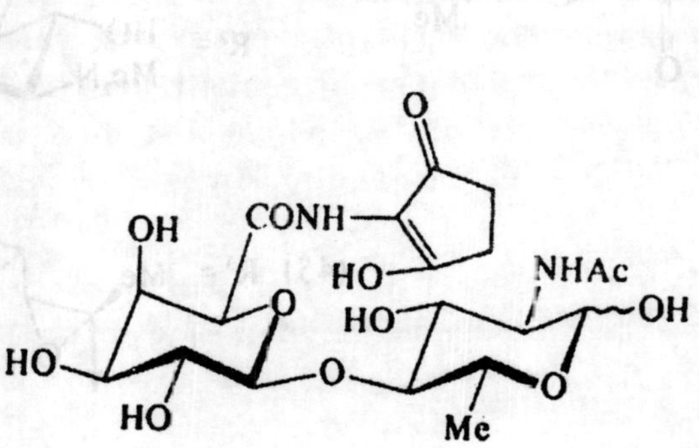

(43)

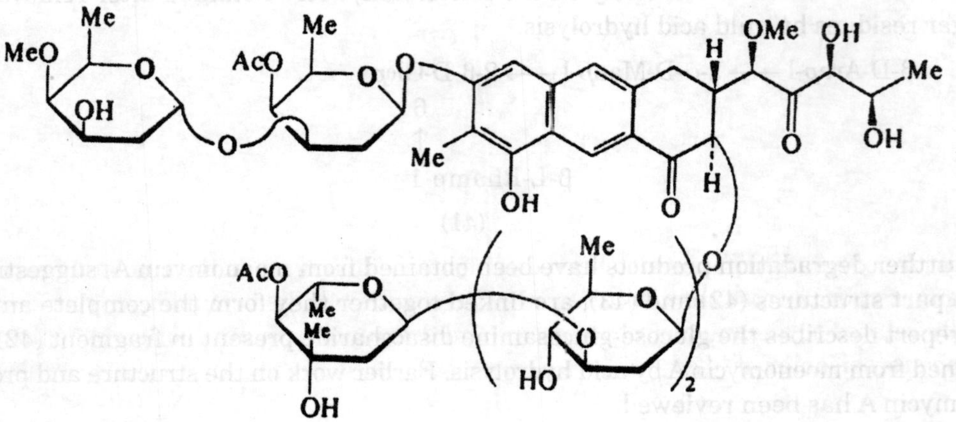

(44)

A combination of [1]H n.m.r. and [13]C n.m.r. techniques has been used to con-firm the gross structure of chromomycin A_3, but with the positions of the acetate and glycosidic groups on the disaccharide non-terminal sugar unit inter-changed to give the revised structure (44).

X-Ray crystal analysis has established the structure of the polyether anti-biotic carriomycin, which contains 2,3,6-trideoxy-4-*O*-methyl-D-*erthro*-hexose, and has shown that hedamycin is the interesting di-*C*-glycoside (45) involving amino and branched-chain sugars, kidamycin being the closely related analogue (46).

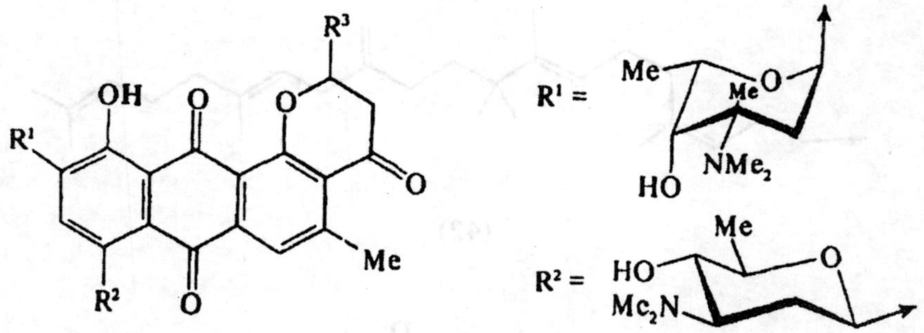

R^1 =

R^2 =

(45) R^3 = Me

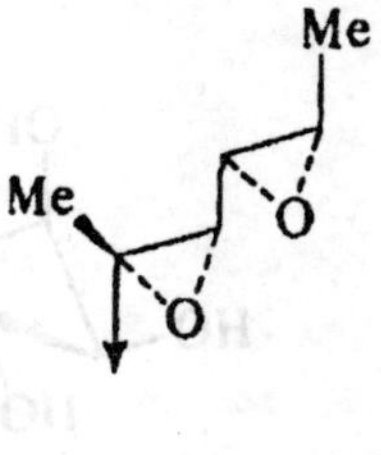

(46) R^3 = Me

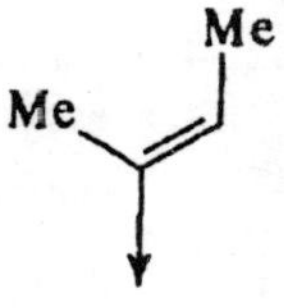

(47)

The structure of rubradirin has also been established. This contains the unusual branched-chain nitro-sugar rubranitrose that is linked as a β-glycoside (47) to a complex aglycone, which is structurally related to ansamycin, novo-biocin, and everninomicin antibiotics; evernitrose is the *L-arabino* stereoisomer of (47).

CHAPTER 7

Biosynthesis of Polysaccharides

The polysaccharides made by bacteria are quite different in function from those of higher plants. They are secreted from the cell to form a layer over the surface of the organism, often of substantial depth in comparison with the cell dimensions, which is believed to have several functions. Because of their position they are characterized as exo-polysaccharides, to distinguish them from any polysaccharides that might be found within the cell. The functions are thought to be mainly protective, either as a general physical barrier preventing access of harmful substances, or more specifically as a way of binding and neutralizing bacteriophage. In appropriate environments they may prevent dehydration. They may also prevent phagocytosis by other micro-organisms or the cells of the immune system.

The capsular polysaccharides are often highly immunogenic, and may have evolved their unusual diversity as a way of avoiding antibody responses. They also have a role in adhesion and penetration of the host, and since in the case of plants this will involve interaction with polysaccharide structures in the cell walls there are clearly possibilities of specific interactions. Plant lectins (glycoproteins) which have specific binding properties with respect to carbohydrate structures may play a part in this, and in the general defences of plants against bacterial infection.

Lectins are thought to be part of the plants' defence against insect and animal depredations, and are known to bind to sites in the gut; their presence predominantly in the seeds is consistent with this but they may also have a role against bacteria and fungi. The secreted polysaccharides can be involved in pathogenicity. *Pseudomonas aeruginosa,* commonly found in respiratory tract infections, produces alginate which contributes to blockage in the respiratory tract, and leads to further infection, while similar blockage of phloem in plants has been described.

BIOTECHNOLOGY

Bacterial products are very important in all aspects of polysaccharide biotechnology. They range from bulk materials such as xanthan, priced at about $14/kg, used mainly in food applications to cyclic dextrans valued at about $50/kg, and used in high-value applications in research and pharmaceuticals. Bacteria are known which produce nearly all the major plant polysaccharides such as glucans, alginate-like materials, and even cellulose — though apparently not pectins — as well as the complex bacterium-specific materials.

Genetic Manipulation of bacteria has been studied for longer and is in general much easier than for higher organisms, so they are an obvious target both for manipulation of biosynthetic pathways and for the expression of heterologous genes to produce especially desirable enzymes. Polysaccharides are not of course under the direct control of genetic material in the way that enzymes are, and must be approached indirectly by manipulating the biosynthetic or degradation pathways by way of the enzymes responsible. We have already seen one example of this, in the control of pectin degradation in higher plants, but there would seem to be great scope for this approach in bacteria as well.

In addition, since there are now fermentation processes. Involving the large scale growth of bacteria all the advances in this that can be achieved by biotechnology will apply. There are limitations. It is unlikely that large-scale fermentation could ever compete with existing processes for pectin manufacture, for example, since processors currently use waste by-products from other processes as their raw material. This is quite apart from the inherent costs of large-scale fermenters in comparison with simple extraction processes. It follows that bacterial products must have particularly useful properties in order to justify their probable greater cost. As we shall see there are a number of such products that can meet this criterion.

BACTERIAL ALGINATES

Although commercial alginates derive from algal sources there is a large potential for producing 'tailor-made' alginates from bacterial sources, especially if advantage is taken of the genetic tools for controlling the production of the enzymes that are responsible for the synthesis and epimerization (conversion of D-mannuronic, 'M', to L-guluronic residues, 'G') of the polymeric alginate chain. There appears to be greater structural diversity (poly-M, poly-G and poly-MG residues) and our understanding of the genes and the enzyme gene products is much greater for bacterial alginate production compared with the case for seaweed. The main alginate-producing bacteria that have been studied are *Pseudomonas aeruginosa* and *Azotobacter vinelandii*. *P. aeruginosa* has been the subject of particular attention because of its association with respiratory disease and is found in patients suffering from cystic fibrosis. A. *vinelandii* appears to be the most promising in terms of industrial production because of its stable output of alginate. *P. aeruginosa* alginate has no poly-G residues (and hence has a low G-content) whereas *A. vinelandii* can, like seaweed alginate, possess all three block sequences (poly-M, poly-G and poly-MG residues). However, all bacterial alginates are *O*-acetylated to varying degrees:

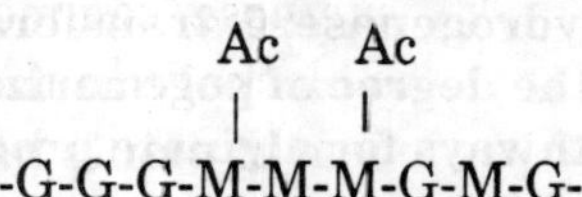

Acetylated mannuronic acids cannot be converted by epimerases from M to G.

Alginate structural enzymes

Starting from fructose-6-phosphate the following enzymes are involved according to the scheme of: hexokinase; phosphomannose isomerase; D-mannose-1-phosphate guanyl transferase; GDP (guanosine diphosphate) mannose dehydrogenase; transferase; acetyl transferase, and mannuronan C-5-epimerase. Such schemes have been worked out for A. *vinelandii, P. aeruginosa* and for the seaweed *Fucus garnweri*. Key structural genes have been identified in a gene cluster for *P. aeruginosa* with algG encoding for the epimerase and algF for the acetylase.

Possibilities

As Ertesvag *et al.* (1996) report, it may be possible to treat alginate with epimerases to increase the poly-G content, producing a firmer polymer. In principle this could be done in vivo by the incorporation and expression of genes from a plasmid in an alginate-producing bacterium. The degree of acetylation is another control point since, as we have already mentioned, an acetylated mannuronic acid residue is immune from epimerization.

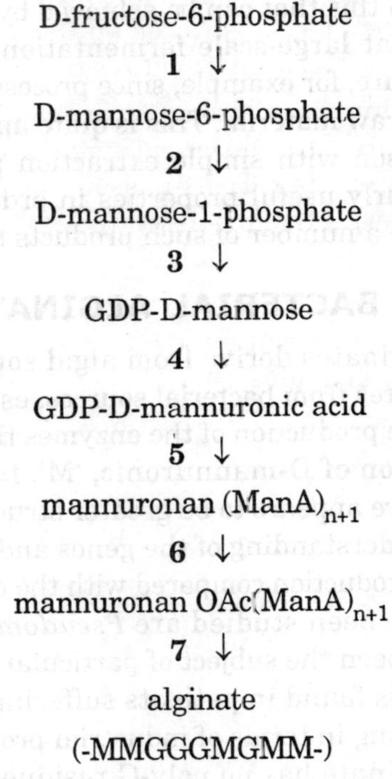

Fig. 5.1. Alginate biosynthesis in *Azotobacter vinelandii* and the principal enzymes involved: 1: hexokinase; 2: phosphomannose isomerase; 3: D-mannose-1 -phosphate guanyl transferase; 4: guanosine diphosphate mannose dehydrogenase; 5: transferase (n is the number of uronic acid residues in the polymer chain, i.e. the degree of polymerization); 6: acetyl transferase; 7: mannuronan C-5-epimerase. Similar pathways for alginate production have been reported in Fucus *gardneri* (a brown alga) and Pseucdomonas *aeruginosa.*

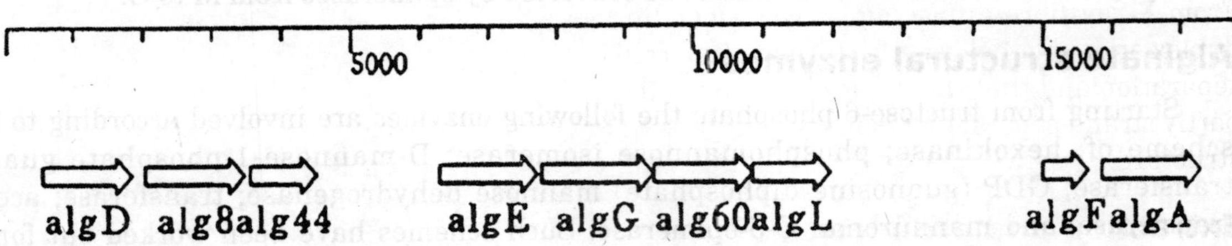

Fig. 5.2. Gene cluster organization encoding most of the alginate structural enzyme genes in *P. aeruginosa*. Scale is in base pairs.

XANTHAN

Xanthan (or, as commonly called, 'xanthan gum') at $14/kg falls into a similar price range to other gums and exudates and has similar functional properties. It is made by the organism *Xanthomonas campestris* (which in its wild existence is responsible for cabbage blight) grown largely on glucose, itself derived from maize starch, which it converts with high efficiency (80 per cent) to the xanthan gum structure. Typically for a fermentation product, the raw material costs are a small part of the total, which are mainly to do with recovering the gum from the culture medium. Production is around 9000 tonnes a year. It has a $\beta(1 \rightarrow 4)$ linked glucan main chain with alternating residues substituted on the 3-position with a trisaccharide chain containing two mannose and one glucuronic acid residue. It is thus a charged polymer.

Some of the mannose residues may also carry acetyl groups. It is useful because it forms relatively rigid-rod-like structures in solution at ambient temperatures, though they convert to the random configuration on heating. These rods are able to align themselves – like agarose and the κ- and ι-carrageenans –with the unsubstituted regions of galactomannans, such as guar and its derivatives and locust bean gum, to produce fairly rigid mixed gels with applications in food manufacture.

The primary structure (as worked out by sequential degradation and methylation analysis) consists of the same backbone as cellulose:

$$\ldots \rightarrow 4)\beta\text{-D-Glc}p\ (1 \rightarrow \ldots$$

with the substitution at C3 of every alternate Glc*p* residue by the negatively charged trisaccharide

$$\beta\text{-D-Man}p\text{-}(1 \rightarrow 4)\ \beta\text{-D-Glc}p\text{A-}(1 \rightarrow 2)\ \alpha\text{-D-Man}p\text{-6-}O\text{-acetyl}\ (1 \rightarrow 3)$$

with some of the terminal β-D-Man*p* residues substituted at positions C4 and C6 by pyruvate ($CH_3.CO.COO^-$). Thus xanthan (texcept under highly acidic conditions) is a highly, charged polyanion. Fig. 7.3 shows the Haworth projection of a (pyruvylated) alternate repeat section of xanthan.

It is not only charge repulsion effects that give xanthan an extended rigid-rod conformation, i.e. a 'zone A'-polysaccharide: X-ray fibre diffraction studies have shown that, like the marine polysaccharides ι- and κ-carrageenan along with agarose, xanthan exists in 'ordered form' as a helix although there is still uncertainty as to whether this helix is a coaxial duplex or just a single chain one (with the projecting trisaccharides folded back along the axis of the helix giving effectively a chain rigidity similar to that of a double helix). Using a combination of electron microscopy (contour chain length, L) and light scattering (weight average molecular weight, M_w), B. T. Stokke and co-workers evaluated a mass per unit length, M_L of (1950 ±250) Danm. From X-ray fibre diffraction, a corresponding value of 950 Da nm^{-1} was obtained, and so a duplex or double-helical structure was inferred in solution. Further studies showed some strand separation and that the duplex structure was more rigid compared with regions that had been partly strand-separated. High temperatures will reversibly melt this helix to give a more random structure.

Extraction and production

The primary laboratory and commercial fermentation medium - as has been known for over fifty years - for *X. campestris* growth and xanthan production is a phosphate-buffered

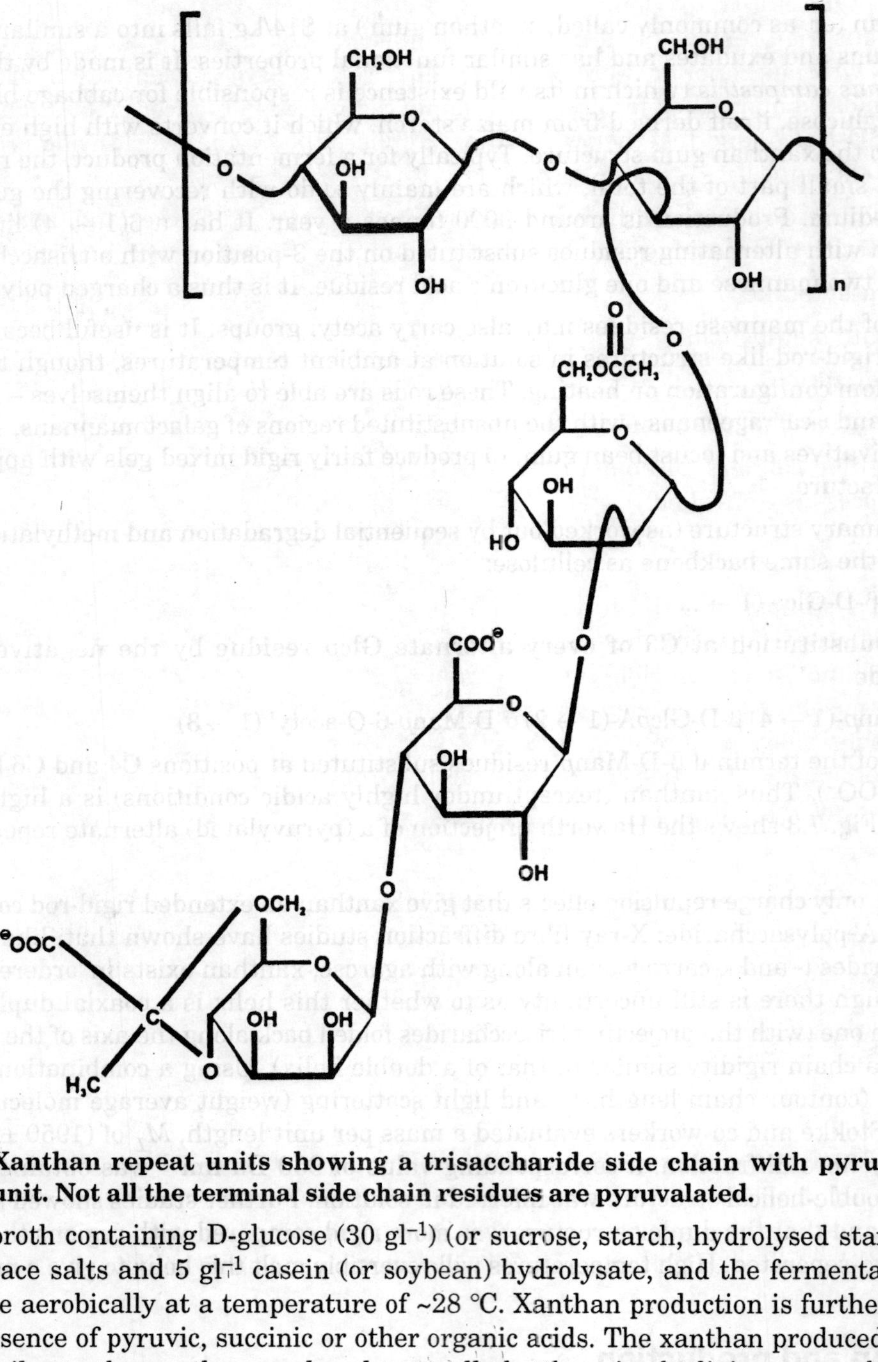

Fig. 7.3. Xanthan repeat units showing a trisaccharide side chain with pyruvalated end mannose unit. Not all the terminal side chain residues are pyruvalated.

(pH ~ 7) broth containing D-glucose (30 gl^{-1}) (or sucrose, starch, hydrolysed starch), NH_4Cl, $MgSO_4$, trace salts and 5 gl^{-1} casein (or soybean) hydrolysate, and the fermentation process takes place aerobically at a temperature of ~28 °C. Xanthan production is further stimulated by the presence of pyruvic, succinic or other organic acids. The xanthan produced in this way is very similar to the xanthan produced naturally by the microbe living on a cabbage. In the commercial process the oxygen uptake from the broth is controlled to a rate of 1 $mmoll^{-1}$ min^{-1}. Treated in this way the bacterium is an extremely efficient enzyme mini-factory converting

>70 per cent of the substrate (D-glucose or related) to polymeric xanthan. The bacterium having done its work is then removed in a rather undignified way by centrifugation and the xanthan precipitated with methanol or 2-propanol at 50 per cent weight concentration. The xanthan slurry is then dried and milled for use. The original commercial producer of xanthan was Kelco Ltd (now MSD-Kelco) and together with other suppliers the annual world-wide production is now over 10000 tonnes.

The biosynthetic process by the bacterium worked out by I. W. Sutherland (1989) and later confirmed by others follows the same basic pattern proposed for other microbial polysaccharides

1. substrate uptake
2. substrate metabolism
3. polymerization
4. modification and extrusion

and involves lipid carriers, although there is still uncertainty over their precise role and how the whole process is controlled.

Properties

Table 7.1 Dsummarizes the fundamental properties of a popular commercial xanthan. Analytical ultracentrifugation, light scattering and viscometric studies using Mark-Houwink, worm-like coil and Wales-van Holde types of analysis are all consistent with an extra-rigid-rod conformation in solution. The weight average M_w of around 6 million daltons for Keltrol-xanthan means that this, along with other commercial xanthans, is one of the largest of the aqueous soluble polysaccharides. Solutions are correspondingly extremely viscous. The intrinsic viscosity

Table 7.1. Fundamental properties of a commercial xanthan : Keltrol

Property	
Polymer type	Polyanionic; complex repeat
Conformation type	Zone A/B (Extra-rigid/rigid rod)
Persistence length, L_p	1500 Å
Solubility	Cold or hot aqueous systems
Molecular weight	$(5.9\pm1.1) \times 10^6$Da
Intrinsic viscosity *[n]*	(7500 ± 2700) ml g^{-1}
Sedimentation coefficient, $s^\circ_{20,w}$	(13.0 ± 0.3) S
Wales–van Holde ratio, k_s/[n]	0.3 ± 0.1
Mark–Houwink *a* coefficient	1.2
Gelation properties	Transient weak gels
Key sites for biomodification	C3 trisaccharide side chains (particularly extent of acetylation of terminal β-D-Man*p*); helical backbone — synergistic interactions with galactomannans

of ~7500 ml g^{-1} is one of the highest known for a polysaccharide, and the dilution solution concentration limit, via the Launay C^* parameter (~[n]/3.3, is very low (~ 0.5 mg ml^{-1}). The very high viscosity at low concentrations (e.g. at 5 mg ml^{-1} a viscosity of -1000 cP has been observed at room temperature) makes it ideal as a thickening and suspending-agent. By itself xanthan, however, only forms transient weak gels since the junction zones are weaker than in those used for networking in carrageenan and agarose.

Modification

The key sites for modification are the first and terminal mannose residues on the trisaccharide side chains (particularly extent of acetylation of the first and pyruvalation of the latter) and the helical backbone by forming non-covalent interactions with galactomannans: these interactions appear to be also affected by the substitutions in the side chains. As to the trisaccharide side chains themselves, there are two approaches for alteration or control: one is changing the physiological conditions of fermentation; the other is the use of different pathovars or strains of *Xanthomonas campestris*.

The pathovars *pv phaseoli* and *oryzae* yield virtually acetyl-free or pyruvate-free xanthan respectively. The other approach has been to look at the genetics of the enzymes controlling the biosynthetic pathway and to attempt to produce and isolate in sufficient quantity genetic mutants deficient or defective in one or more of these enzymes to give 'polytetramer' (i.e. lacking in the terminal mannosyl or pyruvalated mannose group) and 'polytrimer' (lacking in addition the adjacent glucuronic acid residue). Although yields of 50 per cent polytetramer have been found, corresponding attempts for polytrimer and other xanthan variants have thus for been disappointing.

There has been considerable interest in improving the weak gelation characteristics of xanthan by inclusion of galactomannans such as locust bean gum into mixtures. The results are more marked with xanthan whose own gels are weak and transient, but made in the presence of guar or locust bean gum give stronger gels with an optimum mixing ratio ~50:50 by weight. Deacetylating the xanthan side chains seems to enhance these synergistic interactions.

Uses of Xanthan

Xanthan has been food grade approved by the USA Food and Drug Administration for nearly thirty years. This makes it not only attractive as a food product but also for use in packaging material in contact with food and in pharmaceutical and biomedical applications that involve ingestion. Its uses chiefly derive from its solubility in hot and cold water and its very high thickening and suspending potential which in turn derives from the very high viscosity of its suspensions. Despite the high viscosity, xanthan suspensions exhibit high shear thinning which means they also flow easily (i.e. good pourability).

Food Applications

Besides high viscosity, thickening and suspending ability, xanthan suspensions have high acid stability. This makes them highly popular in sauces, syrups, toppings and salad dressings. In drinks, the addition of xanthan together with carboxymethylcellulose adds 'body' to the liquid and assists with uniform distribution of fruit pulp etc. It is also used to add body to dairy products. The high freeze-thaw stability of xanthan suspensions makes them particularly attractive for the frozen food industry. The high suspending and stability properties are also taken advantage of by the animal feed industry for transporting liquid feeds with added vitamins and other supplements that would otherwise sediment out with transport or storage time.

Pharmaceutical and Cosmetic : uses Microencapsulation

Xanthan has recently been added to the list of hydrophilic matrix carriers, along with chitosan, cellulose ethers, modified starches and scleroglucan. Tablets containing 5 per cent xanthan gum were shown under low shear conditions to give successful controlled release of acetaminophen into stomach fluid, and tablets containing 20 per cent xanthan successfully carried a high loading (50 per cent) of the drug theophylline.

The high suspension stability is made use of in pharmaceutical cream formulations and in barium sulphate preparations. This high cream stability is taken advantage of also by the cosmetic industry, including, as with carrageenan, toothpaste technology where the toothpaste will hold its ingredients (high viscosity) and then easily brush onto and off the teeth (high shear thinning). Uniform pigment dispersal, along with other ingredients and long-term stability, make xanthan a good base for shampoos.

Oil Industry

Xanthans are used in a number of aspects of well-bore technology such as (i) a proppant into fissures for rock fracturation, (ii) the suspension and removal of debris from the bore area even in the presence of harsh environments (seawater included) and pipelines and (iii) enhancing oil recovery using the technique of 'polymer flooding'.

Wallpaper Adhesives

The high stability of suspensions makes xanthan an ideal base for suspending adhesive agents for wallpapers.

Textile Industry

Like alginates the suspension stabilizing property of xanthans makes them ideal for producing sharp prints from dyes with a minimum risk of running, for application to textiles and (in conjunction with guar) carpets.

PULLULAN

Pullulan is an α-glucan made up from maltotriose units linked by $\alpha(1 \rightarrow 6)$ bonds. It is obtained from *Aureobasidium pullulans* and is hydrolyzed by pullulanase to yield maltotriose. It is not attacked by digestive enzymes of the human gut, and is used to form films. Production is now substantial, and has found particular application in formulating snack foods in Japan based on cod roe, powdered cheese, and as a packaging film for is water soluble, with molecular weights in the range 5000 Da to 900000 Da, with straight unbranched chains, and behaves as a random coil according to a combination of sedimentation coefficients and intrinsic viscosity measurements. Similarly, light scattering and sedimentation equilibrium measurements show that pullulans above about 50 000 Da molecular weight behave to a close approximation like a classical random coil. It has been proposed as a 'standard polysaccharide' in the sense that it is so near to random coil behaviour, and readily obtainable in very reproducible form, that it could be used for comparative tests with other polysaccharides. The idea is presumably that deviation from pullulan-type behaviour must imply a more complex structure. It might also be used to standardize instruments. The solutions in water are liable to bacterial attack with rapid degradation of the molecular weight and preservatives such as azide may be needed.

GELLAN, XM6 AND CURDLAN

Inspired by the example of xanthan, other bacterial polysaccharides with useful properties have been developed, in the case of gellan as the result of a systematic search for a polysaccharide of the required properties followed by the identification of the organism. Gellan is obtained from cultures of *Pseudomonas elodea,* found growing on the *elodea* plant. It is isolated by ethanol precipitation from the culture medium, and may be partially deacetylated by alkali treatment.

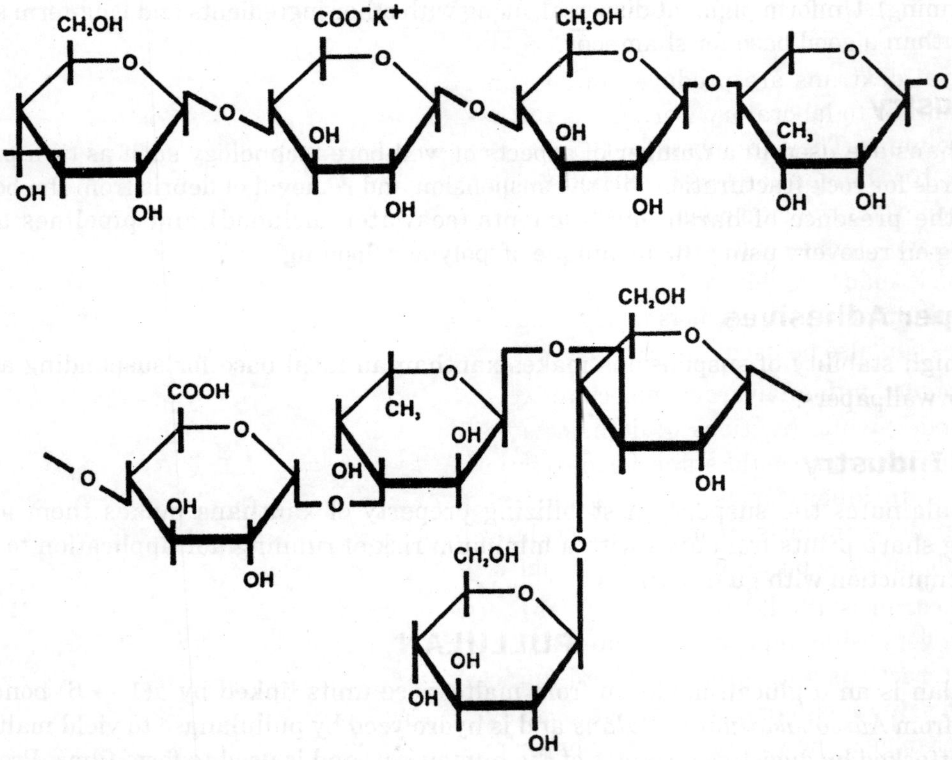

Fig. 7.4. Repeat units, at the top for gellan in deacetylated form and below for the XM6 polysaccharide.

It has a linear structure with a repeat unit of a tetrasaccharide each with one carboxyl group and in the native state one acetyl group. It is therefore sensitive to calcium levels but has rheological properties similar to those of xanthan which has a similar charge density. It was clearly intended to be a xanthan competitor, though before permission for food use was obtained it was promoted as an agar substitute, particularly for use in growth media.

The XM6 repeat unit structure is also shown in Fig. 7.4 and is a polysaccharide made by an *Enterobacter* discovered at Edinburgh University. It lacks acetyl groups but is classed along with xanthan and gellan, and can be induced to gel by adding calcium ions. It is included here as a representative of many similar bacterial polysaccharides, all of which have interesting and potentially useful rheological properties.

It is not commercially available and not approved for food use, and probably never will be. Curdlan is one of two polysaccharides produced by *Alcaligines faecalis* with a repeat unit of three $\beta(1 \rightarrow 3)$ linked D-glucose residues, and unlike the examples above is not charged. Production is now from *Agrobacterium* mutants that make only curdlan in high yield, and is mostly confined to Japan. Curdlan suspensions gel on heating, apparently irreversibly. There are a number of potential applications for it in the food industry, mostly as a replacer for existing gums, and it may be used in Japan in some products. Novel products based on its heat-set ability may also appear.

DEXTRANS

Bacterial dextrans are produced in substantial quantities by *Leuconostoc mesenteroides* and are familiar to laboratory workers as the basis for cross-linked dextran beads used in gel filtration columns. They are mostly $\alpha(1 \rightarrow 6)$D-glucopyranosyl polymers with molecular weights up to $\sim 1 \times 10^6$ Da more or less branched through $(1 \rightarrow 2)$, $(1 \rightarrow 3)$ or $(1 \rightarrow 4)$ links. In most cases the length of the side chains is short, and branched residues vary between 5 and 33 per cent. The major commercial dextran is about 95 per cent $(1 \rightarrow 6)$ linked and 5 per cent $(1 \rightarrow 3)$ linked, and is made by selected strains of *Leuconostoc.* After ethanol precipitation, from the culture medium acid hydrolysis is used to reduce the overall molecular weight, though fungal dextranases can also be used.

The product with an average molecular weight of about 60 000 Da is used in medicine as a blood extender, while fractions of defined molecular weights (e.g. the Pharmacia 'T-' series, where 'T500 Dextran' would stand for dextran of weight average molecular weight 500 kDa) are familiar in laboratories and to some extent, like the pullulan 'P–' series, serve as polysaccharide standards in molecular weight calibrations. They are also used as part of incompatible phase separation systems, usually with polyethylene glycol.

'Blue dextran' is a well-known marker for the void volume for gel filtration studies. Dextrans have found very wide application in laboratory work because they are particularly free from positive interactions with proteins. The interactions can be almost entirely characterized as coexclusion. This has found application, as already noted, in gel filtration media, but cross-linked dextran gels show other effects. For example, they swell and shrink in a way related to the osmotic pressure of the solvent system, and can be used to make miniature osmometers.

CYCLODEXTRINS

Cyclodextrins have attracted more interest than any other bacterial polysaccharide because of their unique structures. The interest has been more scientific than industrial, since xanthans are much more important in commercial terms, but there always seems to be a potential application of the cyclodextrins on the horizon which would lead to a major expansion in demand for them. Formerly called Schardinger dextrins after their discoverer, and still called cycloamyloses and cyclomaltoses here and there, they are cyclic molecules formed by glucopyranose linked by α-D-(1 → 4) linked bonds. They contain six, seven, eight or nine glucopyranose residues. It is impossible to form cyclic structures with less than six residues, while those with more than nine have so far been found to have at most a nine-membered ring with other glucose residues attached as branches through (1 → 6) links.

The structures (for a seven-membered ring, $cGlc_7$ or in common notation $cGlu_7$) are illustrated in Fig. 7.5. The nine-membered ring is uncommon and found only as a minor by-product of enzymatic synthesis of the others. The six-, seven- and eight-membered rings are sometimes called the α-, β- and γ-cyclodextrins, abbreviated as $cGlu_6$, $cGlu_7$, $cGlu_8$. They are water soluble and easily crystallized, and their structures have been determined. The great interest in them is because the cavity is of a size to include many small molecules of interest and is also fairly hydrophobic in character. The outside of the toroidal cone is full of hydrophilic hydroxyl groups.

As might be expected, most of the potential applications involve inserting smaller molecules into the cavity. The dimensions are roughly those of a cylinder with length of 7 Å and diameter of 4.5 Å for the six-membered ring and 8.5 Å for the eight-residue ring. Thus they can accommodate an aromatic ring like benzene, as well as alkyl chains.

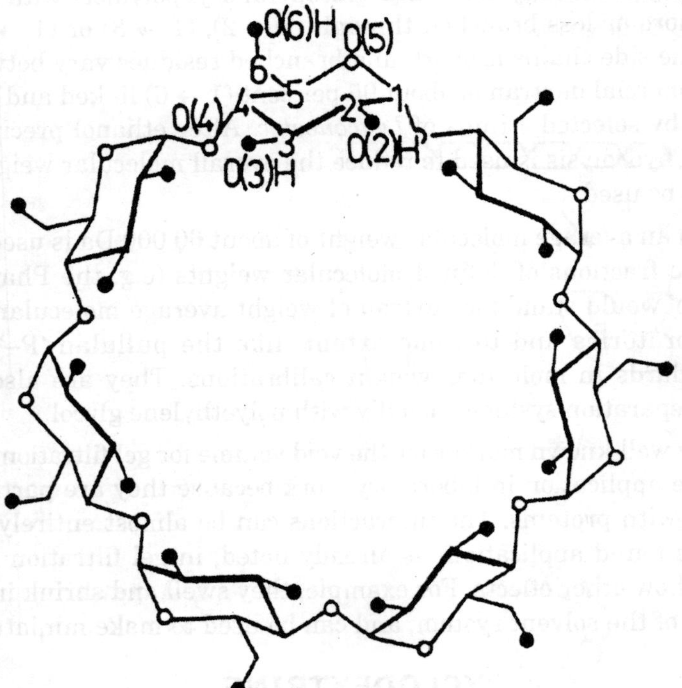

Fig. 5.5. Diagrammatic structure of $cGlu_7$, a cyclodextrin with a seven-residue ring.

Biosynthesis

The enzymes capable of synthesizing cyclic dextrins are exclusively bacterial in origin, but were discovered by observation of their activity of soluble starch *in vitro.* It is far from clear whether cyclic dextrins ever occur intracellularly, or are ever produced by the bacteria in question in the normal course of events. B. *macerans* is the best-known bacterium producing an amylase with glycosyl transferase activity. Seven other species are known, including *B. stearo-thermophilus*, *B. circulans* and *Klebsiella pneumoniae.*

The enzyme is induced when they are grown on starch, and the mechanism is believed to involve fission at the non-reducing ends of the chains, followed by cyclization as suggested in Fig. 7.6. They can also act as general transferases. This is a variation on the usual amylase mechanism where the group which becomes attached to the enzyme is released by hydrolysis, and the glucotransferases show strong homology with the α-amylase family and clearly belong to it. The crystallographic structures of the *B. circulans* and *B. stearothermophilus* enzymes are available. The enzyme can act in the vicinity of branch points, and can produce branched products of the type shown in Fig. 7.8. There is some doubt about the formation of a covalent link between the C1 carbon and the enzyme active centre aspartate residue. Evidence from pancreatic amylase supports the idea but crystallographic data suggest that the barley amylase does not. The distance between the aspartate nucleophile and the C1 carbon is too great. The point is not crucial since there are well-established binding sites for the polysaccharide chain on the enzyme and the exact location of these must play a role in the cyclization reaction. Different bacteria make enzymes with slightly different specificities which have been exploited to make different cyclodextrins.

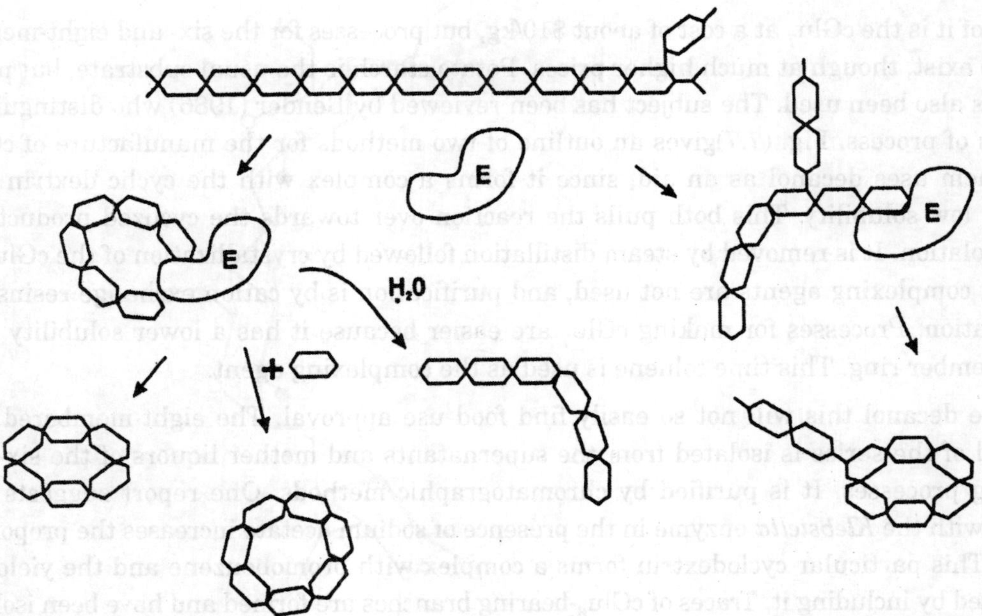

Fig. 7.6. Mode of action of cyclodextrin synthases. They act as amylases with general transferase activity but can also cyclize the product. Branched derivatives can also form as shown.

Thus that from *Bacillus 38-2* produces mainly the $cGlu_7$ while *B. macerans* and *Klebsiella* are used to make the $cGlu_6$ dextrins. However, it is possible to reduce the seven-membered ring to the six-membered by using a mixture with glucose and the transferase enzyme even where it is of the c7 type. There are no enzymes with a $cGlu_8$ preference which occurs as a minor (5 per cent) by-product of the others.

Large scale production

More important from the point of view of production methods is the tendency of the enzyme to make linear fission products under some conditions. Once the enzyme has split the main chain the fate of the groups attached to it depends on the chances of collision with potential receptors and the relative rate of reaction. Thus high substrate concentrations will lead to a better chance of transfer to another sugar residue rather than cyclization (or water) and there is an optimum substrate concentration to produce cyclic fragments. At less than 1 per cent starch, 85 per cent of amylose, 65 per cent of amylopectin and 55 per cent of glycogen are cyclized, while above 5 per cent starch only 35-50 per cent cyclization can be obtained.

Adding other enzymes such as pullulanases or isoamylase, a debranching enzyme, will also shift the ratios. In practice a compromise between the yield and the inconvenience of low concentrations of starch in the form of large vessels and costly water removal leads to a level of about 15 per cent starch although a limited prior hydrolysis to avoid retrogradation is desirable. Total production world wide is now in Excess of 100 tonnes a year, and is exclusively by batch methods in stirred tanks.

Most of it is the $cGlu_7$ at a cost of about \$10/kg, but processes for the six- and eight-member rings also exist, though at much higher prices. Potato starch is the usual substrate, but maize starch has also been used. The subject has been reviewed by Bender (1986) who distinguishes two kinds of process. Fig. (7.7)gives an outline of two methods for the manufacture of $cGlu_6$. One of them uses decanol as an aid, since it forms a complex with the cyclic dextrin with relatively low solubility. This both pulls the reaction over towards the cyclized product and aids its isolation. It is removed by steam distillation followed by crystallization of the $cGlu_6$. In the other, complexing agents are not used, and purification is by cation exchange resins and crystallization. Processes for making $cGlu_7$ are easier because it has a lower solubility than the six-member ring. This time toluene is used as the complexing agent.

Unlike decanol this will not so easily find food use approval. The eight-membered ring compound of the series is isolated from the supernatants and mother liquors of the six- and seven-ring processes. It is purified by chromatographic methods. One report suggests that digestion with the *Klebsiella* enzyme in the presence of sodium acetate increases the proportion of $cGlu_8$. This particular cyclodextrin forms a complex with bromobenzene and the yield can be increased by including it. Traces of $cGlu_6$-bearing branches are formed and have been isolated and crystallized. Although they have slightly different complexing abilities and higher solubilities, they have not been seriously considered for use because of their scarcity.

Decanol complex process
Soluble potato starch
20% in water, 80 °C, 5 minute
↓
Cycloglucan transferase
ex Klebsiella. Decanol (v/v),
40 °C, 24 hours
↓
α-amylase, 65 °C, 6 hours
Decanol complex precipitates
↓
Water suspension, steam distillation
to remove decanol
↓
Filteration, concentration and crystallization
of $cGlu_6$

Non-complexing process
Potato starch, 6% in water
120 °C, 30 minutes
↓
Cycloglucan transferase ex *B. macerans*
40 °C, 40 hours
↓
120 °C, 30 minutes, adjust to pH 5
Glucoamylase, amylase
40 °C, 30 hours
↓
Ion exchange, activated charcoal, concentrate
↓
Cation exchange chromatography
↓
Concentrate and crystallize $cGlu_6$

Fig. 7.7 . Two schemes for the production of $cGlu_6$.

Manufacture of Cycloglucotransferases

The enzyme from *Bacillus 38-2* is isolated partially purified from culture medium by an unusual method. This involves ethanol precipitation at –20 °C, drying and re-solution in water, complexing of the enzyme with maize starch, followed by elution with phosphate buffer and chromatography on gel filtration and DEAE (diethylaminoethyl)-cellulose. As a final stage preparative electrophoresis was used. This cannot be considered to be a serious industrialenzyme isolation, and is much nearer to a small laboratory isolation procedure.

Toluene complex process

Maize starch, 28% in water
80 °C, 10 minutes, 120 °C, 30 minutes
B. subtilis a-amylase
↓
B. macerans cycloglucan transferase
50°C, 30 minutes
Toluene 5% (v/v)
45 °C,105 hours
↓
Toluene distillation
50% water, 50% ethanol
↓
Supernatant, extra toluene
Collect complex
↓
Distill toluene
Crystallize, recrystallize,
active charcoal
↓
Final crystallization of $cGlu_7$

Fig. 7.8. A scheme for the production of $cGlu_7$.

Another method appears to be better suited to larger scale operation. This involves the use of an immobilized $cGlu_6$ column directly on the culture medium, followed by elution with the cyclodextrin. This gave a 100-fold purification and a 90 per cent yield. This is clearly the best method and the one to consider first when making these enzymes. Some selectivity can be obtained with $cGlu_7$ columns for enzymes favouring that product. Surprisingly, considering the nature of the reaction, a transferase immobilized on a vinylpyridine polymer matrix was effective, though it lost 75 per cent of its activity, and apparently requires a very long spacer arm. It could be repeatedly used over two weeks at least. Other transferases have also been successfully immobilized.

Applications

Amylase inhibition : Cyclodextrins are competitive inhibitors of α- and β-amylases, potato phosphorylase and pullulanase. They are similar therefore to a number of other amylase inhibitors that have been found and which are mainly small carbohydrate molecules that interfere with the active site of the enzyme. A different class of non-competitive protein amylase inhibitors are also known to exist. This property immediately raises two questions.

Do the cyclic dextrins present any hazards, perhaps as part of a foodstuff, or possibly by inhalation of a powder in a factory environment? Secondly, is there any possibility of using this property as a component of slimming aids? Feeding trials with rats suggest that $cGlu_7$ is relatively digestible compared with $cGlu_6$ which has been considered as a possible component of a low-calorie diet. It can be regarded as a component of the fibre, on a par with the indigestible part of starch, but considering its high cost and the fact that starch can also provide adequate amounts it is unlikely ever to find applications of this kind. Toxicity trials showed an LD50 of 12 g kg^{-1} in rats for oral intake, which is high, and values around 1 gkg^{-1} for intravenous injection for both $cGlu_6$ and $cGlu_7$. There are unlikely to be handling problems.

Artificial enzymes : The idea of artificial enzymes, designed to have a specific catalytic activity, is rather vague, since it might be thought to apply strictly to peptide chains synthesized *de novo,* albeit with the aid of synthetic cDNA and the cells' protein synthesis system. When this term was introduced this was not feasible, though it is now, and instead it came to mean some sort of polymer, not necessarily of biological origin since polyacrylamide has been involved, that is able to show catalytic activity similar to that shown by *bona fide* enzymes of biological origin. In all discussions cyclic dextrins play a prominent part. If the effectiveness of the 'enzymatic activity, is assessed by comparing the rate of hydrolysis in the presence and absence of the cyclodextrin, then the largest effect is on the hydrolysis of an ester group on E-3-carboxylene 1,2-ferrecene cyclohexane *p*-nitrophenyl ester which is accelerated ~3 200 000 times.

More typical is the hydrolysis of a t-butylphenylacetyl ester, which is accelerated ~250 times. There is a long list of esters, amides and organophosphates which are apparently able to hydrolyze more rapidly, via the formation of a covalent link to the cyclodextrin. They show chiral specificity and other typical kinetic enzyme properties such as competitive inhibition, and are indeed convincing models for enzymes. The effects are not confined to hydrolase-like activity; decarboxylation and oxidation group migration have both been described. The mechanisms are closely similar to those of enzymes themselves. So far applications seem to be confined to laboratory synthesis.

Inclusion complexes : The ability of cyclodextrins to enclose smaller molecules and effectively to sequester them from solution has led to a number of applications. One recent suggestion is as an aid to protein folding. Sodium dodecyl sulphate interacts strongly with proteins, to form a complex with high solubility, but where the peptide chain is generally extended into a rod shape. Removing the SDS is difficult, and the yield of correctly folded protein tends to be poor. When the complex of SDS and the enzymes carbonic anhydrase and citrate synthase was treated with $cGlu_7$, in both cases the SDS was sequestered and the enzymes folded to the correct active form in good yield.

Since present methods involve tedious and lengthy dialysis procedures this could be useful in recovering badly folded proteins from heterologous expression systems. In an application such as this the high cost would be justified. Formation of clathrates can be used to stabilize volatile and otherwise unstable materials. The complexes may also show enhanced solubility in water and a long list of drugs are known to be improved in their handling properties by the formation of inclusion complexes. Improved absorption can result from encapsulation so as to deliver materials to the appropriate part of the intestines.

In food applications flavours and fragrances are more effective when used in the form of complexes. Having said all this it is not easy to point to any particular application, though it is likely that most of the world production goes into a large number of small outlets in pharmaceuticals. One problem is that encapsulation effects can usually be achieved by other, probably cheaper means.

Derivatives of cyclodextrins

Cross linking the cyclodextrins yields some interesting gel filtration matrices. They can be used for chiral resolution, and for the removal of substances such as *naringin* and *limonin* from fruit juice, where they give unwanted bitter flavours. Pairs of cyclodextrins linked by bridges have also been made, with modified complexing properties, but these are only laboratory curiosities so far. Of greater potential use are the cyclodextrins attached to an agarose matrix since these are useful for the isolation of amylases, for which a demand does exist. Modified cyclodextrins can be provided with metal chelating groups, such as diaminoethane, and this extends the range of enzyme-like behaviour.

SCLEROGLUCAN AND SCHIZOPHYLLAN

A consideration of microbial polysaccharides would not be complete without at least a brief consideration of two fungal polysaccharides that are attracting increasing commercial interest: the weak-gelling scleroglucan and the schizophyllan systems. They are both large, neutral polysaccharides of weight average molecular weights, M_w, (as largely established by light scattering techniques) ~ 500000 Da, with a greater diversity being reported for scleroglucan. They both appear from X-ray diffraction to exist as hydrogen-bond stabilized triple helices, with resultant extra-rigid-rod like properties in solution: they have virtually the largest persistence lengths known for polysaccharides: ~1500 Å for schizophyllan and 2000 Å for scleroglucan.

The existence of the triple helix for scleroglucan has been further supported by electron microscopic/light scattering measurements of the mass per unit length, M_L= (2100±200) Da nm^{-1}, along similar lines to those that supported the duplex model for xanthan as discussed above. The Mark-Houwink-viscosity *a* parameter of 1.7 for schizophyllans of M_w <500 000 Da is again amongst the highest known for a polysaccharide and is on the rigid-rod limit. For chains of M_w>500000 Da the α parameter falls to -1.2 and corresponds to slightly more flexibility as the polymer length increases.

Chemically also they are very similar, with a backbone of repeating β(1 → 3) linked glucose residues:

...→ 3) β-D-Glc*p*-(1→...

In scleroglucan every third residue has a β(1 → 6) linked D-glucose branch which protrudes from the triple helix. B. T. Stokke and co-workers have demonstrated using electron microscopy that certain denaturation-renaturation treatments cause the formation of interesting ring structures or 'macrocycles'. In common with other branched glucans they appear to stimulate an immune response against tumour cells and, particularly scleroglucan, have been considered for use in cosmetics (as part of skin and hair products), for application in pesticides (to assist binding to foilage) and, along with xanthan and other polysaccharides, advantage can be taken of their high capacity to bind water and their high heat stability in oil well drilling fluids.

SYNTHETIC POLYSACCHARIDES: USE OF ENZYMES

While chemical synthesis of polysaccharides and oligosaccharides is possible in principle, in practice synthetic strategies are lengthy, complicated and skilled-labour intensive. In practice the main way of achieving this is likely to be by the judicious use of enzymes. The demand for synthetic oligo- and polysaccharides is stimulated mainly by experimental work. It is known that complex oligosaccharides on the surface of cells, usually in the form of conjugates with protein or lipid, are strongly antigenic. This is so whether the cells in question are bacterial (where gram negative organisms have lipopolysaccharide determinants), erythrocytes (since blood groups depend on surface oligosaccharides for their specificity), or transformed tumour cells which have characteristic groups.

In the last case potential diagnostic methods could be developed if suitable ways of detecting these changes were available. It is also possible to target antibodies onto the sites in question leading to potential treatment methods. Interactions between plants and bacteria such as the nitrogen fixers also depend on specific carbohydrate recognition sites. All these areas have stimulated a demand for oligosaccharides of known and often complex structure, in small amounts and where the materials would have high value.

This is quite different from the demand for bulk supplies, such as the 9000 tonnes p.a. of xanthan currently sold, and means that enzyme methods which would be simply not worth consideration can be used, or at least considered. Thus all the known methods of polysaccharide biosynthesis, including those using nucleotide cofactors, have been investigated with this application in mind.

NUCLEOTIDE-DEPENDENT SYNTHASES

The way in which nucleotide-dependent synthesis occurs is well established. Two enzymes are involved, the nucleoside transferase which attaches the nucleotide to the sugar and a glycosyl tranferase which then uses this high-energy intermediate to glycosylate an acceptor molecule, which it often does with very high specificity. It is this specificity that enables the directed synthesis of defined oligosaccharides. Many have been made and clearly within the limitations of the available specificities considerable control can be achieved.

The oligosaccharides are relatively small and the method could not easily be extended to larger molecules. It is possible that chemical methods could be used to link these oligosaccharides to larger chains, but their main use is to see if they block determinant sites for which high molecular weights are not needed. Problems are that the enzymes are not readily available and of course many would be needed so as to exploit their specificity, and they are difficult to store and lose activity. These are all matters that might be improved by genetic engineering approaches, but it is uncertain that the size of the applications would justify it.

Glycosidase Reversal

A second possible approach is to use glycosidases in the reverse of their normal function. In fact all glycosidases are potentially reversible, and after the enzyme substrate complex has been formed can transfer the residue to either water or other receptors such as another sugar, or to an alcohol. It is known for example that during the hydrolysis of lactose by lactases, in a commercial process, one of the problems is that the galactose residue can be transferred to a passing lactose to form a trisaccharide, and even higher polymers can be formed. Eventually they will probably be broken down again, but this may be too slow for the process. They might also be insoluble or attach themselves to some other component. In order to reverse hydrolysis it is necessary to greatly increase the concentration of the monomer, and also keep the water concentration as low as possible.

The action of yeast α-glycosidase on mannose has been investigated. It was possible to get up to about four residues with a variety of acceptor sugars, and a number of other studies using fungal enzymes reached a similar conclusion. Typical conditions were 83 per cent w/w mannose, at 60 °C with α-mannosidase, and only about 4 per cent maltotriose was formed, with correspondingly less of the higher polymers. There have also been attempts to use organic solvents to keep the water level low, without dramatic effect. One particularly interesting attempt made use of aqueous two-phase systems. An outline of the method is given in Fig. 7.9 and, apart from its use in oligosaccharide synthesis, illustrates an application of bacterial dextrans in forming the two-phase system. It is based on the observation that α-mannosidase is heavily concentrated into the dextran-rich phase (this probably indicates a dextran binding site on the mannosidase; proteins in general do not concentrate into one phase in this way).

The *p*-nitrophenol esters of mannose and galactose were the donors and acceptors, and the main advantage of the system was efficient use of the enzyme. Other processes using immobilized enzymes have been described. The main problems with this method are the relatively low specificity of the transfers and the inherent difficulty of obtaining high degrees of polymerization. It may be a method for obtaining certain tri- and tetrasaccharides if a demand should develop, but is likely to be less useful than the third method of oligosaccharide synthesis using sucrose-dependent glycosyl transferases.

Glycosyl Transferases

These enzymes are probably the best prospect for directed polysaccharide synthesis, because they use the energy-rich bond of sucrose (or in some instances other disaccharides) to drive the addition of the residue to the growing chain. The point has already been made when discussing the synthesis of fructans, which are made *in vivo* by enzymes using this reaction. The dextran sucrase of *Leuconostoc mesenteroides* has been extensively investigated to find the scope of the acceptors capable of receiving the transferred group. The group transferred is the glucose half of sucrose, and the newly formed links are almost always α(1 → 6), but a range of receptors can operate.

In addition the newly formed oligosaccharide can itself act as a receptor so that quite long chains can be built up. This is certainly the case with fructans, where the initial receptor is sucrose. Another transferase is found in *Streptococcus mutans,* an organism which produces an α(1 → 3) linked glucan with some α(1 → 6) links. This is insoluble and may be involved in adhesion of the bacteria to teeth. They also produce a soluble dextran with α(1 → 6) links but also 27 per cent α(1 → 3) links, which depends on a different transferase. Another enzyme from *Leuconostoc mesenteroides* synthesizes an *alternan,* with alternating (1 → 3) and (1 → 6) linkages, once again using sucrose as the donor. Table 7.2 shows the result of substituting different receptors and indicates the range of sugars that can act in this capacity.

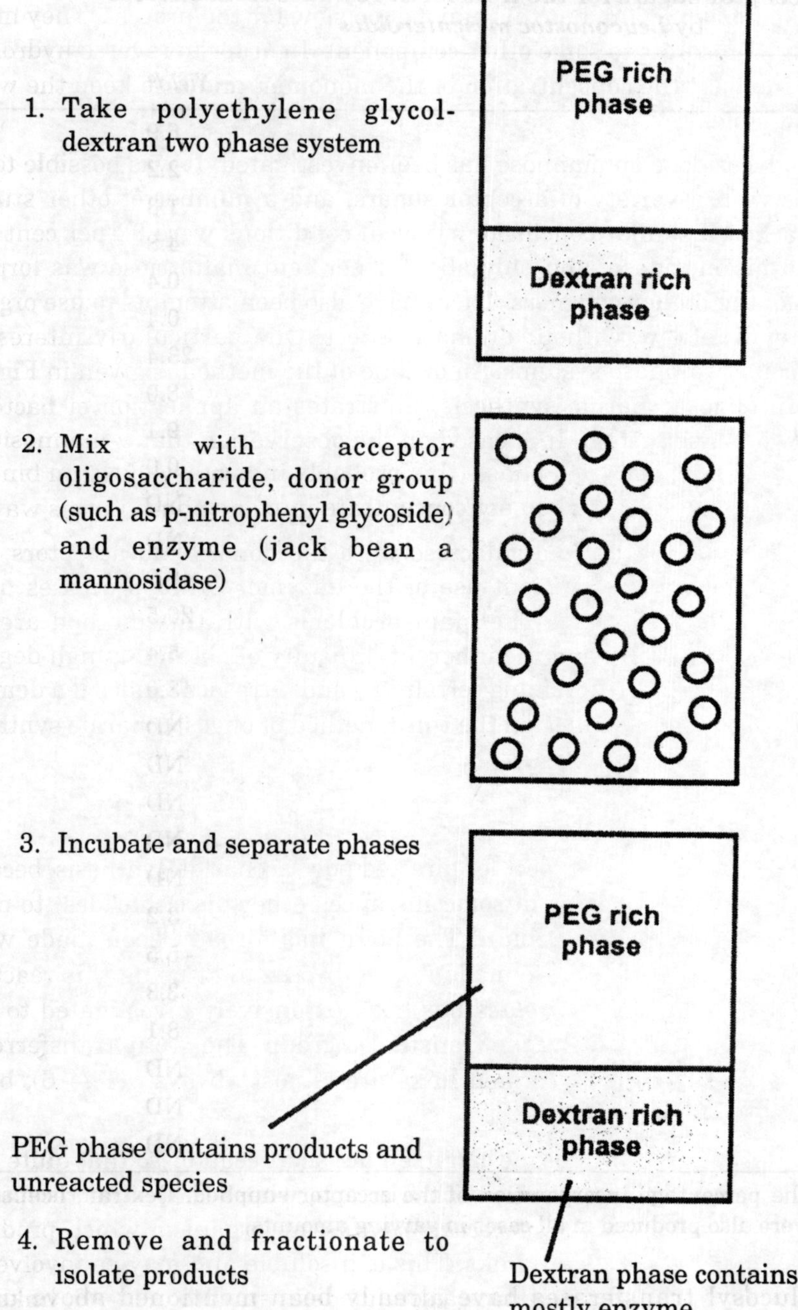

Fig. 7.9. Use of a polyethylene glycol (PEC) dextran two-phase system for the synthesis of oligosaccharides. The basic idea is that the α-mannosidase used is mainly contained within the dextran-rich phase, which can be repeatedly treated with reactants in the PEG-rich phase, with the loss of relatively little enzyme in each cycle.

Table 7.2 Acceptor sugars for the transfer of residues from sucrose by *Leuconostoc mesenteroides*

Acceptor sugar	*Yield*[a]
D-Glucose	6.9
D-Mannose	2.2
D-Galactose	1.3
D-Fructose	4.8
'L-Sorbose	0.4
D-Xylose	0.4
Methyl α-glucoside	23.4
Methyl β-glucoside	8.0
1 ,5-Anhydro-D-glucitol	9.1
α,α-Trehalose	0.0
α,β-Trehalose	ND
α,α-Trehalose	ND
Maltose	48.0
Nigerose	20.5
Cellobiose	5.4
Isomaltose	43.0
Isomaltulose	ND
Gentiobiose	ND
Melibiose	ND
Sophorose	ND
Kojibiose	ND
Laminaribiose	3.2
Turanose	- 5.5
Raffinose	3.3
Lactose	8.1
Lactulose	ND
Leucrose	ND
Theanderose	ND

[a]Yields are expressed as the percentage incorporation of the acceptor supplied. Dextran, isomaltose, leucrose, and free glucose were also produced in all cases in varying amounts.
ND, not determined.

The cyclodextrin glucosyl transferases have already been mentioned above under cyclodextrin. They can however use a wide varietyj of sugars as receptors, and some that can, and some that cannot, are listed in Table 7.3. In one of the most promising uses of this enzyme a maltohexose was transferred to a cellopentose to create an eleven-residue chain. Clearly these enzymes can use quite large chains, but there seems to have been no systematic attempt so far to explore the limits of size. As indicated above, most of the demand for synthetic oligosaccharides so far is for relatively small ones, which can block determinant sites on cell surfaces, and there is no serious demand for, for example, 37-residue linear dextrans which can probably be obtained in other ways - with some difficulty.

Applications in Animal Feeds

Bacteria attach themselves to surfaces by interaction of surface lectins, usually glycoproteins, with the surface carbohydrates of the cells they are hoping to invade. This is particularly true of pathogens such as *Salmonella, E. coli* and *Vibrio cholera* attacking the gut epithelia. It has been found that mannans can block this interaction by themselves binding to the bacterial lectins, and oligosaccharides known to be rich in mannans derived from yeast cell walls have been proposed as additives to chicken, pig and rabbit feed. It is claimed that lower levels of pathogen and better performance result from this. It is also known that mannans can enhance the immune response which may also help to prevent infection. Barley diets benefit by the use of β-glucanases in chicken feeds, apparently by an indirect effect on the viscosity of the digesta, which leads to improved absorption of virtually all the nutrients. It is also possible that digestion provides extra absorbable sugars, but this is less significant.

Table 7.3: Acceptor and non-acceptor molecules for the cyclodextrin glucosyl transferse of *Bacillus* sp.

Acceptors	*Non-acceptors*
D-Glucose	D-Fructose
L-Sorbose	D-Mannose
D-Xylose	D-Glucosamine
D-Galactose	N-Acetyl-D-glucosamine
2-Deoxy-D-glucose	L-Rhamnose
3- O-Methyl-D-glucose	D-Allose
6-Deoxy-D-glucose	D-Quinovose
Methyl- α-glucoside	D-Glucuronic acid
Methyl-β-glucoside	L-Arabinose
Phenyl-α-glucose	D-Ribose
Phenyl-β-glucoside	α-D-Glucose-1 -phosphate
Maltose	D-Glucose-6-phosphate
Nigerose	Trehalose
Sophorose	2-Deoxy-D-arabinohexose
Sucrose	myo-Inositol
Maltulose	Glucitol
Palatinose	Xylitol
Cellobiose	Glycerol
p-Nitrophenylmaltoside	
6^2- α-Maltosylmaltose	
6^3- α-Glucosylmaltotriose	

Similar effects can be seen with pigs, but the main use of this approach is to reduce the differences that may otherwise be found between different batches of barley, since these vary considerably in β-glucan content.

Pentosanases have also been tried. Pentosans have a general reputation as anti-nutritional factors, and hydrolysis does improve digestibility, though exactly why is not clear. Xylanases and glucanases also have effects in wheat diets, and given the probability that all the activities are present to some extent in commercial enzyme preparations it is not easy to say precisely which enzyme is the main one.

The effect is most likely to be indirect via changes in digesta viscosity. Attempts to use cellulases have been limited by the poor performance of available enzymes in breaking down cell wall structures, while amylases have not found much application in the feed industry. This may be because the animals have no difficulty in degrading starch in any case. For the time being the main commercial enzyme use is confined to β-glucanases and pentosanases in chick feed.

CHAPTER

8 Glycolipids

Recent advances in the synthesis of disaccharides, glycosphingolipids, and oligosaccharides have been described. Progress in synthesizing human blood group antigenic determinants has been reviewed.

Polysaccharides

Poly(urea-urethane)s containing 2-acetamido-2-deoxy-D-glucosyl residues in the main chain have been prepared by direct addition polymerization of 2-acetamido-2-deoxy-D-glucose and di-isocyanates. The polymers were soluble in polar solvents and exhibited good reactivity toward acetylation and very high water absorption. Several copolymers of styrene and 2,3 :4,5-di-O-isopropylidene-l-vinylbenzyl-β-D-fructopyranose have been prepared by free-radical initiation. The specific rotation of the copolymers was linearly related to their molar compositions, which also affected the glass transition temperatures. 1,6-Anhydro-2,4-di-*O*-benzyl-3-*O*-but-2-enyl-β-D-glucopyranose has been copolymerized with 1,6-anhydro-2,3,4-tri-O-benzyl-β-D-glucopyranose. Crotyl groups were removed quantitatively and side chains introduced by reaction of the hydroxylated polymers with 2,3,4-tri-*O*-(4-methyl-benzyl)-6-*O*-(*N*-phenyl-carbamoyl)-1-*O*-tosyl-D-glucopyranose.

Decarbanilation and debenzylation yielded a family of stereoregular dextrans with randomly distributed 3-*O*-(α-*D*-glucopyranosyl) side chains, and these dextrans have been compared with naturally occurring microbial dextrans.

Poly(5,6-anhydro-1,2-*O*-isopropylidene-α-D-glucofuranose) has been obtained by anionic ring-opening polymerization of 5,6-anhydro-1,2-O-isopropylidene-α-D-glucofuranose. Phosphorus pentafluoride catalysis of the ring-opening polymerization of 1,6-anhydro-2,3,4-tri-*O*-benzyl-β-D-glucopyranose has been investigated using ^{31}P and ^{19}F n.m.r. spectroscopy, and a polymerization mechanism proposed.

The cationic ring-opening polymerization of 3,5-anhydro-1,2-*O*-isopropylidene-α-D-xylofuranose has been reported. Novel polymers which show a high degree of water absorption have been synthesized by direct addition polymerization of α, α-trehalose with di-isocyanates. Polycondensation of (1) yields a stereoregular polymer (DP 20) identified as a glucan with alternating (1 → 4)-α-D and (1 → 6)-β-D linkages. 2,3,6,2′,3′,4′-Hexa-*O*-acetyl-α-D-maltosyl bromide has been polymerized to give a polymer containing maltosyl structures.

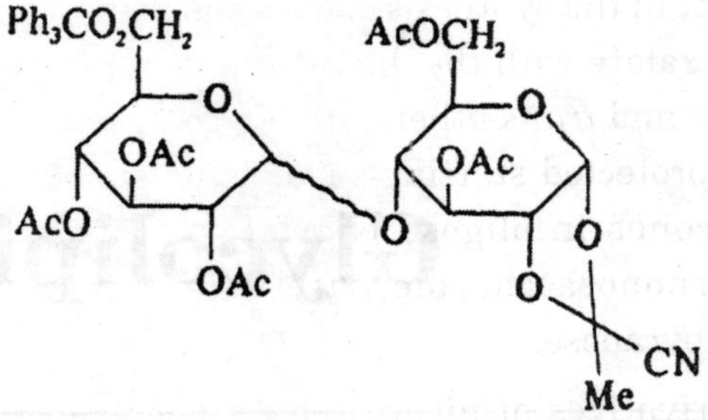

(1)

Oligosaccharides

A new method of glycosylation involving anomeric hydroxy-group activation by iminium salts has been investigated. Starting from a free hemiacetalic hydroxy group, glycosylation is performed by means of the DMF-carbonyl chloride Vilsmeier-Haack reagent and silver tosylate. The 1,2-*trans*-octa-acetates of cellobiose, lactose, and maltose have been converted into β-glycosides of 8-ethoxycarbonyloctanol in good yield by a single step reaction catalysed by tin(IV) chloride in dichloromethane at –10°. β-Glycosides were initially obtained *via* 1,2-orthoacetate intermediates, but *in situ* anomerization occurred during reaction at room temperature and provided a preparative route to the α-glycosides.

Hydrazide derivatives of the disaccharide glycosides were prepared for conversion into artificial carbohydrate antigens. By reaction of the acetylated glycosyl bromides of lactose, cellobiose, and maltose with silver trifluoromethanesulphonate in NN-tetramethylurea, considerably better yields of 1,2-trans-glycosides of 8-ethoxycarbonyloctanol are obtained than by conventional Koenigs-Knorr and Helferich conditions.

Use of trimethylpyridine as proton acceptor provided good yields of the corresponding 1,2-orthoacetates, which were rearranged to the 1,2-*trans*-glycosides using tin(IV) chloride. A facile synthesis of [3,6-di-*O*-acetyl-4-*O*-(chloroacetyl)-1,2-dideoxy-α-D-glucopyrano]-[2,1-*d*]-2-methyl-2-oxazoline (2) has been reported. The compound is a valuable intermediate for the sequential synthesis of complex sugar molecules.

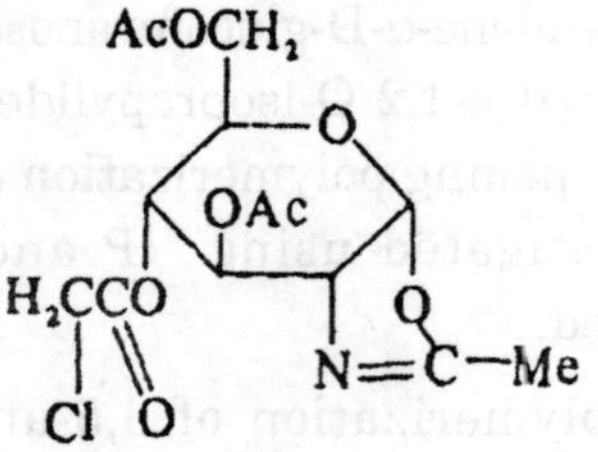

(2)

A two-step procedure for the preparation of 4-nitrophenyl-β-D-mannopyrano-side from D-mannose involves formation of 2,3 : 4,6-di-*O*-cyclohexylidene-α-D-mannopyranose. Treatment with triphenylphosphine, 4-nitrophenol, and diethyl diazocarbonate, followed by mild acid hydrolysis, yields the crystalline glycoside.

Significant improvement in the synthesis of disaccharides by cycloaddition may be achieved by reaction of diethyl mesoxalate with the dienyl ethers of protected sugars. This more active dienophile adds to both *cis*- and *trans*-ethers, unlike compounds hitherto reported, and more efficient utilization of the protected starting sugar is possible. The utility of secondary trityl ethers of sugars as aglycones in oligosaccharide synthesis has been demonstrated by glycosylation of various rnonosaccharide trityl ethers with 3,4,6-tri-*O*-acetyl-1,2-*O*-(1-cyanoethylidene)-α-D-glucopyranose.

Syntheses of deoxy-derivatives of nigerose have been reported. A convenient synthetic method for the preparation of laminaribioside (3) and some of its derivatives, *e.g.* methyl α-laminaribioside (4) and methyl (5), benzyl (6), and phenyl (7) β-laminaribioside, involves condensation of 2,3,4,6-tetra-*O*-acetyl-α-D-glucopyranosyl bromide (8) with methyl 2-*O*-benzoyl-4,6-*O*-benzylidene-α-D-glucopyranoside (9). Methyl 3-*O*-β-D-quinovosyl-α:-D-olivoside has been synthesized from laminaribiose and characterized as its peracetate.

Using the Helferich modification of the Koenigs-Knorr reaction, methyl 4,6-*O*-benzylidene-α-D-mannopyranoside has been treated with tetra-*O*-acetyl-α-D-mannopyranosyl bromide, yielding methyl 3-*O*-α-D-mannopyranosyl-α-D-mannopyranoside. Benzyl 2,4-di-*O*-benzyl-α-D-mannopyranoside has been condensed with 2-*O*-acetyl-3,4,6-tri-*O*-benzyl-α-D-mannosyl chloride in toluene-nitromethane, using a mixture of mercury (II) bromide-mercury(II) cyanide as promoter, in another modification of the Koenigs-Knorr reaction which gives high yield of α-D-mannosides. After deacetylation and catalytic hydrogenation 3,6-di-*O*-(α-D-mannopyranosyl)-α-D-mannose was obtained.

(3) $R^1, R^2 = H, OH; R^3 = H$
(4) $R^1 = R^3 = H, R^2 = OMe$
(5) $R^1 = OMe, R^2 = R^3 = H$
(6) $R^1 = OBn, R^2 = R^3 = H$
(7) $R^1 = OPh, R^2 = R^3 = H$

(8) (9)

Hepta-*O*-acetyl-α-kojibiosyl bromide has been converted into the β-chloride form by treatment with tetraethylammonium chloride, and then condensed immediately with benzyl 2,3,4-tri-*O*-benzyl-β-D-glucopyranoside in an $AgClO_4$-Ag_2CO_3 catalysed reaction. *O*-α-D-Glucopyranosyl-(1 → 2)-*O*-α-D-gluco-pyranosyl-(1 → 6)-D-glucose was obtained on hydrogenolysis and Zemplen deacetylation. 2-[4-(4-Toluenesulphonamido)phenyl] ethyl 2,3,4-tri-*O*-benzyl-α-D-gluco-pyranoside has been condensed with 2,3,4-tri-*O*-benzyl-6-*O*-(*N*-phenyl-carbamoyl)-1-*O*-tosyl-D-glucopyranose to yield 2,4-(4-toluene-sulphonamido)-phenyl ethyl 2,3,4,2′, 3′, 4′-hexa-*O*-benzyl-6′-*O*-(N-phenylcarbamoyl)-α-isomaltoside.

The disaccharide was decarbanilated in ethanol with sodium ethoxide. The sequence of coupling with the 1-*O*-tosyl-D-glucose derivative followed by decarbanilation was repeated to form the tri- and tetra-saccharide derivatives. The synthesis of 3-amino-3-deoxy-α-D-glucopyranosyl α-D-glucopyranoside (3-amino-3-deoxy-α,α-trehalose) from 4,6 : 4′,6′-di-*O*-benzylidene-α,α-trehalose has been reported. Syntheses have been devised for a number of disaccharides of interest as reference compounds in the study of polysaccharide and glycoprotein structure.

Condensation of benzyl 2-acetamido-2-deoxy-3,6-di-*O*-benzyl-α-D-glucopyrano-side with 2,3,4,6 -tetra-*O*-benzyl-1 -*O*-(*N*-methyl)acetimidoyl-β-D-glucopyranose followed by catalytic hydrogenation yielded crystalline 2-acetamido-2-deoxy-4-*O*-α-D-glucopyranosyl-α-D-glucopyranose (*N*-acetyl-maltosamine). An alternative route was also reported. 2-Acetamido-2-deoxy-4-*O*-β-D-glucopyranosyl-α-D-glucopyranose has been synthesized by condensation of benzyl 2-acetamido-3,6-di-*O*-benzyl-2-deoxy-α-D-glucopyranoside with 2,3,4,6-tetra-*O*-acetyl-α-D-glucopyranosyl bromide in the presence of mercury(II) bromide, followed by deacetylation and catalytic hydrogenolysis of the condensation product.

A simple synthetic method for the preparation of 2-acetamido-2-deoxy-4-*O*-β-D-galactopyranosyl-D-glucose (Scheme 1) and -D-mannose from 3 -*O*-β-D-galactopyranosyl-D-arabinose has been reported. Using this method the products are 4 to 1 in favour of the 2-epimer (10), while previous methods (Kiliani reaction or nitromethane addition) favour the other epimer. 2-Acetamido-2-deoxy-3-*O*-, -4-*O*-, and -6-*O*-α-D-mannopyranosyl-D-glucoses have been prepared by condensation of tetra-*O*-acetyl-α-D-mannopyranosyl bromide with benzyl 2-acetamido-2-deoxy-4,6-*O*-(4-methoxybenzylidene)-α-D-glucopyranoside, benzyl 2-acetamido-3,6-di-*O*-acetyl-2-deoxy-α-D-glucopyranoside, and 2-acetamido-1,3,4-tri-*O*-acetyl-2-deoxy-β-D-glucopyranose, respectively. 4-Methoxy-benzylidene groups were then removed with 80% acetic acid at 100 °C, *O*-acetyl groups by methanolic triethylamine or ammonia, and *O*-benzyl by hydrogenolysis.

The anomeric configuration of the α-D-mannopyranosyl linkage in the disaccharides was confirmed on the basis of their susceptibility to hydrolysis by the α-D-mannosidase from *Aspergillus niger.* 2-*O*-β-D-Galactopyranosyl-D-galactose has been synthesized for the first time. The product obtained after reaction of methyl 4,6-*O*-benzylidene-α-D-galactopyranoside with 2,3,4,6-tetra-*O*-acetyl-α-D-galactopyranosyl bromide, de-protection and separation from contaminants yielded D-galactose only on hydrolysis with crystalline (3-D-galactosidase from *Escherichia coli.* 6-*O*-Acetyl-2-azido-3,4-di-*O*-benzyl-2-deoxy-β-D-glucopyranosyl chloride reacts with 2,3,4,6-tetra-*O*-benzyl-α-D-glucopyranose or 2,3,4,6-tetra-*O*-acetyl-α-D-mannopyranose with good stereoselectivity to yield α,α-trehalosamine or α,α-mannotrehalosamine (Scheme 2).

Several other α,α-(1 → 1)-linked disaccharides of the trehalosamine type and containing additional azido-groups were also synthesized. Modification of mannobiose (4-*O*-α-D-mannopyranosyl-D-mannose) yielded the di-iodo compounds (11) and (12), which may be reducec with nickel(II) chloride-sodium borohydride to yield the glycosides (13) and (14) of 4-*O*-(β-D-rhamnopyranosyl)-α-D-rhamnopyranose or 4-*O*-(β-D-rhamnopyranosyl)-α–D-olivo-pyranose, respectively.

Synthesis of (1 → 4)-linked disaccharide glycosides of the 2,3-anhydro-4-*O*-(hex-2-enopyranosyl)-α-D-allopyranoside type has been reported (Scheme 3).

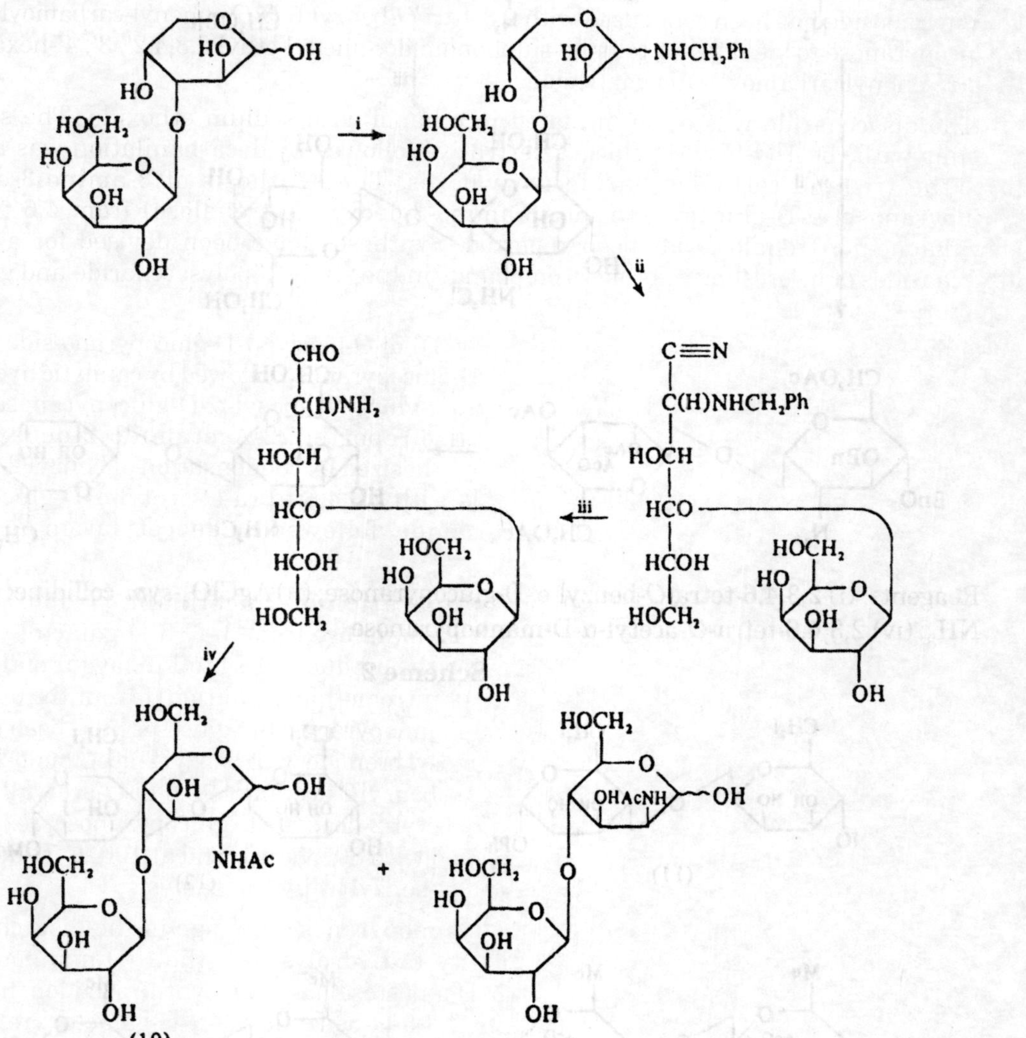

(10)

Reagents: (i) $PhCH_2NH_2$; (ii) HCN; (iii) H_2-Pd; (iv) Ac_2O

Scheme 1

1,2,3,6,2′,3′-Hexa-*O*-acetyl-β-lactose has been synthesized from lactose and treated with 2,3,4-tri-*O*-acetyl-α-L-fucopyranosyl bromide in a Koenigs-Knorr condensation. After Zemplen deacetylation *O*-β-L-fucopyranosyl-(1 → 3)-*O*-β-D-galactopyranosyl-(1 → 4)-D-glucose was obtained. Preparation of 2,4,6-tri-*O*-methyl-D-galactose by hydrolysis of the trisaccharide and from methyl α-D-galactopyranoside was also described.

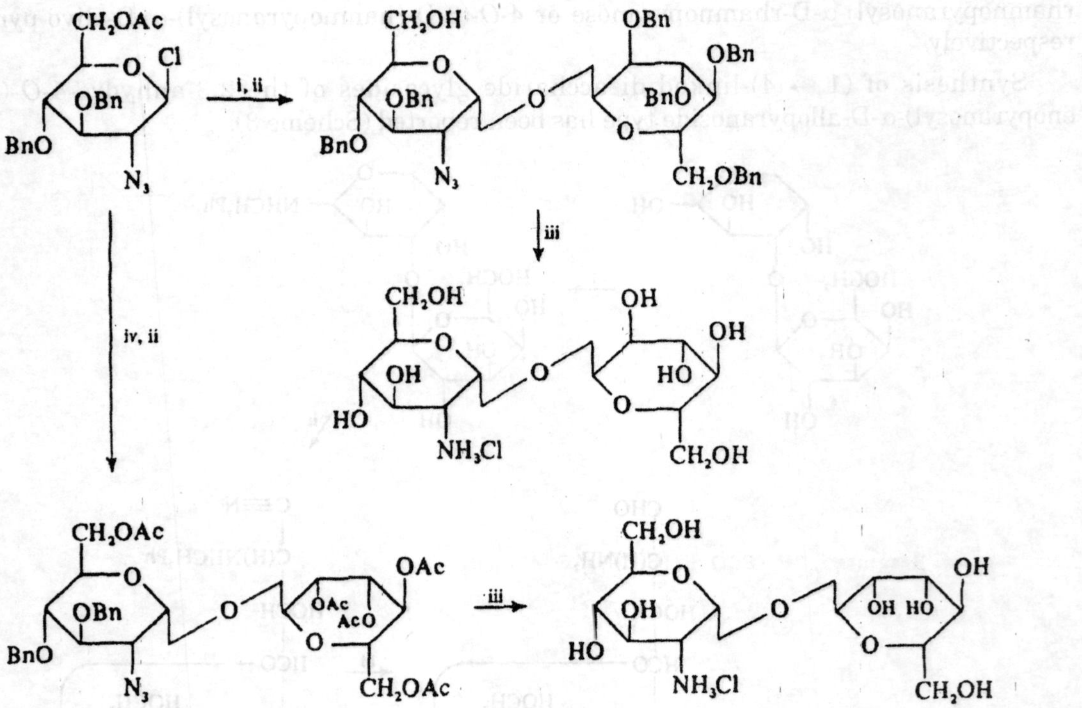

Reagents: (i) 2,3,4,6-tetra-*O*-benzyl-α-D-glucopyranose; (ii) $AgClO_4$-*sym*-collidine; (iii) Na-liq. NH_3; (iv) 2,3,4,6-tetra-*O*-acetyl-α-D-mannopyranose

Scheme 2

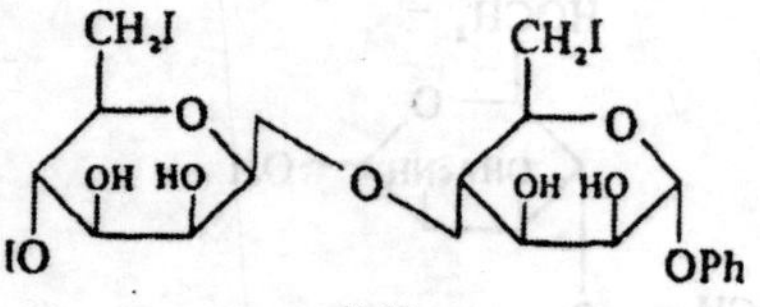

(11) 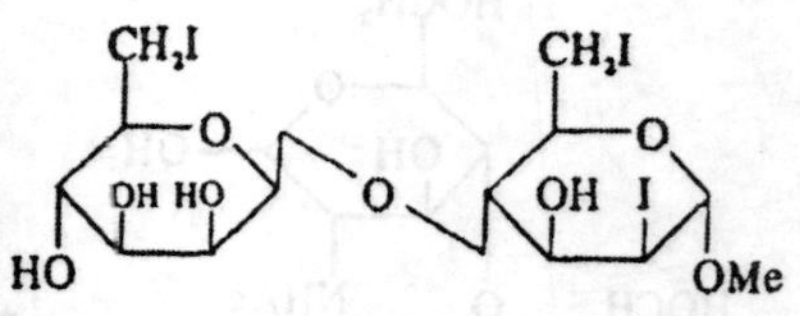 (12)

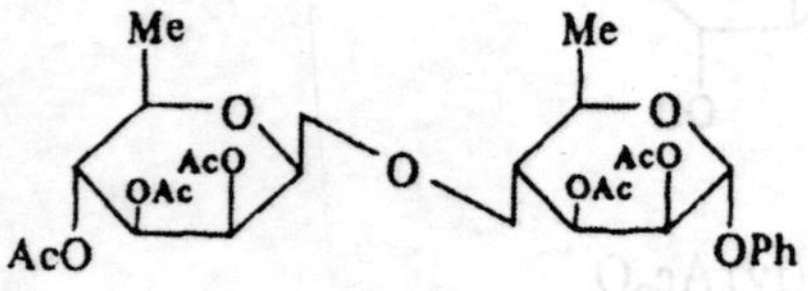

(13) 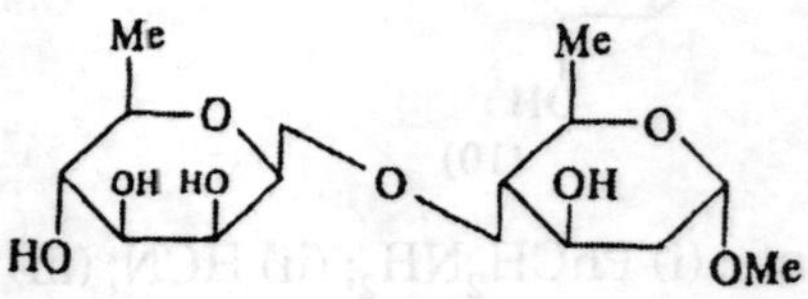 (14)

The synthesis of methyl *O*-β-D-galactopyranosyl-(1 → 6)-*O*-α-D-galactopyranosyl-(1 → 6)-*O*-β-D-galactopyranosyl-(1 → 6)-*O*-β-D-glucopyranoside from stachyose has been reported.

The 1,2′ : 1′,2-dianhydride of 3,4-di-*O*-acetyl-β-L-rhamnopyranose and methyl 3,4-di-*O*-acetyl-α-D-galactopyranuronate (15) has been synthesized from 2-*O*-(α-D-galactopyranosyluronic acid)-L-rhamnopyranose (16), demonstrating that the dianhydride is a cyclization product of (16).

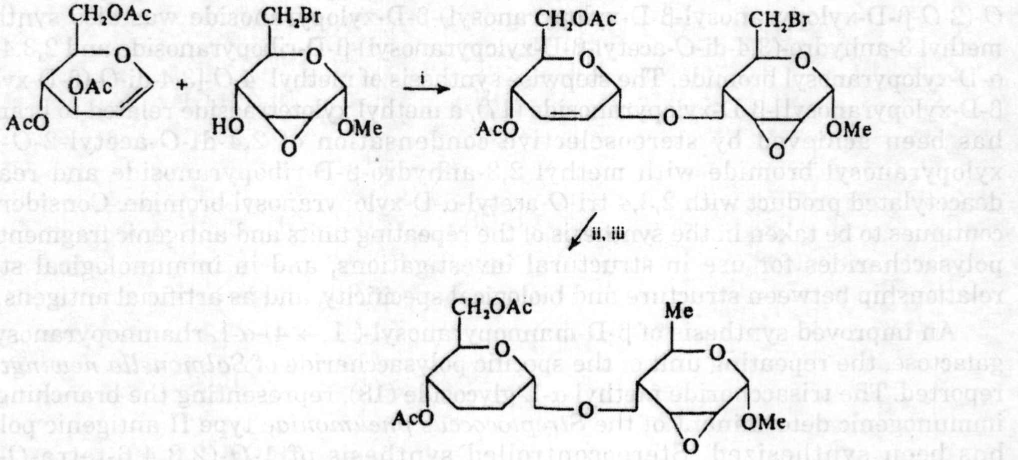

Reagents: i, $BF_3 \cdot Et_2O$; ii, Pd–C–$BaCO_3$–H_2; iii, Raney Ni–H_2

Scheme 3

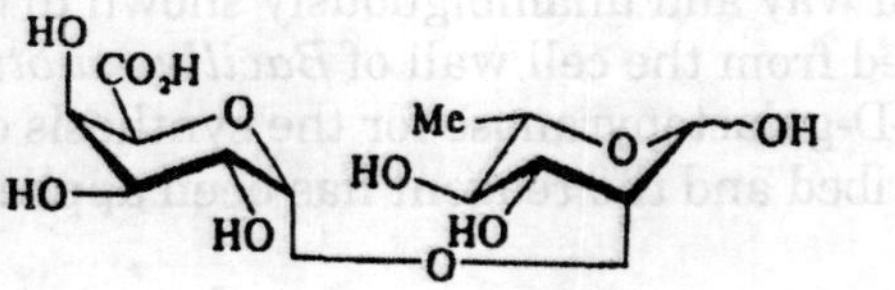

(16)

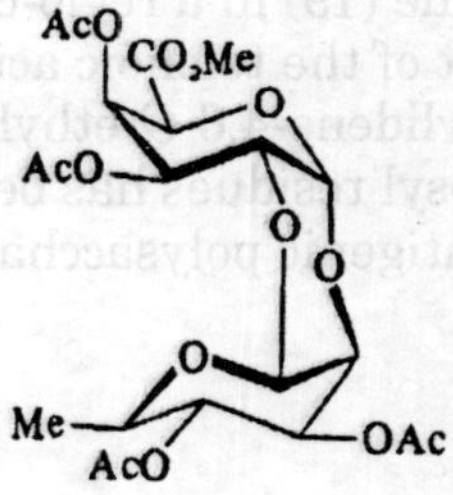

(15)

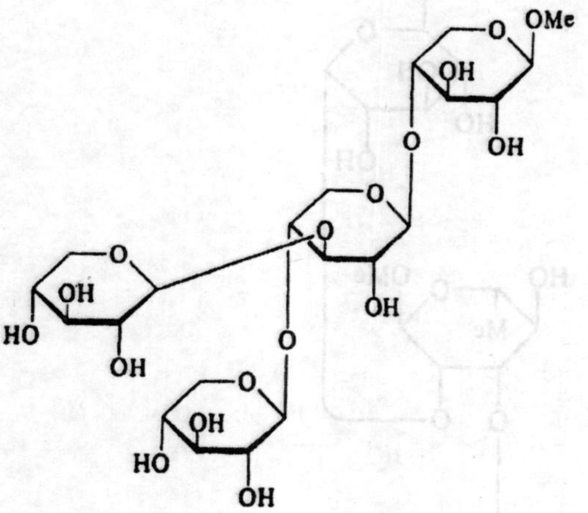

(17)

Methyl 2-*O*-β-D-xylopyranosyl-D-galactopyranuronate, a pseudoaldobiouronic ester, has been synthesized by a route in which the key step was the condensation of methyl (benzyl 3,4-*O*-isopropylidene-β-D-galactopyranosid)uronate with 2,3,4-tri-*O*-acetyl-α-D-xylopyranosyl bromide, followed by removal of the protecting groups.

The reaction of methyl 3-*O*-benzyl-β-D-xylopyranoside with 2,3,4-tri-*O*-acetyl-α-D-xylopyranosyl bromide in acetonitrile in the presence of mercury(II) cyanide yields crystalline methyl 3-*O*-benzyl-2,4-di-*O*-(2,3,4-tri-*O*-acetyl-β-D-xylopyranosyl)-β-D-xylopyranoside. Debenzylation and deacetylation gave 2,4-di-*O*-β-D-xylopyranosyl-β-D-xylopyranoside. Methyl 4-*O*-(2-*O*-β-D-xylopyranosyl-β-D-xylopyranosyl)-β-D-xylopyranoside was also synthesized from methyl 3-anhydro-(3,4-di-*O*-acetyl-β-D-xylopyranosyl)-β-D-ribopyranoside and 2,3,4-tri-*O*-acetyl-α-D-xylopyranosyl bromide. The stepwise synthesis of methyl 4-*O*-[3,4-di-*O*-(β-D-xylopyranosyl)-β-D-xylopyranosyl]-β-D-xylopyranoside (17), a methyl xylotetraoside related to branched xylans, has been achieved by stereoselective condensation of 2,4-di-*O*-acetyl-2-*O*-benzyl-α-D-xylopyranosyl bromide with methyl 2,3-anhydro-β-D-ribopyranoside and reaction of the deacetylated product with 2,3,4-tri-*O*-acetyl-α-D-xylopyranosyl bromide. Considerable interest continues to be taken in the synthesis of the repeating units and antigenic fragments of bacterial polysaccharides for use in structural investigations, and in immunological studies of the relationship between structure and biological specificity, and as artificial antigens.

An improved synthesis of β-D-mannopyranosyl-(1 → 4)-α-L-rhamnopyranosyl-(1 → 3)-D-galactose, the repeating unit of the specific polysaccharide of *Salmonella newington,* has been reported. The trisaccharide methyl α-L-glycoside (18), representing the branching point of the immunogenic determinant of the *Streptococcus pneumoniae* type II antigenic polysaccharide, has been synthesized. Stereocontrolled synthesis of 1-*O*-(2,3,4,6-tetra-*O*-acetyl-β-D-glucopyranosyl)-2,3-*O*-isopropylidene-D-glycerol has been achieved in good yield by use of the modified orthoester method.

The compound was then transformed into 1-deoxy-3-*O*-phosphono-D-glycerol-l-yl β-D-glucopyranoside (19) in a regio-controlled way and unambiguously shown to be identical to the repeating unit of the teichoic acid isolated from the cell wall of *Bacillus subtilis* var *niger.* Use of 1,2-isopropylidene-4,6-*O*-ethylidene-α-D-galactopyranose for the synthesis of 3-*O*-substituted galactopyranosyl residues has been described and the reagent has been applied to the synthesis of bacterial antigenic polysaccharides.

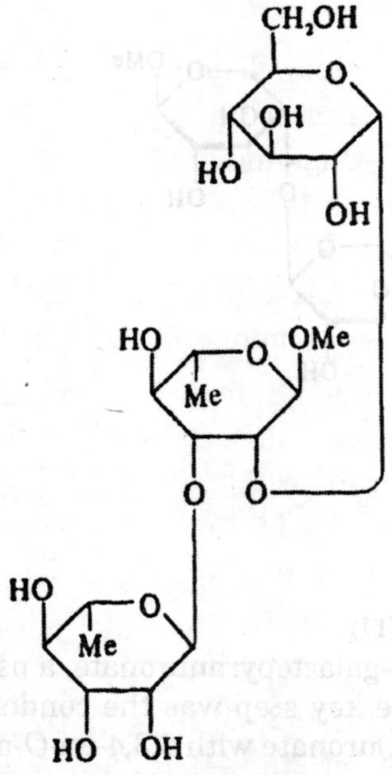

HO–CH₂OH pyranoside structure: CH_2OH, HO, O, HO, O—CH_2, OH, HCOH

α – L-Rha*p*-(1 → 3)-β-D-Glc*p*NAc-(1 → R′)
(20)
α – L-Rha*p*-(1 → 2)-α-L-Rha*p*-(1 → R′)
(21)
β – D-Glc*p*NAc-(1 → 2)-α-L-Rha*p*-(1 → R′)
(22)

where $R' = O(CH_2)_8CO_2Me$

Three disaccharides [(20)-(22)] representing structural elements similar to portions of the repeating unit of the *Shigella flexneri* O-antigen have been synthesized. The glycosylation reactions utilized silver trifluoromethane sulphonate and *NN*-tetramethylurea or *sym*-collidine, conditions which generated the three 1,2-*trans*-linked glycosides in high yield and with good stereospecificity. Utilizing silver trifluoromethanesulphonate-promoted Koenigs-Knorr reaction conditions, the *Shigella flexneri* trisaccharide hapten 8-methoxycarbonyloctyl *O*-α-L-rhamnopyranosyl-(1 → 3)-*O*-(2-acetamido-2-deoxy-β-D-glucopyranosyl)-(1 → 2)-α-L-rhamnopyranoside has been synthesized in good yield. 4-Nitrophenyl 3-*O*-(3, 6-dideoxy-α-D-*xylo*-hexopyranosyl-α-L-rhamnopyranoside has been synthesized from 4-nitrophenyl α-L-rhamnopyranoside.

2-*O*-Benzoyl-3,4,6-tri-*O*-benzyl-1-*O*-tosyl-D-mannopyranose and 2,3,4-tri-*O*-benzyl-6-*O*-(*N*-phenylcarbamoyl)-1-*O*-tosyl-D-glucopyranose have been treated with partially blocked 2-[4-(4-toluene sulphonamido)phenyl] ethyl α-D-manno-and D-gluco-pyranosides. Disaccharides having α-D-mannopyranosyl-(1 → 2)-α-D-mannopyranose, α-D-mannopyranosyl-(1 → 6)-α-D-glucopyranose, α-D-mannopyranosyl-(1 → 6)-α-D-mannopyranose, and α-D-glucopyranosyl-(1 → 6)-α-D-mannopyranose structures, and a trisaccharide having the structure α-D-mannopyranosyl-(1 → 2)-[σ-D-mannopyranosyl-(1 → 6)]-α-D-mannopyranose were synthesized.

The synthesis of blood-group antigenic determinants continues to arouse considerable interest.[2] The blood group B antigenic determinant, α-L-fuco-pyranosyl-(1 → 2)-[α-D-galactopyranosyl-(1 → 3)]-D-galactopyranose (23) has been prepared using the dibutylstannylene derivative (24) which was coupled with 2,3,4,6-tetra-*O*-benzyl-α-D-galactopyranosyl chloride in *NN′ N″* -hexa-methylphosphorotriamide in the presence of lithium iodide. These reaction conditions were found to be superior to those conventionally used for halide-catalysed glycosidation reactions.

The product (25) was condensed with 2,3,4-tri-*O*-benzyl bromide and on deblocking the trisaccharide (23) was obtained. Use of stannylene compounds was not found to result in activation or selectivity in the reactions with benzylated glycosyl halides. The same tri-saccharide has also been synthesized stereospecifically using the imidate procedure. Allyl 3-*O*-benzoyl-4,6-*O*-benzylidene-β-D-galactopyranoside was first α-L-fucosylated by 1-*O*-(*N*-methyl)-acetimidyl-2,3,4-tri-*O*-benzyl-β-L-fucopyranose and then, after *O*-benzoylation, α-D-galactosylated by 1-*O*-(*N*-methyl)-acetimidyl-2,3,4,6-tetra-*O*-benzyl-β-D-galactopyranose.

The resulting trisaccharide was also obtained from allyl 2-*O*-benzoyl-4,6-*O*-benzylidene-β-D-galactopyranoside after α-D-galactosylation, *O*-debenzoylation, and α-L-fucosylation. Deallylation followed by catalytic hydrogenolysis yielded (26).

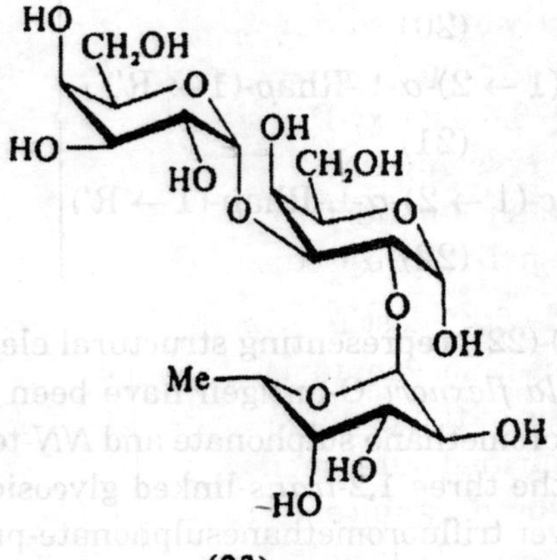

(23)

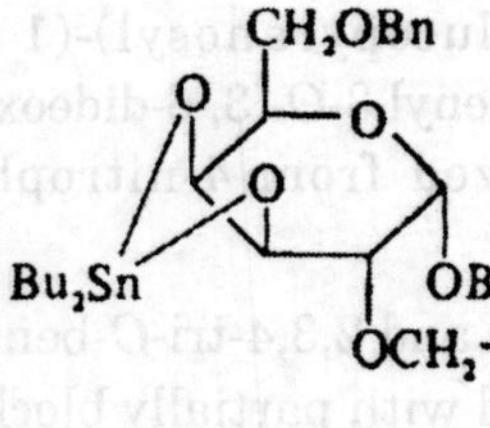

(24)

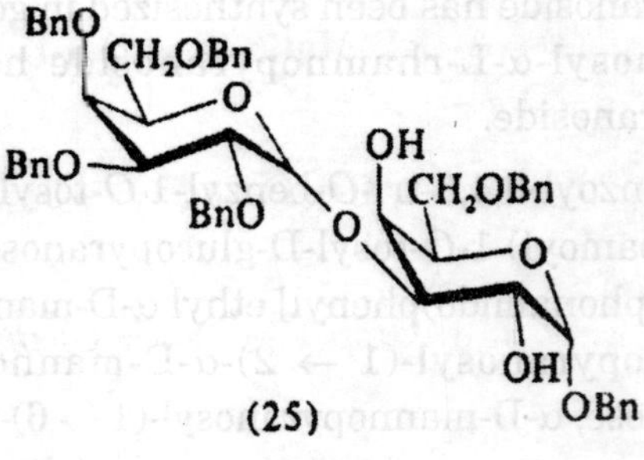

(25)

α-L-Fuc*p*-(1 → 2)-D-Gal*p*
3
↑
1
α-D-Gal*p*

(26)

Oligosaccharide antigenic determinants of blood-group substances A, B, and H have been synthesized by selective glycosidation reactions combined with a series of blocking and deblocking steps. All three oligosaccharides selectively inhibited haemagglutination with the corresponding human antibodies or lectins. The synthesis of *O*-α-L-fucopyranosyl-(1 → 3)-[*O*-β-D-galactopyranosyl-(1 → 4)]-2-acetamido-2-deoxy-D-glucopyranoside, a trisaccharide located at the 'non-reducing end' of numerous oligosaccharides or glycoconjugates of human origin, has been reported.

The imidate procedure for α-glycoside production has been applied successfully to the preparation of 2-acetamido-2-deoxy-4-*O*-(α-L-fucopyranosyl)-3-*O*-(β-D-galactopyranosyl)-D-glucopyranose, 2-acetamido-2-deoxy-3-O-(β-D-galactopyranosyl)-D-glucopyranose, and 2-acetamido-2-deoxy-4-*O*-(α-L-fucopyranosyl)-D-glucopyranose.

Glycoproteins. — Pseudoglycoproteins have been synthesized by treating proteins with carbohydrates, chiefly for use in affinity chromatography. Oligosaccharides coupled to bovine serum albumin for use as immunogens include chito-oligosaccharides (also coupled to cytochrome *c*) lactose covalently attached to 8-ethoxycarbonyloctanol, 4-isothiocyanatophenyl 3-*O*-(3,6-dideoxy-α-D-*xylo*-hexopyranosyl)-α-L-rhamnopyranoside (artificial *Salmonella O*-factor 8), 4-isothiocyanatophenyl-3-*O*-(3,6-dideoxy-D-araibino-hexopyranosyl) α-D-mannopyranoside *(i.e.* 4-isothiocyanatophenyl 3-*O*-α-tyvelopyranosyl-α-D-mannopyranoside, artificial *Salmonella O*-factor 9), synthetic antigenic determinants for yeast D-mannans and a linear (1 → 6)-α-D-gluco-D-mannan, and 2-(4-aminophenyl)ethyl α-isomalto-oligosaccharides.

The last were also coupled with edestin. Artificial *Salmonella* vaccines have been prepared by covalently linking the 2-(4-isothiocyanatophenyl)ethylamine derivative of an octasaccharide obtained by enzymic hydrolysis of the *O*-antigenic polysaccharide chain of *S. typhimurium* to diphtheria toxin, edestin, or purified outer membrane proteins ('porins') from *S. typhimurium.* All vaccines thus prepared induced significant protection against experimental *S. typhimurium* infection in mice. Mannotetraose (27), lacto-*N*-fucopentaose III (28), and lacto-*N*-tetraose (29) have been coupled to edestin, and the specificity of the anti-carbohydrate antibodies that they elicit in rabbits investigated.

α-D-Man*p*-(1 → 3)-α-D-Man*p*-(1 → 2)-α-D-Man*p*(1 → 2)-D-Man

(27)

β-D-Gal*p*-(1 → 4)-α-D-Fuc*p*(1 → 3)-β-D-Glc*pN*Ac-(1 → 3)-β-D-Gal*p*-(1 → 4)-D-Glc

(28)

β-D-Gal*p*(1 → 3)-β-D-Glc*pN*Ac-(1 → 3)-β-D-Gal*p*-(1 → 4)-D-Glc

(29)

D-Mannose, D-glucose, and 2-acetamido-2-deoxy-D-glucose have been coupled to albumin to form pseudoglycoproteins for use in a study of receptor-mediated binding of glycoproteins by alveolar macrophages. Fluoresceinyl derivatives of D-mannose, lactose, 2-acetamido-2-deoxy-α-D-galactose, and L-fucose have been covalently attached to albumin and used for direct visualization of membrane lectins *in situ.* The influence of glycosylation on tissue uptake of enzymes has been investigated.

Lactose and *N*-acetylneuraminyl-lactose have been coupled to *Escherichia coli* L-asparaginase by reductive amination with sodium cyanoborohydride. Qligosaccharides obtained from acetolysis of bakers' yeast D-mannan have been coupled to the cross-linked dimer of ribonuclease A and to human serum albumin by the same method.

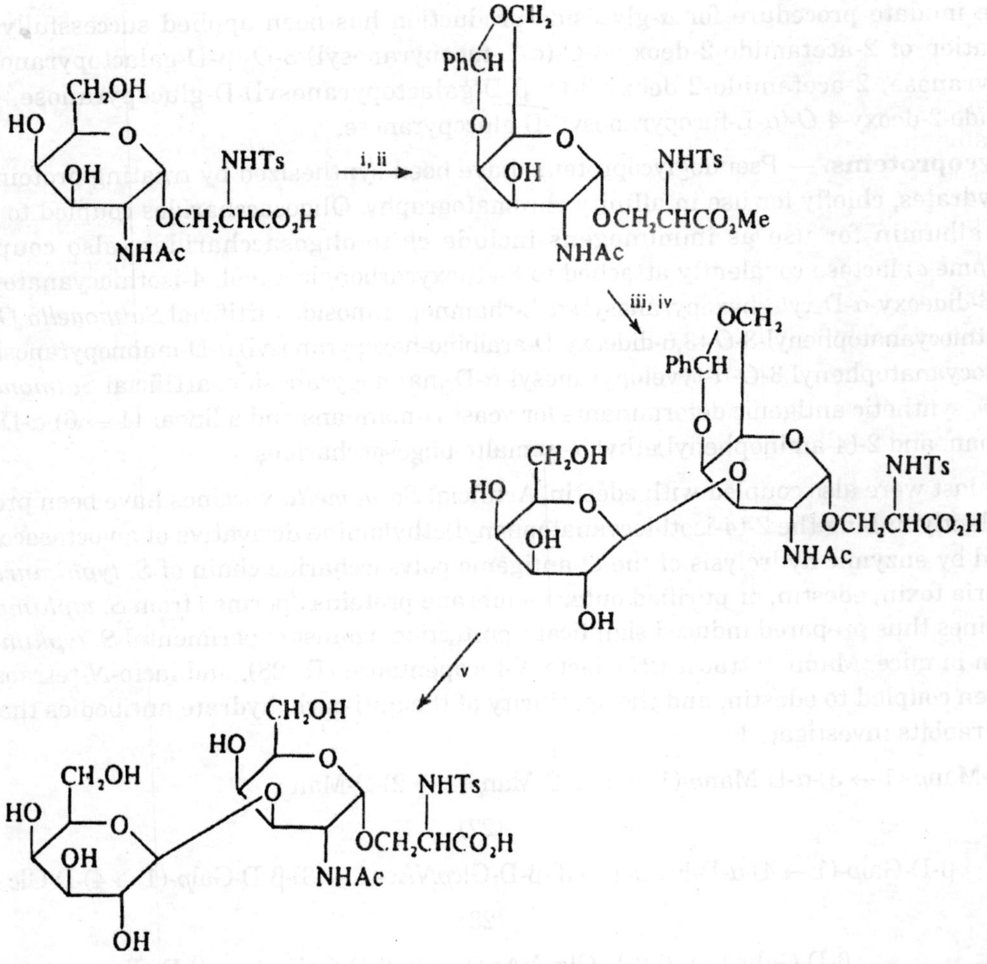

Reagents: (i) methylation; (ii) PhCHO-$ZnCl_2$; (iii) 2,3,4,6-tetra-*O*-acetyl-α-D-galactopyranosyl bromide-$Hg(CN)_2$; (iv) Dowex-1 ion exchange resin, elution with 0.1M HCl-30% aq. C_2H_5OH; (v) approx. 16 h incubation, 16°C

Scheme 4

Glycopeptides

Treatment with sodium in liquid ammonia has been used as a novel method of benzyl- and benzylidene-group deprotection in glycopeptide synthesis.

O^3-{*O*-β-D-Galactopyranosyl-(1 → 3)-*O*-[2-acetamido-2-deoxy-α (and β)-D-galactopyranosyl]}-*N*-tosyl-L-serine have been synthesized by the route shown in Scheme 4. The interaction of the synthetic compounds with D-galactose-binding lectins was investigated.

Two dimers, the galactosyl bromide 2,4,6-tri-*O*-benzoyl-3-*O*-(2,3,4,6-tetra-*O*-benzoyl-β-D-galactopyranosyl)-α-D-galactopyranosyl bromide, and the D-xylosyl-L-serine derivative, O^3-(2,3-di-*O*-benzoyl-β-D-xylopyranosyl)-*N*-carbobenzyloxy-L-serine benzyl ester, have been synthesized and then condensed and the product deblocked to yield (30).

β-D-Glc*p*-(1 → 3)-β-D-Gal*p*-(1 → 4)-β-D-Xyl*p*-(1 → L-Ser

(30)

Syntheses have been reported for a number of glycopeptide derivatives containing the N^4-(2-acetamido-2-deoxy-β-D-glucopyranosyl) hydrogen L-asparagin-ate linkage and having the amino-acid sequences 32-34, 33-35, 33-37, and 33-38 of bovine ribonuclease.

Di- and tri-peptides having an O^3-(2-acetamido-2-deoxy-3,4,6-tri-*O*-acetyl-β-D-glucopyranosyl)-L-serine residue have been synthesized and the mechanism of the β-elimination reaction undergone by the derivatives investigated.

O-(2-Acetamido-2-deoxy-β-D-glucosyl)-(1 → 4)-*N*-acetyl muramoyl-L-alanyl-D-isoglutamine (31), the dipeptide derivative of the disaccharide isolated from lysozyme digests of bacterial cell walls, has been prepared in a fourteen-step sequence from 2-acetamido-2-deoxy-D-glucose. The dipeptide disaccharide derivative (32) isolated from lysostaphin *endo*-β-D-2-acetamido-2-deoxy-glucanase digests of cell walls has also been synthesized (Scheme 5).

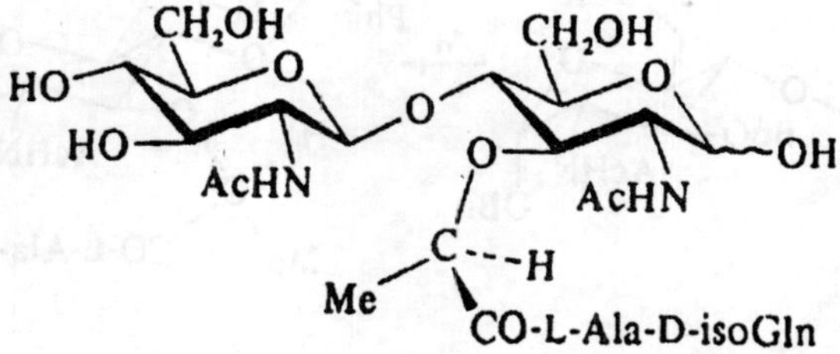

(31)

Glycolipids

A convenient synthesis of D-glucosyl phosphatidyl glycerol *via* a phosphotriester intermediate has been reported (Scheme 6). Use of the pivaloyl (*i.e.*, 2,2-dimethylpropanoyl) group for protection of one of the *sn*-glycerol primary hydroxy-groups was advantageous: it could be introduced selectively, no migration of the group was observed, and its removal could also be accomplished selectively to yield a free primary hydroxy-function.

Starting from 2-acetamido-2-deoxy-4,6-di-*O*-acetyl-3-*O*-(methyl 2,3,4-tri-*O*-acetyl-β-D-glucopyranosyluronate)-α-D-glucopyranose (33), a crystalline intermediate prepared by a conventional sequence of reactions, the total synthesis of P^2-dolichyl-P^1-Z-acetamido-Z-deoxy-3-*O*-(D-glucopyranosyl uronic acid)-D-glucose pyrophosphate (34) has been achieved. One of the key steps involved the transformation of (33) into the methyloxazoline (35) which was then converted into the stable crystalline disaccharide phosphate derivative in approximately 30% yield.

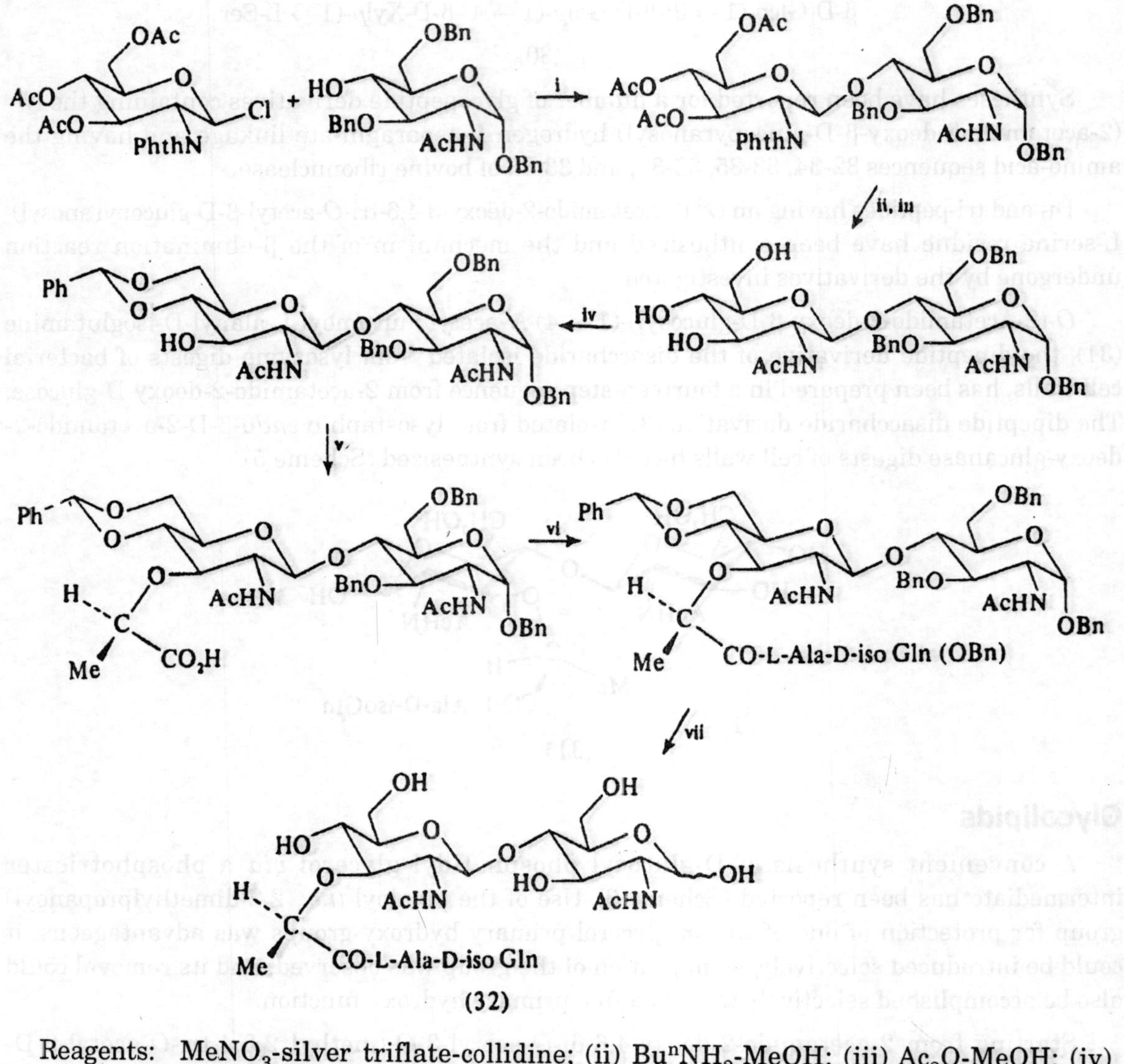

(32)

Reagents: $MeNO_3$-silver triflate-collidine; (ii) Bu^nNH_2-MeOH; (iii) Ac_2O-MeOH; (iv) PhCHO-$ZnCl_2$; (v) (*s*)-2-chloropropionic acid-dioxane-NaH; L-Ala-D-isoGln benzyl ester hydrochloride-*N*-methylmorpholine-isobutyl chloroformate-DMF; (vii) H_2-Pd-$MeCO_2H$

Scheme 5

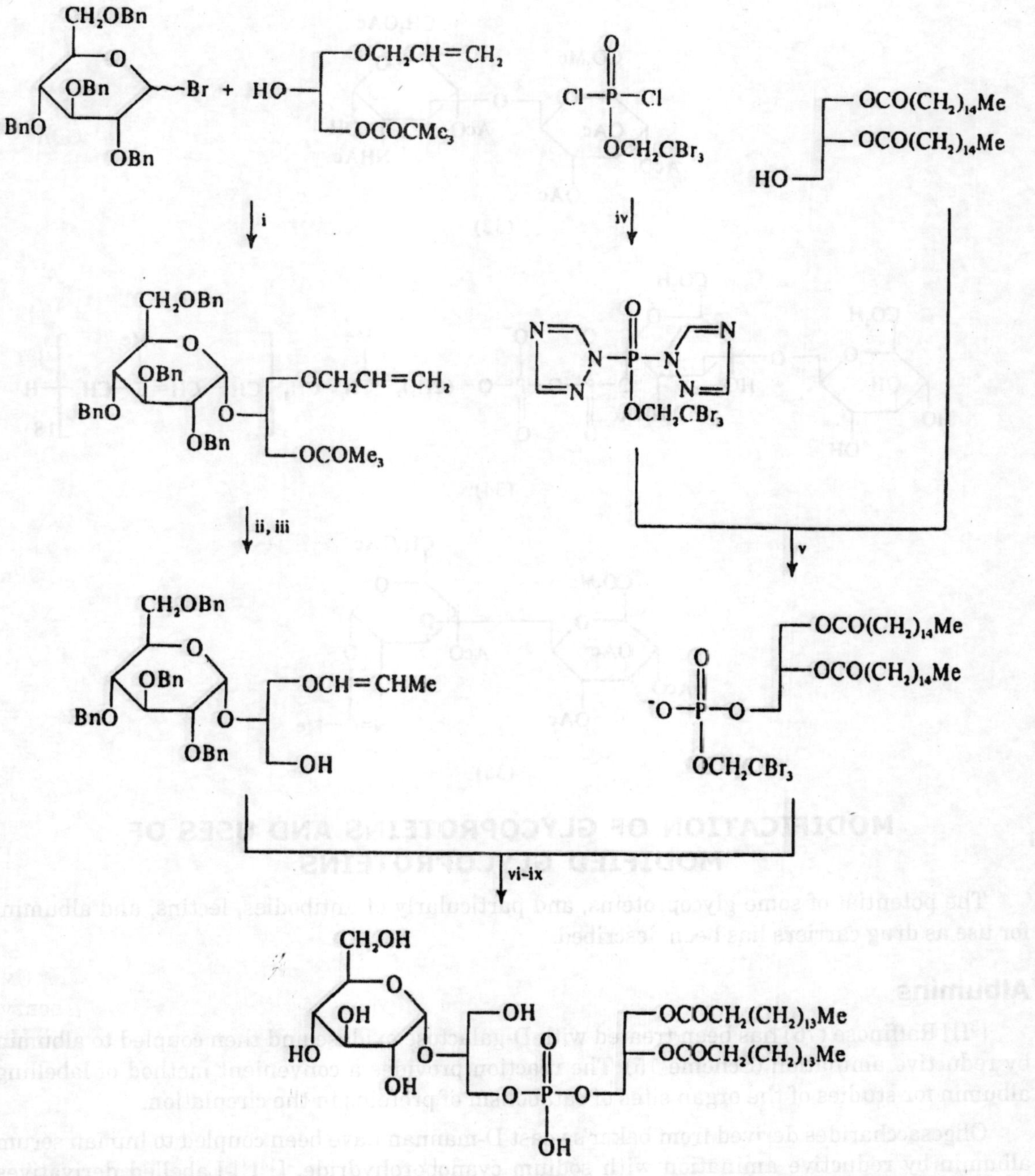

H_2O; (vi) 2, 4, 6-tri-isopropylbenzenesulphonyl-3nitro-1,2,4-triazolide; (vii) $HgCl_2$-HgO; (viii) Zn-pyr-pentane-2, 4-dione; (ix) H_2-Pd-C

Scheme 6

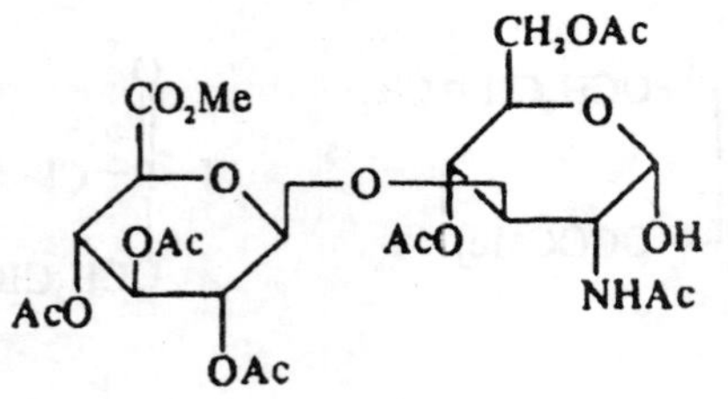

(33)

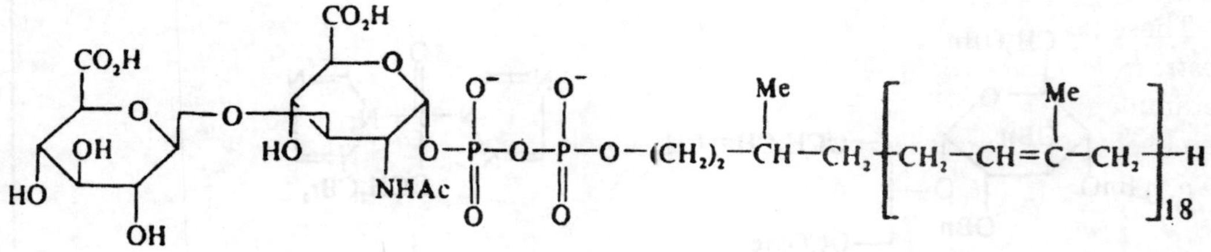

(34)

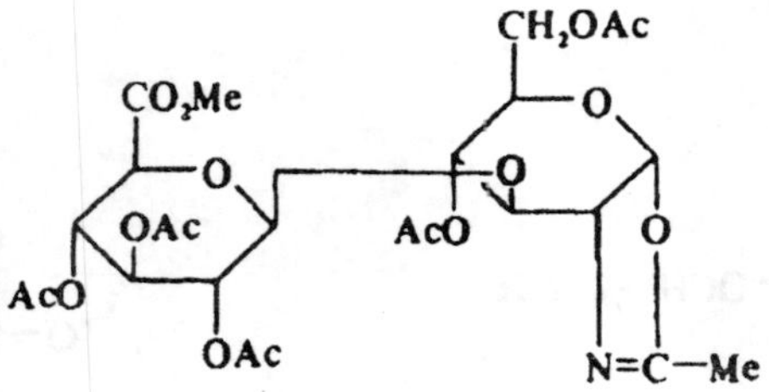

(35)

MODIFICATION OF GLYCOPROTEINS AND USES OF MODIFIED GLYCOPROTEINS

The potential of some glycoproteins, and particularly of antibodies, lectins, and albumin, for use as drug carriers has been described.

Albumins

[^{3}H] Raffinose (76) has been treated with D-galactose oxidase and then coupled to albumin by reductive amination (Scheme 15). The reaction provides a convenient method of labelling albumin for studies of the organ sites of catabolism of proteins in the circulation.

Oligosaccharides derived from baker's yeast D-mannan have been coupled to human serum albumin by reductive amination with sodium cyanoborohydride. [^{14}C]-Labelled derivatives containing two or four D-mannose residues per 10000 mol. wt. were administered intravenously to rats; selective uptake of these derivatives by the endothelial and Kupffer cells of the liver was observed within ten to fifteen minutes.

The extent of hepatic uptake was a function of the number and size of the D-manno-oligosaccharides coupled. Reductive 'manno-samination' may provide a means of directing proteins of potential therapeutic interest towards specific liver cells. Soluble copolymers of albumin and L-glutamate dehydrogenase have been prepared by glutaraldehyde cross-linking.

The kinetic and electron microscopic properties of the soluble derivatives were compared with data available concerning the enzyme immobilized within proteic films.

Crude preparations of gangliosides have been treated with albumin *via* a carbodi-imide mediator. The reaction was stopped by precipitation with ammonium sulphate. Following resuspension of the precipitate, unsubstituted albumin was mixed with the albumin-ganglioside conjugate and the suspension treated with glutaraldehyde and then coated onto cellulose filters.

These 'ganglioside affinity filters' could be used for the identification of toxigenic strains of *Clostridium botulinum* types C and D. Fluoresceinyl derivatives of D-mannose, lactose, 2-acetamido-2-deoxy-α-D-galactopyranose, and L-fucose coupled to bovine serum albumin have been used as cytochemical markers for the direct visualization of membrane lectins *in situ*.

β-D-Glc*p*-(1 → 6)-α-D-Glc*p*-(1 → 2)-β-D-Fru*f*

(76)

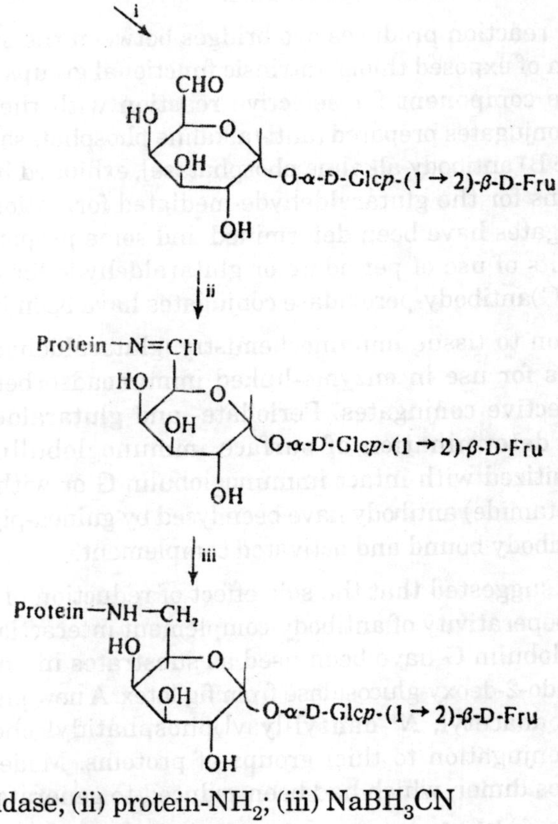

i

CHO
HO
O
OH
O-α-D-Glc*p*-(1 → 2)-β-D-Fru
OH

ii

Protein—N=CH
HO
O
OH
O-α-D-Glc*p*-(1 → 2)-β-D-Fru
OH

iii

Protein—NH—CH_2
HO
O
OH
O-α-D-Glc*p*-(1 → 2)-β-D-Fru
OH

Reagents: (i) D-galactose oxidase; (ii) protein-NH_2; (iii) $NaBH_3CN$

Scheme 15

Antibodies

A new volume of 'Methods in Enzymology' covers the preparation and use of affinity-labelled immunoglobulins. The specialized procedures required for the preparation of ferritin-antibody conjugates have been reviewed and the use of the conjugates for the localization of enzymes in mitochondria! membranes discussed. The procedures for the conjugation of rabbit immunoglobulin G and Fab′ antibodies with β-D-galactosidase using *NN′* -2-phenylene-dimaleimide have been improved by performing conjugation under an atmosphere of nitrogen at 4 °C and pH 6.5 for at least fifteen hours.

The conjugates were almost completely separated from unreacted molecules by gel filtration and there was no significant impairment of β-D-galactosidase activity or the ability of anti-(human immunoglobulin G) antibody to bind to human immunoglobulin G following conjugation. The conjugate preparation, however, was heterogeneous and one-third of each preparation consisted of aggregated conjugates less useful in sandwich enzyme immunoassay than the remaining material. A new method for the stepwise conjugation of antibodies and marker enzymes has been reported. The amino-groups of both proteins are selectively modified by reaction with 3-hydroxy-3-nitromethyl benzimidate hydrochloride, followed by reduction, and then diazotization of one of the components.

The coupling reaction produces azo bridges between the substituent groups and, with the possible exception of exposed thiols, intrinsic functional groups are not involved in the coupling. Activation of one component for selective reaction with the other avoids the formation of homopolymers. Conjugates prepared [anti-(alkaline phosphatase) antibody-lactate dehydrogenase and anti-(aldolase B) antibody-alkaline phosphatase], exhibited high immunoreactivity (80-100%). Optimal conditions for the glutaraldehyde-mediated formation of human immunoglobulin G-peroxidase conjugates have been determined and some properties of the conjugates reported. The relative merits of use of periodate or glutaraldehyde for the preparation of anti-(human immunoglobulin G)antibody-peroxidase conjugates have been investigated.

For application to tissue immunochemistry, glutaraldehyde conjugates were found to be superior, whereas for use in enzyme-linked immunoadsorbent assays, periodate treatment yielded more effective conjugates. Periodate- and glutaraldehyde-conjugates were equally suitable for the determination of surface immunoglobulin G on lymphoid cells. Sheep erythrocytes sensitized with intact immunoglobulin G or with reduced (dithioerythritol) and alkylated (iodoacetamide) antibody have been lysed by guinea-pig serum, indicating that reduced and alkylated antibody bound and activated complement.

Observations suggested that the sole effect of reduction of antibody disulphide bonds was to diminish the cooperativity of antibody-complement interaction. Glycopeptides prepared from human immunoglobulin G have been used as substrates in an investigation of the specificity of a β-D-2-acetamido-2-deoxy-glucosidase from fig latex. A new phospholipid-containing alkylating reagent, *N*-(N^{α}-iodoacetyl, N^{ε}-dansyl-lysyl)phosphatidyl choline ethanolamine, has been synthesized for conjugation to thiol groups of proteins. Model experiments were performed with a Bence-Jones dimer, which had been reduced to generate one thiol group per monomer.

After alkylation with the reagent, the modified protein became spontaneously attached to preformed liposomes and red cell ghosts.

Sea urchin egg tropomyosin has been isolated from an immunoprecipitate formed between anti-(lantern muscle tropomyosin) antibody and a sea urchin egg tropomyosin fraction. The antibody-antigen complex was heated to denature the antibody and the components were then separated.

Fetuin

After introduction of aldehyde groups at C-7 of neuraminic acid residues in fetuin, an amine-containing spectroscopic probe has been attached to the molecule by reductive amination. The spin-label has also been introduced into asialofetuin after modification of the glycoprotein with either D-galactose oxidase or sodium periodate.

[^{3}H]Raffinose (76) has been coupled to fetuin as shown in Scheme 15 for use in investigation of the organ sites of catabolism of proteins in the circulation.

Mucins

Bovine submaxillary mucin and its asialo-derivative have been spin-labelled after introduction of aldehyde groups as described for fetuin.

Phytohaemagglutinins

Cell fractionation by selective agglutination with lectins has been reviewed. The pH dependence of lectin interaction with sugars has been determined by affinity electrophoresis. Binding studies of native lectins and their ferritin conjugates to dispersed pancreatic acinar cells show that the conjugation procedure does not significantly affect the lectin-binding properties. When fibroblasts were treated with di- and mono-valent succinylated concanavalin A, α-D-2-acetamido-2-deoxyglucosidase endocytosis was inhibited, but there was no effect on extracellular β-D-2-acetamido-2-deoxyglucosidase accumulation. Tetravalent concanavalin A and divalent succinylated concanavalin A bind to the surface glycoproteins of macrophages and cause a marked increase in the rate of oxygen consumption owing to activation of the hexose monophosphate shunt.

The metabolic change induced could be reversibly inhibited by methyl a-D-glucopyranoside and was temperature-dependent. The effect of polyvalent lectins, including concanavalin A, di- and mono-valent succinylated concanavalin A, and wheat-germ agglutinin, on the secretion and endocytosis of lysosomal enzymes has been investigated. Concanavalin A has been treated with 3,3′-dimethyldithiobis-propio-imidate and the disulphide bond thus introduced then reduced and treated with ricin to yield a concanavalin A-ricin conjugate, retaining about one third of the inhibitor activity of ricin in a cell-free protein synthesis system.

The hybrid conjugate therefore contains a moiety that can bind to cell membrane receptors and an active fragment of a toxic protein. In order to study the mechanisms involved in the interaction of cell-surface components and effector proteins, a double-labelled dimeric hybrid molecule, consisting of the toxic A sub-unit of ricin and the cell-specific β-sub-unit of human chorionic gonadotrophin, has been synthesized. The biological properties of the conjugate were reported separately.

Horse-radish peroxidase chemically activated with glutaraldehyde and co-valently coupled to wheat-germ agglutinin has been shown to be forty times more sensitive than free horse-radish peroxidase for studies of retrograde axonal transport.

Orosomucoid

A glycopeptide obtained by pronase digestion of orosomucoid has been obtained and the carboxy-group of its terminal *N*-acetylneuraminic acid esterified with a carbodi-imide and then reduced with sodium borohydride. The reduced glycopeptide was shown to be resistant to hydrolysis by neuraminidase and other glycosidases.

Asialo-orosomucoid inhibits the disappearance of human glucocerebrosidase from the rat circulation following intravenous administration.

Ovalbumin

Glycopeptides prepared from ovalbumin have been used as substrates for investigation of the specificity of β-D-2-acetamido-2-deoxy-glucosidase from fig latex.

The nature of the cross-linking of proteins by glutaraldehyde has beer investigated using ovalbumin. Several quaternary pyridinium compounds were isolated from acid hydrolysates of glutaraldehyde cross-linked ovalbumin and their structures were confirmed by synthesis.

Miscellaneous Glycoproteins

Bovine lutrophin, bovine thyrotrophin, and human chorionic gonadotrophin preparations with carbodi-imide-mediated covalently cross-linked sub-units have been isolated in high yield and characterized by c.d. spectroscopy, radioligand receptor assays, and hormone-specific bioassays. The c.d. spectra of the native and cross-linked hormones did not differ much, suggesting that there is little conformational change on cross-linking.

The reaction of 1-ethyl-3(3-dimethylaminopropyl)carbodi-imide with bovine lutrophin has been studied and the reaction conditions have been investigated.

Spectrophotometric measurements, tryptic peptide maps, and sodium dodecyl sulphate-urea polyacrylamide gel electrophoresis banding patterns were obtained and compared for native and cross-linked derivatives of the glycoprotein hormone. Plasma transglutaminase has been shown to catalyse the cross-linking of fibronectin and collagen. The reaction may be of importance in wound healing and tissue repair. Disuccinimidyl suberate, a cross-linking agent, has been synthesized and used to probe the insulin-responsive D-glucose transport of the rat adipocyte.

Incubation of cells with the reagent blocked methyl-α-D-glucopyranoside transport without affecting either basal hexose transport or simple diffusion as indicated by uptake in the presence of cytochalasin B. α_1-Protease inhibitor reacts with proelastase 2 to form a 1:1 molar complex, which is stable to reduction and denaturation with sodium dodecyl sulphate. The application of chemical cross-linking for studies on cell membranes and the identification of surface receptors has been reviewed.

Treatment of human erythrocyte membranes with Cu^{2+}-2-phenanthroline at 0 °C results in the cross-linking of the Band 3 polypeptide to a dimer. The cross-linked protein was digested with proteolytic enzymes and the products were resolved by two-dimensional gel electrophoresis. The thiol groups involved in the cross-linking were identified.

The potential use of diazotized [^{125}I]di-iodosulphanilic acid as a label for cell-surface membranes has been investigated. Colorimetric determination of the compound is possible as it reacts at neutral pH with naphth-1-ol to yield a coloured complex (λ_{max} 430 nm). Results obtained suggest that when used under appropriate conditions di-iodo-sulphanilic acid can serve as a highly selective membrane label with minimal incorporation into intracellular soluble protein. A high molecular weight carbohydrate-rich fragment has been obtained by gel filtration of a collagenase-pronase digest of the bovine glomerular basement membrane. After performic acid oxidation, this fragment became susceptible to further digestion by these proteolytic enzymes and yielded three glycopeptide fractions which were characterized.

Immobilized Derivatives of Glycoproteins.

The modification of glycoproteins to yield biologically active immobilized derivatives continues to attract much attention.

A mechanical device for the continuous purification of biological material using an immunoadsorbent has been described. The use of immobilized lectins in affinity chromatography has been discussed, and the purification of cell-membrane glycoproteins on columns of immobilized lectins reviewed. New techniques of cell fractionation involving use of immobilized lectins or selective agglutination with lectins have been described. Covalent coupling of hog sulphated gastric mucin glycopeptides to agarose has provided a 'universal' affinity chromatographic material for purification of lectins and their ferritin conjugates. The 'spreading-out' of a protein on an affinity matrix at a critically low density, below which intermolecular bridge formation does not occur, prior to reaction with a chemical cross-linker *(e.g.,* glutaraldehyde) has provided an approach which has yielded valuable information about the tertiary and quaternary structure of concanavalin A.

Results were in agreement with evidence available from crystallography and other chemico- and bio-physical techniques. This matrix approach promises to be particularly useful in the structural study of poorly soluble membrane-bound proteins and other proteins which are difficult to crystallize. The immunoglobulin G fraction of rabbit anti-haemocyanin has been specifically adsorbed onto protein A-agarose and then covalently coupled to the matrix using dimethyl suberimidate.

The antibody molecules are thus in the correct orientation for combination with antigen since the association with protein A is *via* the Fc portion of the immunoglobulin G molecules, leaving the Fab combining sites free to interact with antigen. Development of a simple technique has been reported for coating glass cover slips and plastic Petri dishes with various immune complexes. The supports were coated with poly(L-Lys), which was then treated with 2,4-dinitrobenzene sulphonate followed by glutaraldehyde-mediated reaction with the appropriate substance.

Table 6. Modification of glycoproteins by coupling to insoluble matrices and other macromolecules

Glycoprotein	*Macromolecule or matrix coupled and mode of coupling*	*Use of product*
Albumin	Reaction with agarose cyclic imidocarbonate	Investigation of binding of fatty acids, steroids, and drugs to albumin
	Glutaraldehyde-mediated reaction with aminopropyl-silica gel	Purification of goat anti-(human serum albumin) antibody by immunoadsorption
Albumin, glutaraldehyde cross-linked	Carbodi-imide mediated reaction with aminoethyl-agarose	Affinity chromatographic purification of hepatitis B surface antigen from human sera
Anti-(A, B, H, blood group antigen) antibody		Detection by immunoadsorption of ABO blood group specificity on carcinoembryonic antigen preparations
Anti-(AMP-deaminase) antibody		Purification of AMP-deaminases by immunoadsorption
Anti-(adenovirus capsid sub-unit) antibody		Preparation of specific antisera against adenoviruses by 'affinity bead immunization'
Anti-(alkaline phosphatase) antibody	Reaction with agarose cyclic imidocarbonate	Development of an immunoadsorbent method for the continuous purification of human placental alkaline phosphatase
Anti-[α-amylase inhibitor *(Phaseolus vulgaris)*] antibody		Purification by immunoadsorption of α-amylase inhibitor from *P. vulgaris*
Anti-(carboxypeptidase N) antibody		Purification by immunoadsorption of rat serum carboxypeptidase N
Anti-cellulase antibody		Purification by immunoadsorption of *Trichoderma viride* cellulase

Glycoprotein	*Macromolecule or matrix coupled and mode of coupling*	*Use of product*
Anti-digoxin antibody	Microencapsulation in a semipermeable nylon membrane	Reagent for use in the radioimmunoassay of digoxin
Antienterobacterial common antigen) antibody	Carbodi-imide mediated reaction with Affi-gel 10®	Purification by immunoadsorption of entero-bacterial common antigen from *Eskkherichia coli*
Anti-(Factor VIII) antibody		Immunoadsorption of Factor VIII and fibrinogen from normal plasma; calibration of a one-stage Factor VIII assay
Anti-fibronectin antibody		Purification of fibronectin by immunoadsorption
Anti-(α-foetoprotein) antibody		Purification by immunoadsorption of human α-foetoprotein
Anti-(α-D-glucosidase) antibody		Comparison of α-D-glucosidase activities in urine of normal subjects and patients with Pompe's disease
Anti-(β-D-glucuronidase) antibody	Reaction with agairose cyclic imido-carbonate	Isolation by immunoadsorption of β-D-glucuronidase from human plasma
Anti-(glyceraldehyde 3-phosphate dehydrogenase) antibody or Fc fragments thereof		Immobilization by immunoadsorption of glyceraldehyde 3-phosphate dehydrogenase
Anti-(β_2-glycoprotein) antibody		Purification by immunoadsorption of an agglutinin in human sera
Anti-(hepatitis A virus) antibody		Purification by immunoadsorption of hepatitis A virus particles
Anti-(human chorionic gonadotrophin) antibody		Purification by immunoadsorption of human chorionic gonadotrophin-receptor complex
Anti-(human immunoglobulin G) antibody		Study of interactions between immobilized anti-bodies and antigen-peroxidase complex
		Isolation by immunoadsorption of human immunoglobulin G for use in enzyme immunoassay

Glycoprotein	*Macromolecule or matrix coupled and mode of coupling*	*Use of product*
Anti-haemocyanin antibody	Reaction with protein A-derivatized agarose	Purification by immunoadsorption of haemocyanin from *Limulus*
Anti-(human serum) antibody		Isolation by immunoadsorption of Clq (a sub-component of the first component of complement) from human serum
Anti-(infectious bovine rhinotracheitis antigen) antibody		Purification by immunoadsorption of infectious bovine rhinotracheitis antigen
Anti-interferon antibody		Study of nonspecific adsorption of interferon on immobilized serum immunoglobulin
	Reaction with agarose cyclic imidocarbonate	Purification by immunoadsorption of human interferon
Anti-(L-cell colony stimulating factor) antibody		Purification by immunoadsorption of L-cell colony stimulating factor
Anti-legumin antibody		Purification by immunoadsorption of legumin from *Pisum* seeds
Anti-lysozyme antibody		Assessment of immunological reactivity of synthetic lysozyme peptides
Anti-(β_2-microglobulin) antibody		Isolation by immunoadsorption of HLA antigen from human tissue
Anti-myosin antibody		Purification by immunoadsorption of rabbit ventricular myosin
Anti-oestradiol antibody	Adsorption on to poly(vinylidene fluoride) film	Study of oestradiol-binding proteins
Anti-(pregnancy zone protein)-antibody		Purification by immunoadsorption of pregnancy zone protein
Anti-(rabbit immunoglobulin G) antibody	Reaction with agarose cyclic imidocarbonate	Immunoadsorption of anti-(human immunoglobulin G) antibody-β-D-galactosidase conjugates prepared for enzyme immunoassay
Anti-(rabbit immunoglobulin G) antibody $F(ab)_2$ fragments	Reaction with poly(-L-Lys)-coated polyacrylamide beads	Detection of T-cell surface antigens
Anti-(rat cell surface antigen)-anti-body	Reaction with agarose cyclic imidocarbonate	Study by immunoadsorption of tritiated glyco-proteins from the surface of rat lymphoid cells

Glycoprotein	*Macromolecule or matrix coupled and mode of coupling*	*Use of product*
Anti-(soluble egg antigen) antibody		Isolation by immunoadsorption of *Schistosoma mansoni* egg antigen
Anti-tetanolysin antibody		Purification by immunoadsorption of streptolysin O
Antithrombin III) antibody		Separation of heparin into functionally distinct fractions on the basis of immunoreactivity
Anti-(L-thyroxine) antibody	Reaction with N-hydroxy-succinimide-derivatized controlled pore glass	Study of kinetic and thermodynamic properties of antigen-antibody interactions in heterogeneous reaction phases
	Microencapsulation in a semipermeable nylon membrane	Radioimmunoassay determination of free L-thyroxine in serum
Anti-(L-tri-iodothyroxine) antibody	Reaction with *N*-hydroxy-succinimide-derivatized controlled pore glass	Study of kinetic and thermodynamic properties of antigen-antibody interactions in heterogeneous reaction phases
Arachis hypogaea lectin		Affinity chromatography of antigenic determinants of membranes
Bandeiraea simplicifolia lectin		Affinity chromatography of human liver hydrolases
		Affinity chromatography of soluble conjugates of *Salmonella* specific oligosaccharides and bovine serum albumin
Blood group A substance		Affinity chromatography of *Vicia villosa* lectin
Bovine serum albumin		Purification by immunoadsorption of anti-(bovine serum albumin)antibodies; assessment of a modified procedure for preparation of agarose cyclic imidocarbonate
3-(*O*-Carboxymethyl)-oximino-5α-testosterone-bovine serum albumin conjugate		Purification by immunoadsorption of antibodies against the steroid antigen

Glycoprotein	*Macromolecule or matrix coupled and mode of coupling*	*Use of product*
7-(*O*-Carboxymethyl)-oximino-5α-dihydrotestosterone-bovine serum albumin conjugate		Purification by immunoadsorption of antibodies against the steroid antigen
Carcinoembryonic antigen		Analysis of human sera for Carcinoembryonic antigen-binding proteins
		Isolation by immunoadsorption of a carcino-embryonic antigen-binding human immuno-globulin G from normal sera
Concanavalin A		Purification of bovine sperm forward motility protein
		Separation of peroxidase-immunoglobulin G conjugate from unconjugated immunoglobulin G for use in enzyme immunoassay
		Purification of β-D-2-acetamido-2-deoxyhexosidases
		Purification of α-L-fucosidase
		Purification of 2-dansylamido-2-deoxy-D-glucose
		Affinity chromatography of soluble membranc receptors from mouse submaxillary glands
		Affinity chromatographic adsorption of activating factor of lymphocyte adenylate cyclase with consequent reversible deactivation of the enzyme
		Purification of mammalian serum haptoglobin
		Purification of *Schistosoma mansoni* soluble egg antigens
		Affinity chromatography of human alkaline phosphatases
		Purification of Carcinoembryonic antigen
		Purification of D-glucocerebrosidase
		Purification of *Phaseolus vulgaris* phyto-haemagglutinin
		Affinity chromatography of human liver acid hydrolases

Glycoprotein	*Macromolecule or matrix coupled and mode of coupling*	*Use of product*
Glycoprotein		Purification of human liver 'acid' β-D-galactosidases
		Separation of cellulase and 'cellobiase' from *Trichoderma viride*
		Purification of a myeloma immunoglobulin M and its tryptic Fab and $(Fc)_5$ fragments
		Fractionation of soluble brain glycoproteins
		Purification of rat kidney glycoproteins for production of heterologous antisera which induce congenital abnormalities
		Affinity chromatographic separation of integral membrane glycoproteins E1 and E2 of Semliki Forest virus
		Separation of A- and B-type glycopeptides of Semliki Forest virus
		Purification of human placental α-D-galactosidase
		Affinity chromatography of antigenic fractions of human seminal plasma
		Affinity chromatography of glycopeptides obtained by pronase digestion of immunoglobulin G from a patient with multiple myeloma
	Reaction with agarose cyclic imidocarbonate	Purification of 2-deoxy-2-sulphamido-D-glucose sulphohydrolase
		Affinity chromatography of glycoproteins in purified myelin
		Affinity chromatographic purification of a D-galacto-D-mannan antigen from *Aspergillus fumigatus*
		Studies of the interaction of concanavalin A with tryptic fragments from plasma fibronectin

Glycoprotein	Macromolecule or matrix coupled and mode of coupling	Use of product
		Affinity chromatography of acid ribonuclease from HeLa cell lysosomes; demonstration that the enzyme is a glycoprotein
		Purification of human liver 'acid' α-D-glucosidase
		Purification of two forms of α-D-2-acetamido-2-deoxygalactosidase
		Study of specific binding of [^{35}S]-labelled proteoglycans from secretions of human skin fibroblasts
		Separation of two molecular variants of α-foetoprotein
		Purification of bull seminal plasma hyaluronidase
		Purification of ribonuclease B
		Purification of biliary glycoprotein I
		Separation of human amylase isoenzymes
		Purification of cathepsin D from rat spleen
		Purification of a cathepsin E-like acid proteinase from rat spleen
		Study of inhibition of binding of immobilized concanavalin A to insulin receptors on isolated intact rat fat cells by inhibitors of concanavalin A
		'Anchoring units' for enzymes reversibly immobilized on the support for use in an analytical flow system
		Comparison of 4-methylumbelliferyl-β-D-galactosidase and G_{M1} β-D-galactosidase levels in normals and patients with I-cell disease
		Adsorption of *N*-[^{14}C]methyl ribonuclease B for use as substrate in a semi-quantitative assay of *endo*-β-D-2-acetamido-2-deoxyglucanase
		Isolation of antibodies specific for the carbohydrate-binding site of concanavalin A
	Glutaraldehyde-cross-linked onto Sephadex®	Investigation of the tertiary and quaternary protein structure of concanavalin A

Glycoprotein	*Macromolecule or matrix coupled and mode of coupling*	*Use of product*
	Reaction with N-hydroxysuccinimide ester of cross-linked agarose	Support matrix for the immobilization of enzymes
Concanavalin A, succinylated	Reaction with agarose cyclic imidocarbonate	Adsorption of β-D-fructofuranosidase for use in enzyme immunoassay
Cowper's gland mucous, desialylised		Purification of β-D-2-acetamido-2-deoxyhexosidase from rabbit sperm cytoplasmic droplets
2,4-Dinitrophenylated albumin		Isolation by immunoadsorption of anti-(2,4-dinitrophenyl) antibody preparations
Factor VIII		Purification of low molecular weight Factor VIII
Factor X	Reaction with agarose cyclic imidocarbonate	Separation of heparin into functionally distinct fractions; investigation of mechanism of action of heparin
Fetuin		Purification of neuraminidase from *Streptomyces griseus*
Fetuin, desialylised		Purification of β-D-2-acetamido-2-deoxyhexosidase from rabbit sperm cytoplasmic droplets
Haptoglobin		Purification of the lymphocytosis-promoting factor haemagglutinin of *Bordetella pertussis*
Hog sulphated gastric mucin glycopeptides		Purification of lectins by a general affinity chromatographic method
Immunoglobulin A	Adsorption onto polystyrene	Reagent for enzyme-linked immunoassay
Immunoglobulin E		Immunological study of surface glycoproteins and the immunoglobulin E receptor from rat basophilic leukaemia cells
		Purification by immunoadsorption of a cell-surface receptor for immunoglobulin E from rat basophilic leukaemia cells

Glycoprotein	*Macromolecule or matrix coupled and mode of coupling*	*Use of product*
Immunoglobulin G	Reaction with agarose cyclic imidocarbonate	Reagent for enzyme-linked immunoassay
		Immunoadsorption of anti-(human immunoglobulin G) antibody-β-D-galactosidase conjugate prepared for enzyme-linked immunoassay
		Purification by immunoadsorption of Clr, an activated component of the first component of the complement system
	Adsorption onto polystyrene latex beads	Investigation of the interaction between antibody and immobilized antigen using density-gradient centrifugation
	Adsorption onto polystyrene	Reagent for enzyme-linked immunoassay
Immunoglobulin G, Fc fragments	Reaction with agarose cyclic imidocarbonate	Isolation by immunoadsorption of immunoglobulin G Fc fragment-binding glycoproteins from human blood platelets
Immunoglobulin M	Adsorption onto polystyrene	Reagent for enzyme immunoassay
Influenza A virus	Adsorption onto polystyrene	Reagent for enzyme immunoassay
α-Lactalbumin *Lens culinaris* lectin	Reaction with agarose cyclic imidocarbonate	Purification by immunoadsorption of anti-carbohydrate antibodies
		Purification of lactose synthase from pig thyroid
		Investigation of mitogenic response of lymphocytes
		Immunoadsorption of surface proteins and glycoproteins of rat basophilic leukaemia cells
		Purification of a hyaluronidase from *Streptococcus pyogenes*
		Purification of an organ-specific tumour-associated antigen from rat mammary carcinoma
		Affinity chromatography of surface glycoproteins and immunoglobulin *E* receptors from rat basophilic leukaemia cells
	Reaction with agarose cyclic imidocarhonatc	Affinity chromatography of membrane immunoglobulins M and D from murine lymphocytes; study of their affinities for the lectin

Glycoprotein	*Macromolecule or matrix coupled and mode of coupling*	*Use of product*
		Affinity chromatography of glycoprotein fractions from long-term cultured human lymphocytes
		Purification of *Trypanosoma cruzi* cell-surface protein
		Affinity chromatography of tritiated glycoproteins from the surface of rat lymphoid cells
		Purification of the major surface coat glycoprotein of *Trypanosoma brucei brucei*
		'Anchoring units' for enzymes reversibly immobilized on the support for use in an analytical flow system
	Carbodi-imide-mediated reaction with gelatin-coated tubes and petri dishes	Affinity fractionation of human peripheral blood lymphocytes
Lotus tetragonolobus lectin		Affinity chromatography of glycoproteins in purified myelin
Melanoma cell membranes		Isolation by immunoadsorption of immunoglobulins from sera of patients with melanoma, sarcoma, and carcinoma
Ovomucoid	Reaction with agarose cyclic imidocarbonate	Purification of a lectin from rice bran
Prealbumin		Investigation of interaction between prealbumin and retinol-binding proteins
Prothrombin		Purification of staphylocoagulase
Ricinus communis lectin		Affinity chromatography of human liver hydrolases
		Purification of a myeloma immunoglobulin M and its tryptic Fab and $(Fc)_5$ fragments
		Affinity chromatography of glycoproteins in purified myelin

Glycoprotein	*Macromolecule or matrix coupled and mode of coupling*	*Use of product*
		Affinity chromatography of major cell-surface glycoproteins from normal and diabetic rat liver membranes
		Affinity chromatography of tyrosine tRNA's
	Reaction with epichlorohydrin-activated agarose	Affinity chromatographic purification of porcine thyroglobulin
	Reaction with dextran cyclic imidocarbonate	Isolation of serum inhibitors of desialylised glycoprotein-binding to hepatocyte membranes
	Glutaraldehyde-mediated reaction with agarose	Support for lysozyme immobilization
Rubella virus	Adsorption onto polystyrene	Reagent for enzyme-linked immunoassay
Soybean agglutinin	Reaction with agarose cyclic imidocarbonate	Affinity chromatography of human liver hydrolases
Soybean trypsin inhibitor	Carbodi-imide-mediated reaction with an alkyl spacer arm previously treated with 1-carbonyldi-imidazole-activated agarose	Affinity chromatographic purification of trypsin
	Glutaraldehyde-mediated reaction with sub-micron ferrite particles	Recovery of trypsin from product liquors by bioaffinity adsorption
Tamm-Horsfall glycoprotein		Affinity chromatographic separation of *Phaseolus vulgaris* leucoagglutinin and haemagglutinin lectins
Tetragonolobus purpurea lectin		Affinity chromatography of human liver hydrolases
Thrombin		Fractionation of low molecular weight porcine heparin into 'active' and 'relatively inactive' forms
		Affinity chromatography of heparin
Thyroglobulin		Affinity chromatography of rabbit anti-thyroglobulin antibodies
		Purification of *Phaseolus vulgaris* isolectins
Transplantation antigens		Purification of *in vivo* activity of an H-Z allo-antiserum

Glycoprotein	*Macromolecule or matrix coupled and mode of coupling*	*Use of product*
Tridacna maxima lectin	Reaction with agarose cyclic imidocarbonate	Affinity chromatographic purification of an L-arabino-D-galactan protein from *Gladiolus*
Ulex europaens I lectin		Affinity chromatography of human liver hydrolases
Urinary glycoprotein		Purification of human urinary glycoprotein by sub-unit exchange chromatography
Vicia faba lectin		Affinity chromatography of human serum proteins
Vicia villosa lectin		Selective immunoaffinity fractionation of murine cytotoxic T lymphocytes
Viscum album lectin		Affinity chromatography of human serum proteins
Wheat-germ agglutinin		Affinity chromatography of human liver hydrolases
		Affinity chromatographic investigation of interactions between wheat-germ agglutinin and sialoglycoproteins
		Affinity chromatography of glycoproteins; development of improved coupling procedure for wheat-germ agglutinin

Immobilization of Cells

The immobilization of cellular material for use both in affinity chromatography and for the production of enzymatically active preparations continues to receive considerable attention, as has been described in a general review and in a discussion of the transformation of organic compounds by immobilized microbial cells. *A chromobacter butyri* cells entrapped in polyacrylamide gel have been used in column form for the continuous high yield production of D-glucose 6-phosphate from D-glucose and metaphosphate.

The potential of active immobilized cells as catalysts for the production of hydrogen from water as an energy source has been investigated using *Anabaena cylindrica* cells adsorbed onto glass beads.

Arthrobacter simplex cells coprecipitated with calcium alginate gel have been used for the production of prednisolone from cortisol. The conversion of hydrocortisone into prednisolone has been carried out using *Arthrobacter simplex* cells entrapped in a polypropylene glycol-polyethyleneglycol copolymer, a maleic polybutadiene gel, or in urethane prepolymers. Mass-transfer effects on the rate of isomerization of D-glucose into D-fructose catalysed by commercially available immobilized *Arthrobacter* cells have been investigated and a theoretical treatment proposed. The nitrogenase activity of *Azotobacter vinelandii* cells immobilized by adsorption onto anionic exchange cellulose has been investigated under batch and continuous flow conditions. *Tiacillus* cells copolymerized in acrylamide have been used for the production of bacitracin. *Brevibacterium ammoniagenes* cells immobilized by copolymerization with acrylamide have been used in column form for the continuous production in high yield of pure nicotinamide adenine dinucleotide phosphate, for the continuous production of L-malic acid from fumaric acid, and for study of the NAD-kinase activity of the immobilized cells. The continuous production of L-malic acid has also been achieved with *Brevibacterium flavum* cells immobilized by gelation in K-carrageenan. *Catharanthus roseus, Digitata lanata,* and *Morinda citrifolia* plant cells have been immobilized in calcium alginate gels and applied to the production and transformation of natural products. Chloroplasts of *Brassica campestris* L copolymerized with acrylamide have been used to investigate carbon dioxide fixation by the immobilized cells, whose fixation activity was 65% that of free cells but which were more stable than free cells to alkaline conditions and high temperature.

The properties of spinach chloroplasts adsorbed onto diethylaminoethyl-cellulose have been reported and the immobilized chloroplasts used for the continuous photoreduction of 2,6-dichloroindophenol as an indicator of photo-system II activity in the chloroplasts. The formation of ATP from ADP and orthophosphate *via* the cyclic photophosphorylation system has been achieved using chloroplasts copolymerized with polyvinylalcohol onto an electrode.[665] *Citrobacter freundii* cells immobilized by co-casting with collagen have been used in a flow type sensor for cephalosporins.

The properties of *Escherichia coli* cells entrapped in κ-carrageenan and then glutaraldehyde-cross-linked have been investigated and used for the production of L-aspartic acid. L-Tryptophan has been synthesized by *E. coli* cells entrapped in polyacrylamide. A correction to a previously published paper on the synthesis of L-tryptophan by immobilized *E. coli* cells has been noted.

Steroid conversions have been carried out with *Nocardia rhodocrous* cells immobilized by entrapment in hydrophobic poly(propylene glycol) and poly(urethane) polymers, in hydrophilic poly(urethane) polymers, and in hydrophilic or lipophilic gels.

Penicillium cells immobilized by copolymerization with acrylamide, by co-casting with collagen followed by glutaraldehyde-cross-linking, and by coprecipitation with calcium alginate have been used for the production of penicillin G from D-glucose.

The analytical determination of D-glucose in molasses has been achieved using a microbial electrode consisting of *Pseudomonas fluorescens* cells co-cast with collagen onto an oxygen electrode.

Immobilized *Saccharomyces cerevisiae* cells have been used for the continuous production of glutathione, and for the continuous regeneration of ATP from adenosine. *Sparaduxus* cells have been immobilized onto glutaraldehyde-activated glycyl, β-alanyl, and 6-aminohexanoyl derivatives of 1,6-diaminohexane-derivatized hydroxyalkyl methacrylate gel. The effect of the length of the spacer arm and the manner of its activation was investigated.

The physical properties and characteristics of calcium alginate gels for cell immobilization have been investigated using *Saccharomyces uvarum* cells.

The properties of *Streptomyces phaeochromogenes* cells immobilized by copolymerization with 2-hydroxyethyl methacrylate gels have been reported. The radiation-induced polymerization was performed at low temperatures to avoid radiation damage to the cellular enzymes.

Immobilized *Trichosporon brassicae* cells have been used for the amperometric determination of acetic acid. Some properties of *Zygosaccharomyces lactis* cells immobilized on a modified hydroxyalkyl methacrylate gel have been reported.

Rat-liver mitochondria have been adsorbed onto alkylsilylated Porasil beads for the simulation of dynamic cellular environments in flow experiments. The immobilized mitochondria were functionally indistinguishable from mitochondria in suspension.

CHAPTER 9

Metabolism of Carbohydrates

The general principles of biosynthesis, as well as the pathways of formation of major carbohydrate and lipid precursors, are considered. Also described are the processes of glucorieogertesis, the synthesis of glucose 6-phosphate and fructose 6-phosphate from free glucose, and typical polymerization pathways for formation of polysaccharides. In this chapter, additional aspects of the metabolism of monosaccharides, oligosaccharides, polysaccharides, glycoproteins, and glycolipids are considered. These are metabolic transformations that affect the physical properties of cell surfaces and body fluids. They are essential to signaling between cells, to establishment of the immunological identity of individuals, and to the development of strong cell wall materials. Some of the differences in carbohydrates found in bacteria, fungi, green plants, and mammals are considered.

INTERCONVERSIONS OF MONOSACCHARIDES

Chemical interconversions between compounds are easiest at the level of oxidation of carbohydrates. Consequently, many reactions by which one sugar can be changed into another are known. Most of the transformations take place in the "jugar nucleotide derivatives". The first of this group of compounds to be recognized was **uridine diphosphate glucose** (UDPG), which was discovered around 1950 by L. F. Leloir during his investigation of the metabolism of galactose 1-*P*. The fact that interconversions of hexoses take place largely at the sugar nucleotide level was unknown at the time. Leloir's studies led to the characterization of both UDP-glucose and UDP-galactose. Summarizes pathways by which glucose 6-phosphate or fructose 6-phosphate can be converted into many of the other sugars found in living things.

Galactose and mannose can also be interconverted with the other sugars. A kinase forms **mannose 6-phosphate** which equilibrates with fructose 6-phosphate. Galactokinase converts free galactose to **galactose 1-phosphate,** which can be isomerized to glucose 1-phosphate by the reactions. Fructose, an important human dietary constituent derived largely by hydrolysis of sucrose, can also be formed in tissues via the **sorbitol pathway**. Fructose can be phosphorylated to fructose 1-phosphate by liver **fructokinase.** We have no mutase able to convert fructose 1-*P* to fructose 6-*P*, but a special aldolase cleaves fructose 1-*P* to dihydroxyacetone phosphate and free glyceralclehyde. Lack of this aldolase leads to occasionally observed cases of fructose intolerance. The glyceraldehyde formed from fructose can be metabolized by reduction to glycerol followed by phosphorylation (glycerol kinase) and reoxidation to dihydroxyacetone phosphate.

Some phosphorylation of fructose 1-*P* to fructose 1,6-P_2 apparently also occurs. Interconversion of ribose 5-*P* and other sugar phosphates is a central part of the pentose phosphate pathway. Free ribose can be phosphorylated by a **ribokinase.** Oxidation of UDP-glucose in two steps by yields **UDP-glucuronic** acid, which can be epimerized to **UDP-galacturonic** acid. Likewise, **guanosine disphosphate-mannose** (GDP-mannose) is oxidized to **GDP-mannuronic acid,** which undergoes 4-epimerization to **GDP-guluronic acid.** Looking again at the top of the scheme, notice that UDP-D-glucuronic acid may be epimerized at the 5 position to **GDP-L-iduronic acid.** However, the iduronic acid residues in dermatan sulfate arise by inversion at C-5 of D-glucuronic acid residues in the polymer. The mechanism of these reactions, like that of the decarboxylation of UDP-glucuronic acid to UDP-xylose, apparently have not been well investigated. Notice that glucuronic acid is abbreviated GlcA, in accord with IUB recommendations. However, many authors use GlcUA, ManUA, etc., for the uronic acids.

The Metabolism of Galactose

The reactions of galactose have attracted biochemists' interest because of the occurrence of the rare (30 cases/million births) hereditary disorder **galactosemia.** When this defect is present, the body cannot transform galactose into glucose metabolites but reduces it to the sugar alcohol **galactitol** or oxidizes it to **galactonate,** both products being excreted in the urine. Unfortunately, severe gastrointestinal troubles often appear within a few days or weeks of birth. Growth is slow and cataracts develop in the eyes, probably as a result of the accumulating galactitol. Death may come quickly from liver damage. Fortunately, galactose-free diets can be prepared for young infants, and if the disease is diagnosed promptly the most serious damage can be avoided. However, it has not been possible to prevent long-term effects that include speech difficulties, learning disabilities, and ovarian dysfunction. In some less seriously affected galactosemic patients **galactokinase,** is absent, but it is more often **galactose-l-phosphate uridyltransferase,** that is missing or inactive. This enzyme transforms galactose 1-*P* to UDP-galactose by displacing glucose 1-*P* from UDP-glucose. The UDP-galactose is then isomerized by the NAD^+-dependent **UDP-Gal 4-epimerase**. Absence of this enzyme also causes galactosemia.

The overall effect of the reactions, is to transform galactose into glucose 1-*P*. At the same time, the 4-epimer-ase can operate in the reverse direction to convert UDP-glucose to UDP-galactose, when the latter is needed for biosynthesis. Another enzyme important to galactose metabolism, at least in *E. coli,* is **galactose mutarotase.** Cleavage of lactose by β-galactosidase produces β-D-galactose which must be converted to the α-anomer by the mutarotase before it can be acted upon by galactokinase. Galactose is present in most glycoproteins and glycolipids in the pyranose ring form. However, in bacterial O-antigens, in cell walls of mycobacteria and fungi, and in some protozoa glactose occurs in the furanose form. The precursor is UDP-Gal*f*, which is formed from UDP-Gal*p* by **UDP-Gal*p* mutase.**

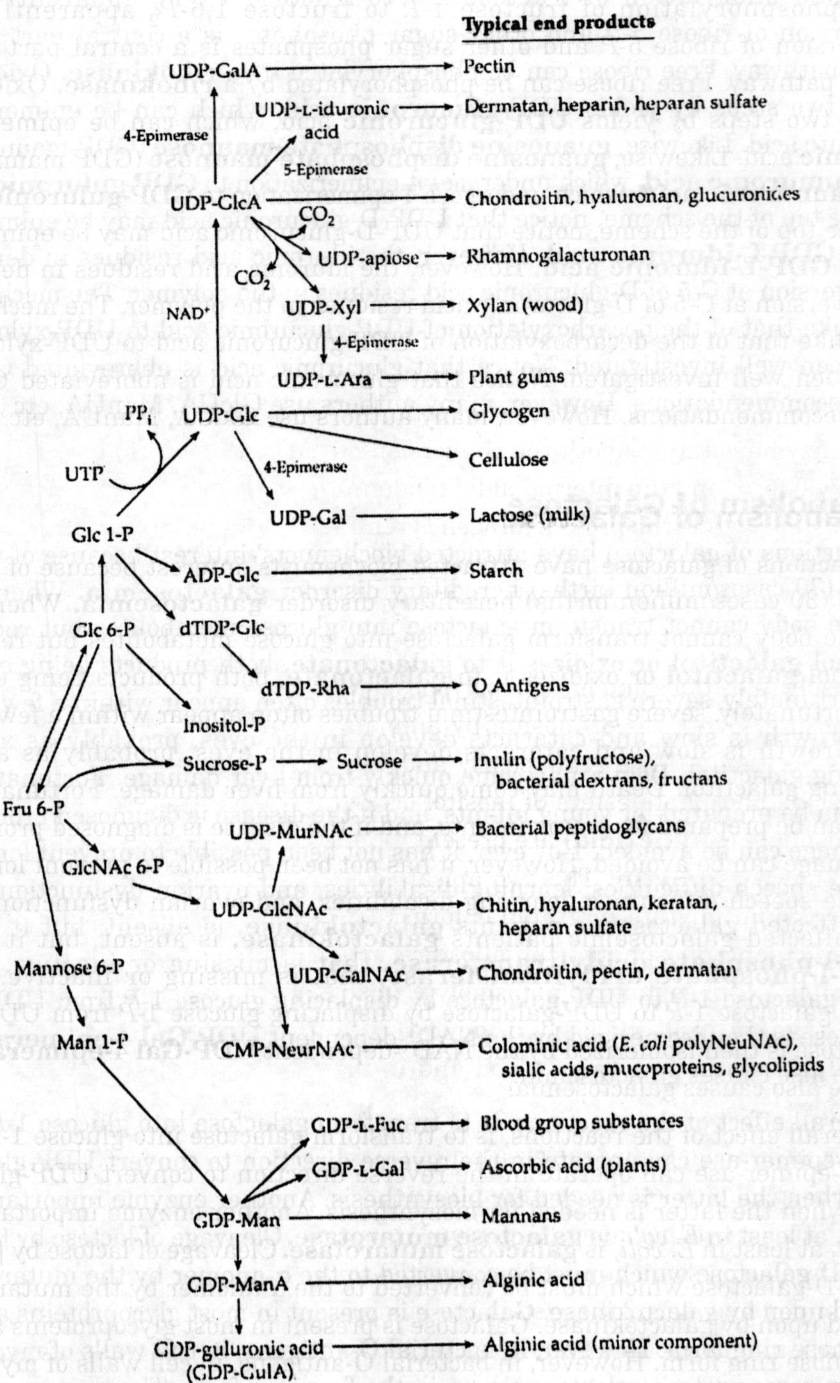

Fig. 9.1. Some routes of interconversion of monosaccharides and of polymerization of the activated glycosyl units.

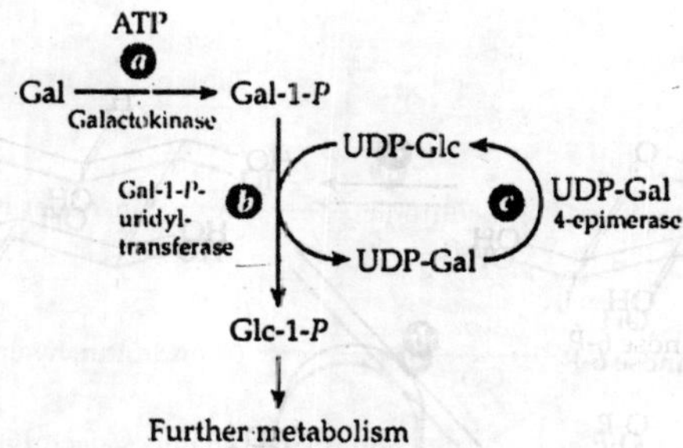

Inositol

Related to the monosaccharides is the hexahydroxy-cyclohexane ***myo*-inositol.** This **cyclitol,** which is apparently present universally within cells, can be formed from glucose-6-*P* according to Eq. 9.2. using a synthase that contains bound NAD^+. In addition to the two redox steps, this enzyme catalyzes both the conversion of the β anomer of glucose 6-*P* to the open-chain aldehyde form and the internal aldol condensation. The pro-R hydrogen at C-*6* of glucose 1-*P* is lost in step *b* while the pro-S hydrogen is retained. The ring numbering system is different for glucose and for the inositols, C-5 of glucose 1-*P* becoming C-2 of i.-*myo*-inositol. Since *myo*-inositol contains a plane of symmetry D- and L- forms are identical. However, they are numbered differently.

The phospho-inositides and inositol polyphosphates are customarily numbered as derivatives of D-myo-inositol. Synthesis of inositol by animals is limited and myo-inositol is sometimes classified as a vitamin. Mice grow poorly and lose some of their hair if deprived of dietary inositol. Various phosphate esters of inositol occur in nature. For example, large amounts of the hexaphosphate **(phytic acid)** are present in grains, usually as the calcium or mixed Ca^{2+}-Mg^{2+} salts known as **phytin.** The two apical cells of the 28-cell larvae of mesozoa, contain enough magnesium phytate in granular form to account for up to half of the weight of the larvae. Inositol pentaphosphate is an allosteric activator for hemoglobin in birds and turtles. Di-myo-inositol-1,1'-phosphate is an osmolyte in some hypothermophilic archaea. Insitol is a component of **galactinol,** the β glycoside of D-galactose with inositol. Galactinol, as well as free inositol, circulates in human blood and in plants and may be a precursor of cell wall polysaccharides.

However, in our own bodies the greatest importance for inositol doubtless lies in the inositol-containing phospholipids known as **phosphoinositides**. Their function m generation of "second messengers" for various hormones. A person typically ingests daily about one gram of inositol, some in the free form, some as phosphoinositides, and some as phytin. As much as four grams of inositol per day may be synthesized in the kidneys. Breast milk is rich in inositol and dietary supplementation with inositol has increased survival of premature infants with respiratory distress syndrome. The action of insulin is reported to be improved by administration of D-chiro-inositol to women with polycystic ovary syndrome.

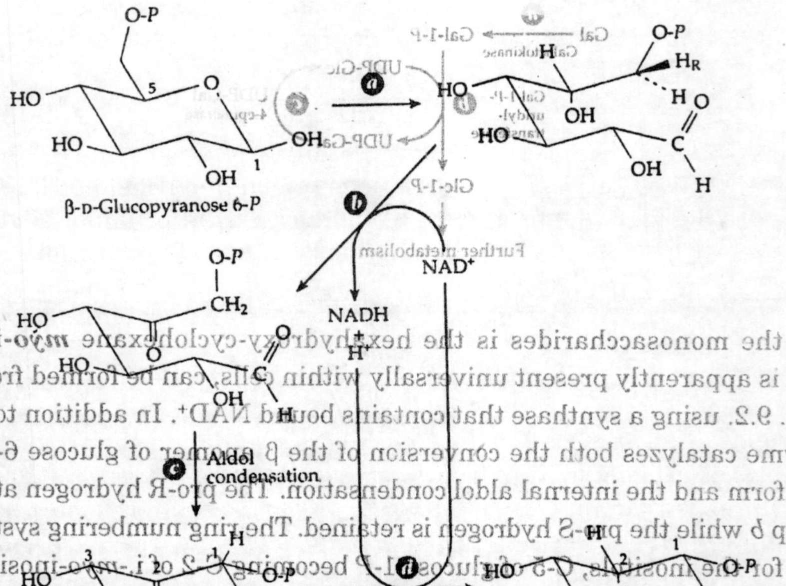

D-Glucuronic Acid, Ascorbic Acid, and Xylitol

In bacteria, as well as in animal kidneys, inositol may be converted to D-glucuronic acid, with the aid of an oxygenase. Free glucuronic acid may also be formed by animals from glucose or from UDP-glucose. Within the animal body glucuronic acid can be reduced with NADH, Fig. 9.2. to yield **L-gulonic acid,** an aldonic acid that could also be formed by oxidation at the alde-hyde end of the sugar **gulose.** Because C-6 of the glucuronic acid has become C-1 of gulonic acid, the latter belongs to the L family of sugars. Gulonic acid can be converted to a cyclic lactone (step *b)* which, in a two-step process involving dehydrogenation and enolization (steps

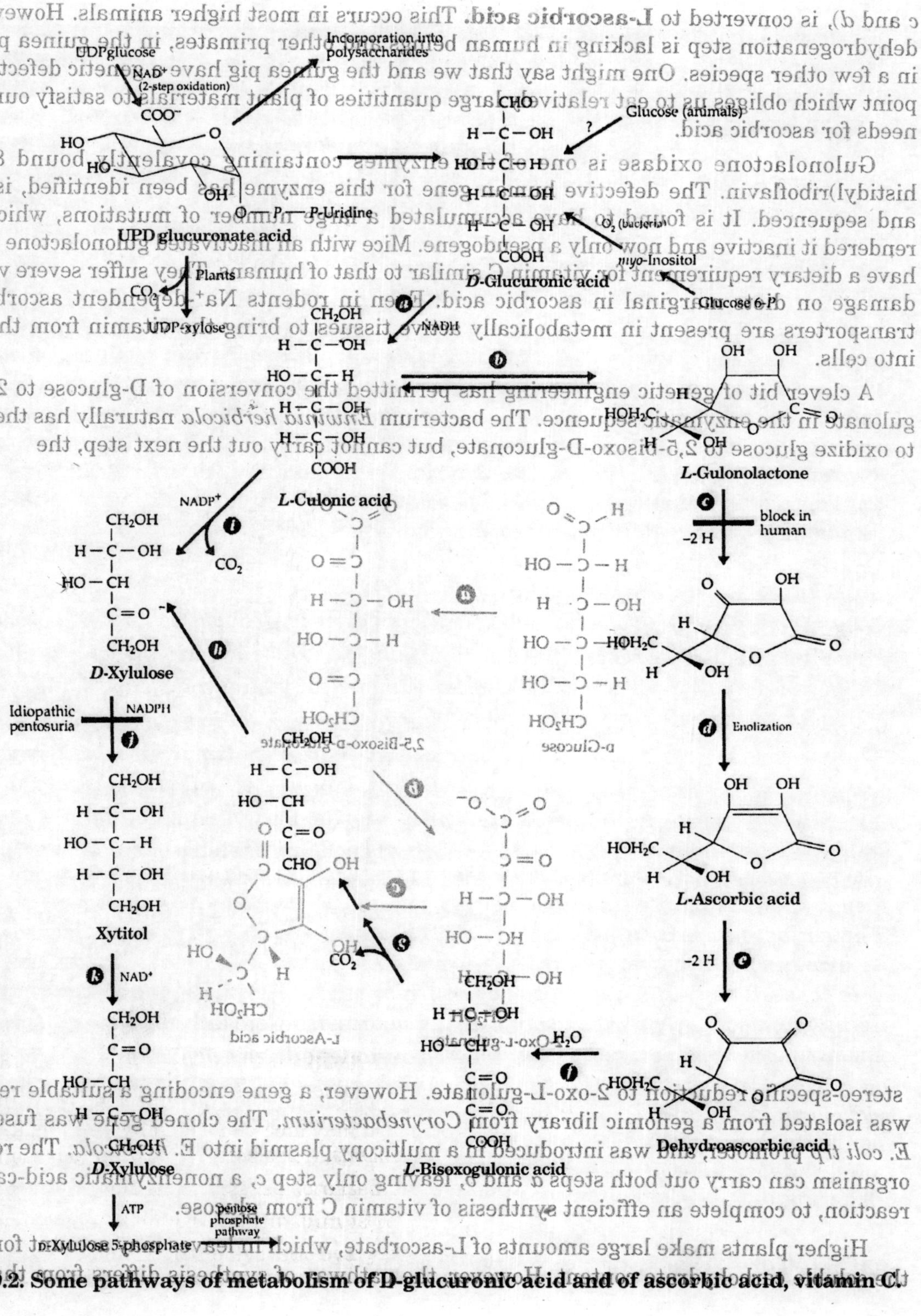

Fig. 9.2. Some pathways of metabolism of D-glucuronic acid and of ascorbic acid, vitamin C.

c and *d*), is converted to **L-ascorbic acid.** This occurs in most higher animals. However, the dehydrogenation step is lacking in human beings and other primates, in the guinea pig, and in a few other species. One might say that we and the guinea pig have a genetic defect at this point which obliges us to eat relatively large quantities of plant materials to satisfy our bodily needs for ascorbic acid.

Gulonolactone oxidase is one of the enzymes containing covalently bound 8α-(N^1-histidyl)riboflavin. The defective human gene for this enzyme has been identified, isolated, and sequenced. It is found to have accumulated a large number of mutations, which have rendered it inactive and now only a pseudogene. Mice with an inactivated gulonolactone oxidase have a dietary requirement for vitamin C similar to that of humans. They suffer severe vascular damage on diets marginal in ascorbic acid. Even in rodents Na^+-dependent ascorbic acid transporters are present in metabolically active tissues to bring the vitamin from the blood into cells.

A clever bit of genetic engineering has permitted the conversion of D-glucose to 2-oxo-L-gulonate in the enzymatic sequence. The bacterium *Enwinia herbicola* naturally has the ability to oxidize glucose to 2,5-bisoxo-D-gluconate, but cannot carry out the next step, the

D-Glucose —(a)→ 2,5-Bisoxo-D-gluconate —(b)→ 2-Oxo-L-gulonate —(c)→ L-Ascorbic acid

stereo-specific reduction to 2-oxo-L-gulonate. However, a gene encoding a suitable reductase was isolated from a genomic library from *Corynebacterium.* The cloned gene was fused to an *E. coli trp* promoter, and was introduced in a multicopy plasmid into E. *herbicola.* The resultant organism can carry out both steps *a* and *b*, leaving only step *c*, a nonenzymatic acid-catalyzed reaction, to complete an efficient synthesis of vitamin C from glucose.

Higher plants make large amounts of L-ascorbate, which in leaves may account for 10% of the soluble carbohydrate content. However, the pathway of synthesis differs from that. Both

D-mannose and L-galactose are efficient precursors. The pathway in which starts with in Fig 9.2. GDP-D-mannose and utilizes known enzymatic processes, has been suggested. The GDP-D-mannose-3, 5-epimerase is a well documented but poorly understood enzyme. Multistep mechanisms related to that of UDP-glucose 4-epimerase, can be envisioned.

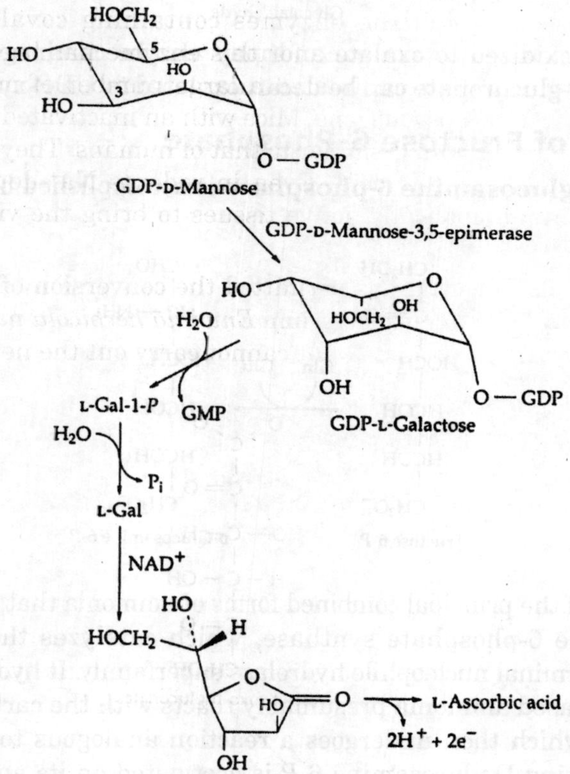

GDP-D-Mannose

GDP-D-Mannose-3,5-epimerase

GDP-L-Galactose

H_2O → GMP

L-Gal-1-*P*

H_2O → P_i

L-Gal

NAD^+

→ L-Ascorbic acid

$2H^+ + 2e^-$

Ascorbic acid is readily oxidized to dehydro-ascorbic acid, which may be hydrolyzed to L-bisoxogulonic acid (step *f*). The latter, after decarboxylation and reduction, is converted to L-xylulose (steps *g* and *h)*, a compound that can also be formed by a standard oxidation and decarboxylation sequence on L-gulonic acid (step *i*). Reduction of xylulose to xylitol and oxidation of the latter with NAD^+ (steps *j* and *k)* produces D-xylulose, which can be phosphorylated with ATP and enter the pentose phosphate pathways. A metabolic variation produces a condition called **idiopathic pentosuria.** Affected individuals cannot reduce xylulose to xylitol and, hence, excrete large amounts of the pentose into the urine, especially if the diet is rich in glucuronic acid.

The "defect" seems to be harmless, but the sugar in the urine can cause the condition to be mistaken for diabetes mellitus. Xylitol is as sweet as sucrose and has been used as a food additive. Because it does not induce formation of dental plaque, it is used as *a* replacement for sucrose in chewing gum. It appeared to be an ideal sugar substitute for diabetics. However, despite the fact that it is already naturally present in the body, ingestion of large amounts of xylitol causes bladder tumors as well as oxalate stones in rats and mice. Its use has, therefore, been largely discontinued. A possible source of the problem may lie in the conversion by fructokinase of some of the xylitol to D-xylulose 1-*P*, which can be cleaved by the xylulose 1-*P* aldolase to dihydroxyace-tone *P* and glycolaldehyde.

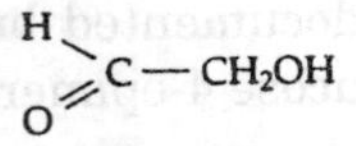

$H(O{=})C-CH_2OH$

Glycolaldehyde

The latter can be oxidized to oxalate and may also be carcinogenic. As indicated in the upper left corner, UDP-glucuronate can be decarboxylated to UDP-xylose.

Transformations of Fructose 6-Phosphate

Biosynthesis of **D-glucosamine 6-phosphate** is accomplished by reaction of fructose 6-*P* with glutamine:

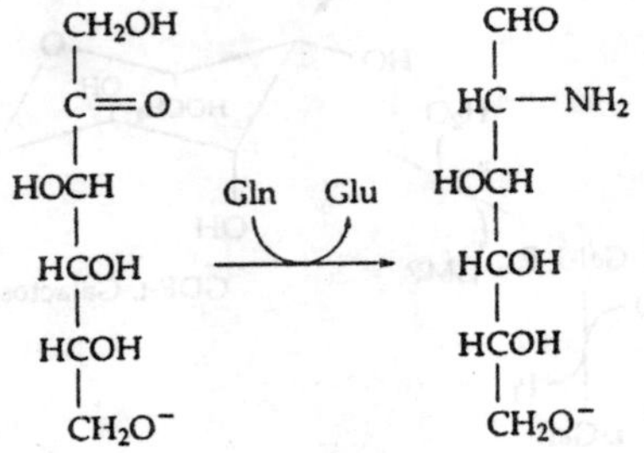

$CH_2OH-C{=}O-HOCH-HCOH-HCOH-CH_2O^-$ (Fructose 6-*P*) $\xrightarrow{\text{Gln} \rightarrow \text{Glu}}$ $CHO-HC(NH_2)-HOCH-HCOH-HCOH-CH_2O^-$ (D-Glucosamine 6-*P*)

Fructose 6-*P* D-Glucosamine 6-*P*

Glutamine is one of the principal combined forms of ammonia that is transported throughout the body. Glucosamine 6-phosphate synthase, which catalyzes the reaction, is an amido-transferase of the N-terminal nucleophile hydrolase superfamily. It hydrolyzes the amide linkage of glutamine. The released ammonia presumably reacts with the carbonyl group of fructose 6-*P* to form an imine, which then undergoes a reaction analogous to that catalyzed by sugar isomerases. The resulting D-glucosamine 6-*P* is acetylated on its amino group by transfer of an acetyl group, and a mutase moves the phospho group to form N-acetylglucosamine 1-*P*. In *E. coli* acetylation occurs on GlcN 1-*P* and is catalyzed by a bifunctional enzyme that also has mutase activity. The resulting N-acetylglucosamine 1-*P* is converted to UDP-N-acetylglucosamine (UDP-GlcNAc) with cleavage of UTP to inorganic pyrophosphate as in the synthesis of UDP-glucose.

Cells of E. *coli* are also able to catabolize glucosamine 6-phosphate. A **deaminase** with many properties similar to those of GlcN 6-*P* synthase, catalyzes a reaction resembling the reverse, of but releasing NH_3. One of the compounds formed from UDP-GlcNAc is **UDP-N-acetylmuramic acid.** The initial step in its synthesis is an unusual type of displacement reaction on the α-carbon of PEP by the 3-hydroxyl group of the sugar. Inorganic phosphate is displaced with formation of an enolpyruvyl derivative of UDP-GlcNAc. This derivative is then reduced by NADPH. A second sugar nucleotide formed from UDP-GlcNAc is **UDP-*N*-acetylgalactosamine** (UDP-GalNAc), which may be created by the same 4-epimerase that generates UDP-Gal. (Eq. 9.1.) Some animal tissues such as kidney and liver also have a **GalNAc kinase** that may salvage, for reuse, GalNAc that arises from the degradation of complex polysaccharides.Bacteria may dehydrogenate UDP-GalNAc to UDP-N-acetylgalactosaminuric acid (UDP-GalNAcA).

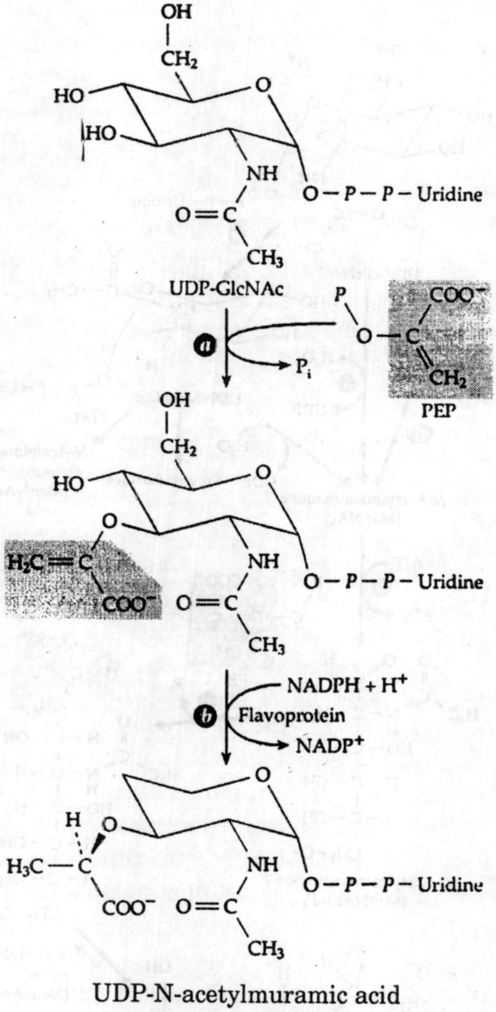

OH
CH_2
HO O
HO
NH
O − P − P − Uridine
O=C
CH_3
UDP-GlcNAc
P
COO^-
O − C
a
P_i
CH_2
PEP
OH
CH_2
HO O
O
$H_2C=C$
NH
O − P − P − Uridine
COO^- O=C
CH_3
$NADPH + H^+$
b Flavoprotein
$NADP^+$
O
H O
H_3C-C
NH
O − P − P − Uridine
COO^- O=C
CH_3

UDP-N-acetylmuramic acid
(UDP-MurNAc)

UDP-GlcNAc can be converted to UDP-*N*-acetyl-mannosamine (UDP-ManNAc) with concurrent elimination of UDP. This unusual epimerization occurs without creation of an adjacent carbonyl group that would activate the 2-H for removal as a proton. As indicated by the small arrows, the UDP is evidently eliminated. In a bacterial enzyme it remains in the E-S complex and is returned after a conformational change involving the acetamido group. This allows the transient C1-C2 double bond to be protonated from the opposite side.

In bacteria the UDP-ManNAc may be dehydrogenated to UDP-*N*-acetylmannos-aminuronic acid (ManNAcA). Both ManNAc and ManNAcA are components of bacterial capsules. In mammals the epimerase, probably utilizes a similar chemical mechanism but eliminates UDP and replaces it with HO^- to give free *N*-acetylmannosamine, which is then phosphor-ylated on the 6-hydroxyl. ManNAc may also be formed from free GlcNAc by another 2-epimerase (step a").

Extending a Sugar Chain with Phosphenolpyruvate (PEP)

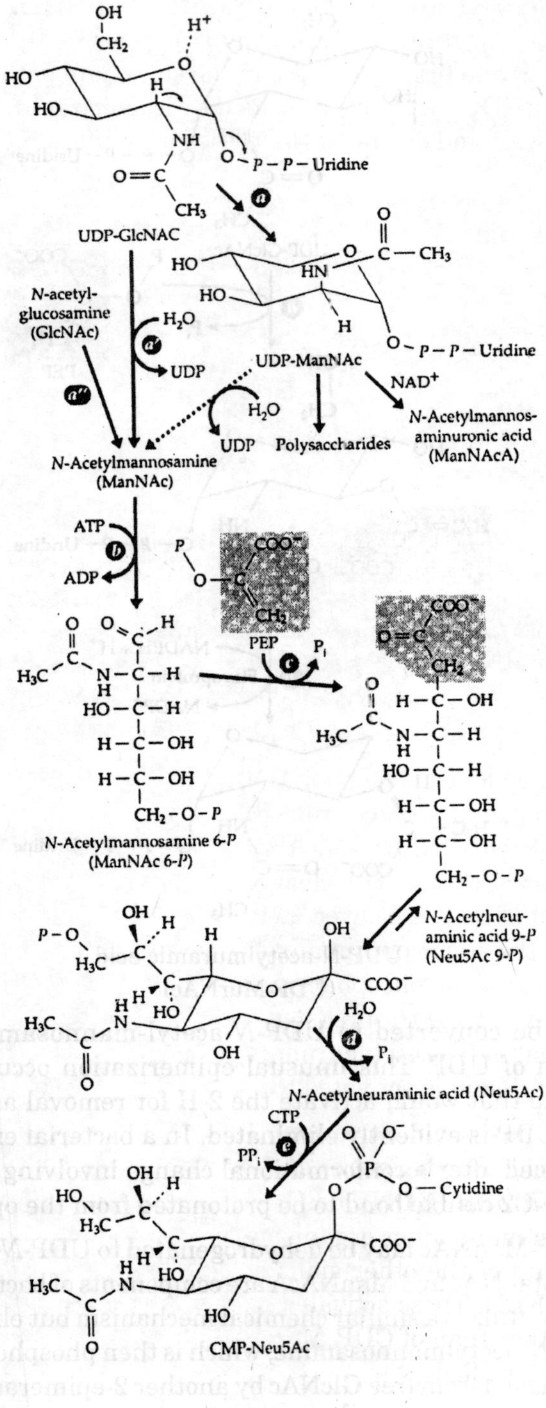

The six-carbon chain of ManNAc *6-P* can be extended by three carbon atoms using an a ldol-type condensation with a three-carbon fragment from to steop c to give ***N*-acetylneuraminic acid** (sialic acid). The nine-carbon chain of this molecule can cyclize to form a pair of anomers with 6-membered rings. In a similar manner, arabinose 5-*P* is converted to the 8-carbon **3-deoxy-D-manno-octulosonic acid (KDO)**, a component of the lipopolysaccharide of gram-negative bacteria, and D-Erythrose 4-*P* is converted to 3-deoxy-D-*arabino*-heptulosonate 7-*P*, the first metabolite in the shikimate pathway of aromatic synthesis. The arabinose-*P* used for KDO synthesis is formed by isomerization of D-ribulose 5-*P* from the pentose phosphate pathway, and erythrose 4-*P* arises from the same pathway.

Arabinose 5-*P* PEP KDO-8-*P*

The mechanism of the aldol condenstion that forms these sugars is somewhat unexpected. A reactive enolate anion can be formed from PEP by hydrolytic attack on the phospho group with cleavage of the O-P bond. However, in reactions such and also in EPSP synthase the initial condensation does not involve O-P cleavage. NMR studies of the action of KDO synthase reveal that the C-O bond of PEP is cleaved as is indicated. The *si* face of PEP faces the *re* face of the carbonyl group of the sugar phosphate. A carbanionic center is generated at C-3 of PEP with possible participation of the phosphate oxygen as well as electrostatic stabilization of the carbocation formed in step *a.* Ring closure (step *b)* occurs with loss of Pi. The immediate product of the aldol condensation, is N-acetylneuraminic acid 9-phosphate, which is cleaved through phosphatase action (step *d)* and is activated to the CMP derivative by reaction with CTP.

Further alterations may occur. For example, CMP-Neu5Ac is hydroxylated to form CMP-*N*-glycolylneuraminic acid. Furthermore, an additional type of sialic acid, 2-oxo-3-deoxy-D-*glycero*-D-*galacto-nononic acid* (**KDN**), has been found in human developmentally regulated glycoproteins and also in many other organisms. It has an –OH group in the 5-position rather than the acetamido group of the other sialic acids. Like NeuNAc it is activated by reaction with CTP forming CMP-KDN. These activated monosaccharides differ from most others in being derivatives of a CMP rather than of CDP. More than 40 different naturally occuring variations of sialic acid have been identified.

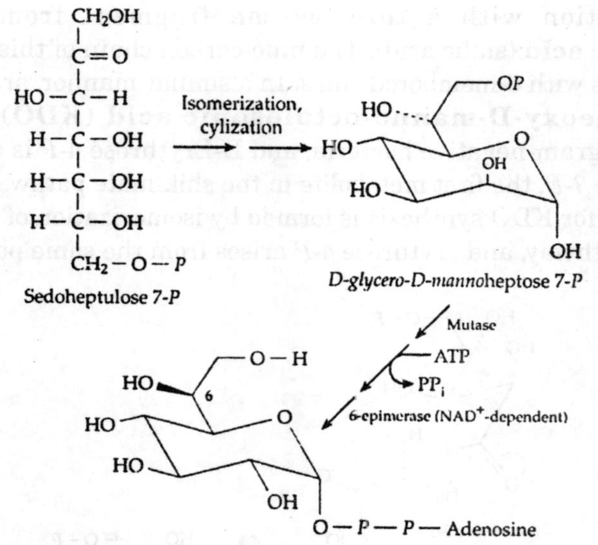

Isomerization, cylization

Sedoheptulose 7-*P*

*D-glycero-D-manno*heptose 7-*P*

Mutase

ATP

PP_i

6-epimerase (NAD^+-dependent)

O—*P*—*P*—Adenosine

In a similar fashion, KDO is converted to the β-linked **CMP-KDO,** which is incorporated into lipid A as shown in Fig 9.10. The ADP derivative of the **L-*glycero*-D-*manno*-heptose**, which is also present in the lipopolysaccharide of gramnegative bacteria, is formed from sedoheptulose 7-*P* in a five-step process.

Synthesis of Deoxy Sugars

Metabolism of sugars often involves dehydration to α,β-unsaturated carbonyl compounds. An example is the formation of 2-oxo-3-deoxy derivatives of sugar acids. Sometimes a carbonyl group is created by oxidation of an —OH group, apparently for the sole purpose of promoting dehydration. For example, the biosynthesis of L-rhamnose from D-glucose is a multistep process that takes place while the sugars are attached to deoxythymidine diphosphate. Introduction of the carbonyl group by dehydrogenation with tightly bound NAD^+, is followed by dehydration. To complete the sequence, the double bond formed by dehydration is reduced (step *c*) by the NADH produced in step *a.* A separate enzyme, a 3,5-epimerase catalyzes inversion at both C-3 and C-5 (step *d*). Finally, a third enzyme is needed for a second reduction (step *e*) using NADPH. The biosynthesis of **GDP-L-fucose** from GDP-D-mannose occurs by a parallel sequence. The metabolism of free L-fucose (6-deoxy-L-galactose), which is present in the diet and is also generated by degradation of glycoproteins, resembles the Entner-Doudoroff pathway of glucose metabolism.

Similar degradative pathways act on D-arabinose and L-galactose. Bacterial surface polysaccharides contain a variety of dideoxy sugars. The four 3,6-dideoxy sugars **D-paratose** (3,6-dideoxy-D-glucose), **D-abequose** (3,6-dideoxy-D-galactose), **D-tyvelose** (3,6-dideoxy-D-mannose), and **L-ascarylose** (3,6-dideoxy-L-mannose), whose structures are shown, arise from CDP-glucose. This substrate is first converted, in reactions parallel to the first three steps, of to 4-oxo-6-deoxy-CDP-glucose which reacts in two steps with pyridoxamine 5'-phosphate (PMP) and NADH. This unusual reaction is catalyzed by

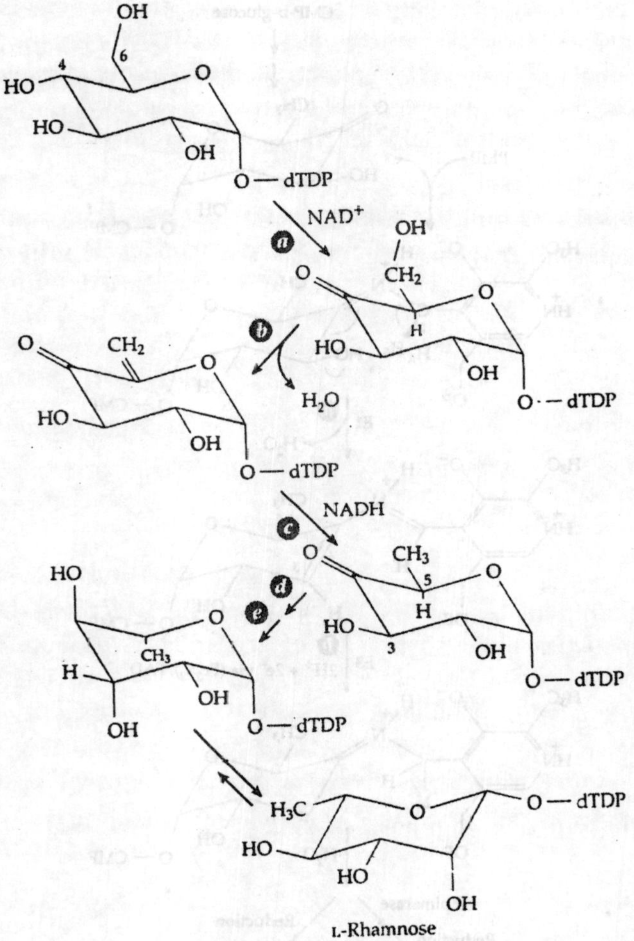

OH
6
4
HO
O
HO
OH
O — dTDP
NAD+
a
OH
CH2
O
O
b
H
O
CH2
O
HO
H2O
OH
O·— dTDP
HO
OH
O — dTDP
NADH
c
O
CH3
HO
5
O
d
O
e
H
HO
CH3
3
OH
H
O — dTDP
OH
OH
O — dTDP
H3C
O
O — dTDP
HO
HO
OH
L-Rhamnose

a two-enzyme complex. The first component, E1, catalyzes the formation of a Schiff base of the substrate with PMP and a transamination, which also accomplishes dehydration, to give an unsaturated sugar ring. The protein also contains an Fe_2S_2 center suggesting a possible one-electron transfer. The second component, E_3, contains both an Fe_2S_2 plant type ferredoxin center and bound FAD.

Observation by EPR spectroscopy revealed accumulation of an organic free radical that may be an intermediate. Hydrolysis, epimerization at C-5, and reduction yields L-ascarylose. A similar reaction sequence without the last epimerization would yield D-abequose. CDP-D-tyvelose arises by C-2 epimerization of CDP-D-paratose. Other unusual sugars are formed from intermediates. One is a **3-amino-3,4,6-trideoxyhexose** in which the amino group has been provided by transamination. The unusual sulfur-containing sugar **6-sulfoquinovose** is present in the sulfolipid of chloroplasts. A possible biosynthetic sequence begins with transamination of cysteic acid to 3-sulfopyruvate, reduction of the latter to sulfolactaldehyde, and aldol condensation with dihydroxyacetone-*P*. However, biosynthesis in chloroplasts appears

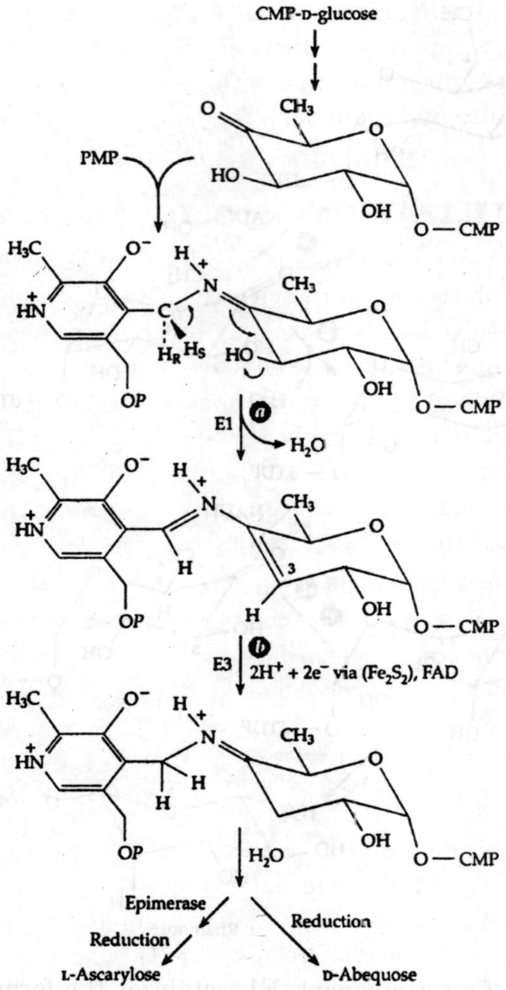

CMP-D-glucose

PMP

E1 (a) H_2O

E3 (b) $2H^+ + 2e^-$ via (Fe_2S_2), FAD

H_2O

Epimerase

Reduction

Reduction

L-Ascarylose

D-Abequose

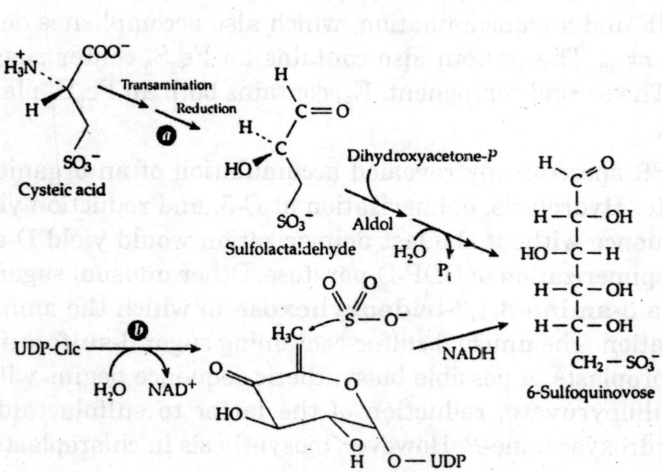

Cysteic acid

Transamination

Reduction

(a)

Sulfolactaldehyde

Dihydroxyacetone-P

Aldol

H_2O

P_i

UDP-Glc

(b)

H_2O NAD^+

NADH

6-Sulfoquinovose

to start with action of a 4,6-dehydratase on UDP-glucose followed by addition of sulfite and reduction. The sulfite is formed by reduction of sulfate via adenylyl sulfate. However, biosynthesis in chloroplasts appears to start with action of a 4,6-dehydratase on UDP-glucose followed by addition of sulfite and reduction. The sulfite is formed by reduction of sulfate via adenylyl sulfate.

SYNTHESIS AND UTILIZATION OF OLIGOSACCHARIDES

Our most common food sugar **sucrose** is formed in all green plants and nowhere else. It is made both in the chloroplasts and in the vicinity of other starch deposits. It serves both as a transport sugar and, dissolved within vacuoles, as an energy store. Sucrose is very soluble in water and is chemically inert because the hemiacetal groups of both sugar rings are blocked. However, sucrose is thermodynamically reactive, the glucosyl group having a group transfer potential of 29.3 kJ mol^{-1}. It is extremely sensitive toward hydrolysis catalyzed by acid. Transport of sugar in the form of a disaccharide provides an advantage to plants in that the disaccharide has a lower osmotic pressure than would the same amount of sugar in monosaccharide form. Biosynthesis of sucrose utilizes both UDP-glucose and fructose 6-*P*.

Reaction of UDP-glucose with fructose can also occur to give sucrose directly. Because this reaction is reversible, sucrose serves as a source of UDP-glucose for synthesis of cellulose and other polysaccharides in plants. Metabolism of sucrose in the animal body begins with the action of **sucrase** (invertase), which hydrolyzes the disaccharide to fructose and glucose. The same

$$\text{UDP–Glc} \xrightarrow{\text{Fru-6–}P} \text{Sucrose 6–}P \xrightarrow[\text{P}_i]{\text{H}_2\text{O}} \text{Sucrose}$$

enzyme is also found in higher plants and fungi. Mammalian sucrase is one of several carbohydrases that are anchored to the external surfaces of the microvilli of the small intestines. Sucrose is bound tightly but noncovalently to **isomaltase,** which hydrolyzes the α-1,6-linked isomaltose and related oligosaccharides. A nonpolar N-terminal segment of the isomaltase anchors the pair of enzymes to the micro-villus membrane. The two-protein complex arises naturally because the two enzymes are synthesized as a single polypeptide, which is cleaved by intestinal proteases.

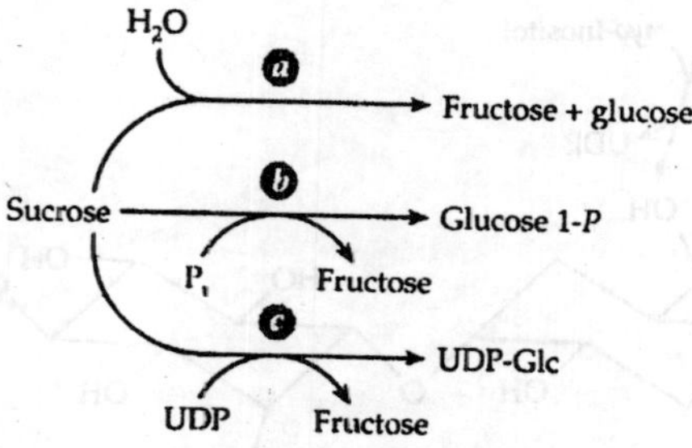

Because of the relatively high group transfer potential of either the glucosyl or fructosyl parts, sucrose is a substrate for glucosyltransferases such as sucrose phosphorylase, associated discussion). In certain bacteria this reaction makes available the activated glucose 1-*P* which may enter catabolic pathways directly. Cleavage of sucrose for biosynthetic purposes can occur by reaction a-14, step c, which yields UDP-glucose in a single step. A disaccharide with many of the same properties as sucrose is **trehalose,** which consists of two α-glucopyranose units in 1,1 linkage.

The biosynthetic pathway from UDP-glucose and glucose 6-*P* parallels that for synthesis of sucrose. In *E. coli* the genes for the needed glucosyl-transferase and phosphatase are part of a single operon. Its transcription is controlled in part by glucosemediated catabolite repression and also by a represser of the Lac family. The represser is allosterically activated by trehalose 6-*P*, the intermediate in the synthesis. Trehalose formation in bacteria, fungi, plants, and microscopic animals is strongly induced during conditions of high osmolality. Both trehalose and maltose can also be taken up via an ABC type transporter. **Lactose,** the characteristic sugar of milk, is formed by transfer of a galactosyl unit from UDP-galactose directly to glucose. The similar transfer of a galactosyl unit to *N*-acetylglucosamine to form *N*-acetyllactosamine occurs in many animal tissues.

An interesting regulatory mechanism is involved. The transferase catalyzing, reaction *b*, forms a complex with **α-lactalbumin** to become **lactose synthase**, the enzyme that catalyzes reaction *a*. Lactalbumin was identified as a milk constituent long before its role as a regulatory protein was recognized. A very common biochemical problem is intolerance to lactose. This results from the inability of the intestinal mucosa to make enough **lactase** to hydrolyze the sugar to its monosaccharide components galactose and glucose. Among most of the peoples of the earth only infants have a high lactase level, and the use of milk as a food for adults often leads to a severe diarrhea. The same is true for most animals. In fact, baby seals and walruses, which drink lactose-free milk, become very ill if fed cow's milk.

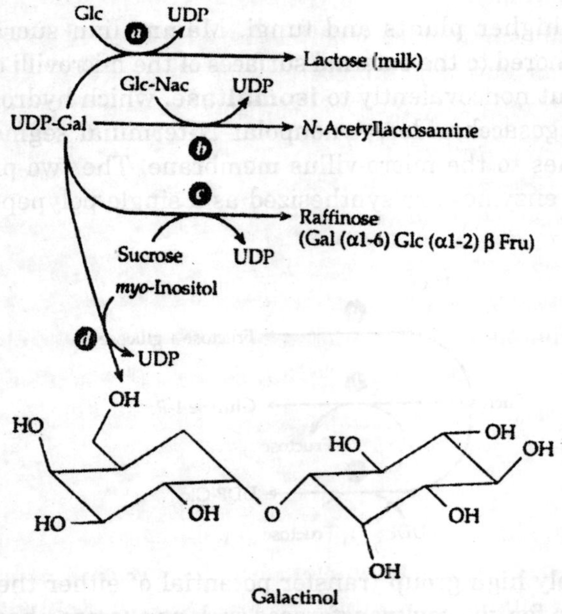

The plant trisaccharide **raffinose** arises from UDP-galactose by transfer of a galactosyl unit onto the 6-hydroxyl of the glucose ring of sucrose. Transfer of a galactosyl unit onto *myo*-inositol, produces **galactinol,** whose occurrence is widespread within the plant kingdom. Galactinol, in turn, can serve as a donor of activated galactosyl groups. Thus, many plants contain **stachyose** and higher homologs, all of which are formed by transfer of additional α-D-galactosyl units onto the 6-hydroxyl of the galactose unit of raffinose. These sugars appear to serve as antifreeze agents in the plants. The concentration of stachyose in soy beans can be as high as that of sucrose.

Some seeds, *e.g.*, those of maize, are coated with a glassy sugar mixture of sucrose and raffinose in a ratio of ~3:l. Besides the oligosaccharides, living organisms form a great variety of glycosides that contain nonsugar components. Among these are the **glucuronides** (glucosiduronides), excretion products found in urine and derived by displacement of UDP from UDP-glucuronic acid by such compounds as phenol, benzoic acid, and sterols. Phenol is converted to phenyl glucuronide, while benzoic acid (also excreted in part as hippuric acid, yields an ester by the same type of displacement reaction. Many other aromatic or aliphatic compounds containing —OH, —SH, —NH_2, or —COO^- groups also form glucuronides.

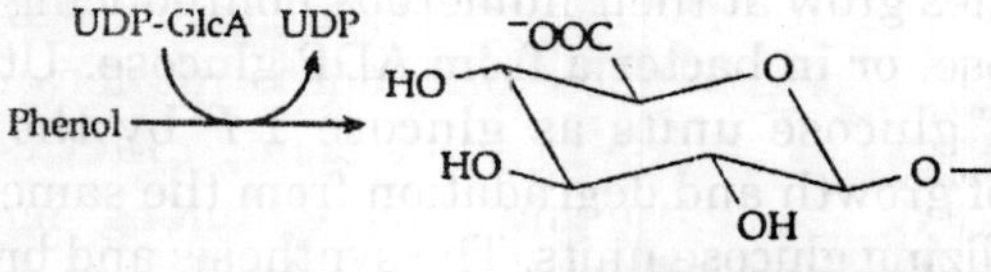

UDP-GlcA UDP
Phenol
^-OOC
HO
HO
O
OH
Phenyl glucuronide

Among these is bilirubin. UDP-glucuron-osyltransferases responsible for their synthesis are present in liver microsomes.

Among the many glycosides and glycosylamines made by plants are the anthocyanin and flavonoid pigments of flowers, cyanogenic glycosides such as amygdalin, and antibiotics.

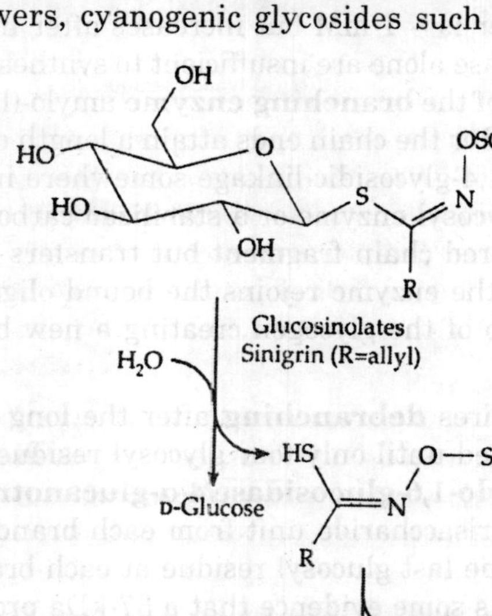

OH
HO
O
OSO_3^-
S
N
HO
OH
R
Glucosinolates
Sinigrin (R=allyl)
H_2O
HS
O — SO_3^-
D-Glucose
N
R
HSO_4^-
R — N = C = S

. Some are characteristic of certain families of plants. For example, more than 100 β-thioglucosides known as **glucosinolides** are found in the Cruciferae (cabbages, mustard, rapeseed). The compounds impart the distinctive flavors and aromas of the plants. However, some are toxic and may cause goiter or liver damage. The enzyme **myrosinase** hydrolyzes these compounds releasing isothiocyanates, thiocyanates, and nitriles. L-Ascorbate acts as a cofactor for this enzyme, evidently providing a catalytic base.

SYNTHESIS AND DEGRADATION OF POLYSACCHARIDES

Polysaccharides are all formed by transfer of glycosyl groups onto initiating molecules or onto growing polymer chains. The initiating molecule is usually a glycoprotein. However, let us direct our attention first to the growth of polysaccharide chains. The glycosyl are transferred by the action of glycosyl-transferases from substrates such as UDP-glucose, other sugar nucleotides, and sometimes sucrose. The glycosyltransferases act by mechanisms discussed, and are usually specific with respect both to substrate structure and to the type of linkage formed.

Glycogen and Starch

The bushlike glycogen molecules grow at their numerous nonreducing ends by the transfer of glucosyl units from UDP-glucose, or in bacteria from ADP-glucose. Utilization of glycogen by the cell involves removal of glucose units as glucose 1-*P* by the action of glycogen phosphorylase. The combination of growth and degradation from the same chain ends provides a means of rapidly storing and utilizing glucose units. The synthesis and breakdown of glycogen in mammalian muscle, involves one of the first studied and best known metabolic control systems. Various aspects have been discussed. The mechanism and regulatory features have been described. An important recent development is the observation of glycogen concentrations in human muscles *in vivo* with ^{13}C NMR. This can be coupled with observation of glucose 6-*P* by ^{31}P NMR.

The concentration of the latter is ~ 1 mM but increases after intense exercise. Glycogen phosphorylase and glycogen synthase alone are insufficient to synthesize and degrade glycogen. Synthesis also requires the action of the **branching enzyme** amylo-(l,4 → 1,6-transglycosylase, an enzyme with dual specificity. After the chain ends attain a length of about ten glucose units, the branching enzyme attacks a 1,4-glycosidic linkage somewhere in the chain. Acting much as does a hydrolase, it forms a glycosyl enzyme or a stabilized carbocation intermediate. The enzyme does not release the severed chain fragment but transfers it to another nearby site on the glycogen molecule. There the enzyme rejoins the bound oligosaccharide chain that it carries to a free 6-hydroxyl group of the glycogen creating a new branch attached in α-1,6-linkage.

Degradation of glycogen requires **debranching** after the long nonreducing ends of the polysaccharide have been shortened until only four glycosyl residues remain at each branch point. This is accomplished by **amylo-1,6-glucosidase/4-α-glucanotransferase.** This 165-kDa bifunctional enzyme transfers a trisaccharide unit from each branch end to the main chain and also removes hydrolytically the last glucosyl residue at each branch point. How are new glycogen molecules made? There is some evidence that a 37-kDa protein primer **glycogenin** is needed to initiate their formation. Thus, glycogen synthesis may be analogous to that of the glycosaminoglycans.

Muscle glycogenin is a self-glycosylating protein, which catalyzes attachment of ~7 to 11 glucose units in α-1,4 linkage to the hydroxyl group of Tyr 194. The glucose units are added one at a time and when the chain is long enough it becomes a substrate for glycogen synthase. The role of glycogenin in liver has been harder to demonstrate, but a second glycogenin gene, which is expressed in liver, has been identified. Genes for several glycogenins or glycogenin-like proteins have been identified in yeast, *Caenorhabditis elegans,* and *Ambidopsis.* In contrast to animals, bacteria such as *E. coli* synthesize glycogen via ADP-glucose rather than UDP-glucose. ADP-glucose is also the glucosyl donor for synthesis of starch in plants. The first step in the biosynthesis is catalyzed by the enzyme ADP-glucose pyrophosphorylase (named for the reverse reaction).

Glucose 1-*P* + ATP → ADP-glucose + PP;

In bacteria this enzyme is usually inhibited by AMP and ADP and activated by glycolytic intermediates such as fructose 1,6-P_2, fructose 6-*P*, or pyruvate. In higher plants, green algae, and cyanobacteria the enzyme is usually activated by 3-phosphoglycerate, a product of photosynthetic CO_2 fixation, and is inhibited by inorganic phosphate (Pi). In eukaryotic plants starch is deposited within chloroplasts or in the cytoplasm as granules in a specifically differentiated and physically fragile plastid, the **amyloplast.** Within the granules the starch is deposited in layers ~ 9 nm in thickness. About two-thirds of the thickness consists of nearly crystalline arrays, probably of double helical amylopectin side chains, with "amorphous" segments between the layers.

In maize there are at least five starch synthases, one of which forms the straight chain amylose. There are also at least three branching enzymes and two or three debranching enzymes. As in the synthesis of glycogen the molecules of amylopectin may grow at the many nonreducing ends. A current model, which is related to the broom-like cluster model of French, is shown. The branches are thought to arise, in part, by transglycosylation within the double helical strands. After branching the two chains remain in a double helix but the cut chain can now grow. Only double helical parts of strands pack well in the crystalline layer. A recent suggestion is that debranching enzymes then trim the molecule, removing singlestranded regions. The location (within the granule) of amylose, which makes up 15-30% by weight of many starches, is uncertain. It may fill in the amorphous layers. It may be cut and provide primer pieces for new amylopectin molecules.

Another possibility is that it grows by an insertion mechanism such as that portrayed for cellulose, and is extruded inward from the membrane of the amyloplast. This mechanism might explain a puzzling question about starch. The branched amylopectin presumably grows in much the same way as does glycogen. A branching enzyme transfers part of the growing glycan chain to the —CH_2—OH group of a glucose uint in an adjacent polysaccharide chain that lies parallel to the first, possibly in a double *helix.* Since amylose and amylopectin are intimately intermixed in the starch granules, it seems strange that the branching enzyme never transfers a branch to molecules of the straight-chain amylose. However, if the linear amylose chains are oriented in the opposite direction from the amylopectin chains, the nonreducing ends of the amylose molecules would be located toward the center of the starch granule.

Growth could occur by an insertion mechanism at the reducing ends and the ends could move out continually with the amyloplast membrane as the granule grows. Recent evidence from ^{14}C labeling indicates that both amylose and amylopectin too may grow by insertion at the reducing end of glucose units from ADP-glucose. Branching could occur to give the structure of Fig. 9.3. Starch synthesis in leaves occurs by day but at night the starch is degraded by amylases, a-glucosidases, and starch phosphorylase. Both the starch synthases and catabolic enzymes are present within the amyloplasts where they may be associated with regulatory proteins of the 14-3-3 class.

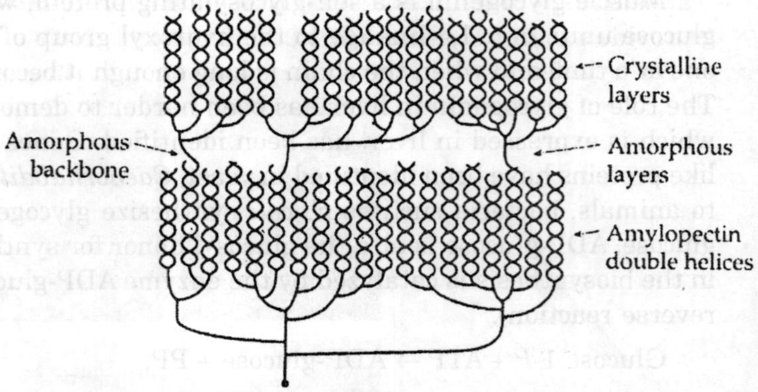

Fig. 9.3. Proposed structure of a molecule of amylopectin in a starch granule. The highly branched molecule lies within 9 nm thick layers, about 2/3 of which contains parallel double helices of the kind shown in a semicrystalline array. The branches are concentrated in the amorphous region. Some starch granules contain no amylose, but it may constitute up to 30% by weight of the starch. It may be found in part in the amorphous bands and in part intertwined with the amylopectin.

Digestion of dietary glycogen and starch in the human body begins with the salivary and pancreatic amylases, which cleave α-1,4 linkages at random. It continues with a **glucoamylase** found in the brush border membranes of the small intestine where it occurs as a complex with **maltase.** Carbohydrases are discussed, elsewhere.

Cellulose, Chitin, and Related Glycans

Cellulose synthases transfer glucosyl units from UDP-glucose, while chitin synthases utilize UDP-*N*-acetylglucosamine. Not only green plants but some fungi and a few bacteria form cellulose. The ameba *Dictyostelium discoideum* also coats its spores with cellulose. Electron microscopic investigations suggest that both in bacteria and in plants multienzyme aggregates located at the plasma membrane synthesize many polymer chains side by side to generate hydrogen-bonded microfibrils which are extruded through the membrane.

Both green plants and fungi also form important β-1,3-linked glycans. The bacterial cellulose synthase from *Acetobacter xylinum* can be solubilized with detergents, and the resulting enzyme generates characteristic 1.7 nm cellulose fibrils, from UDP-glucose. These are similar, but not identical, to the fibrils of cellulose I produced by intact bacteria. Each native fibril appears as a left-handed helix which may contain about nine parallel chains in a crystalline array. Three of these helices appear to coil together, to form a larger 3.7-nm left-handed helical fibril. Similar fibrils are formed by plants. In both bacteria and plants the cellulose I fibrils that are formed are highly crystalline, contain parallel polysaccharide chains, and have the tensile strength of steel.

Electron micrographs show that the cell envelope of *A. xylinum* contains 5-80 pores, through which the cellulose is extruded, lying along the long axis of the cell. The biosynthetic enzymes are probably bound to the plasma membrane. Similar, but more labile, cellulose synthases are present in green plants. In *Arabidopsis* there are ten genes. The encoded cellulose synthases appear to be organized as rosettes on some cell surfaces. The rosettes may be assembled to provide parallel synthesis of ~ 36 individual cellulose chains needed to form a fibril.

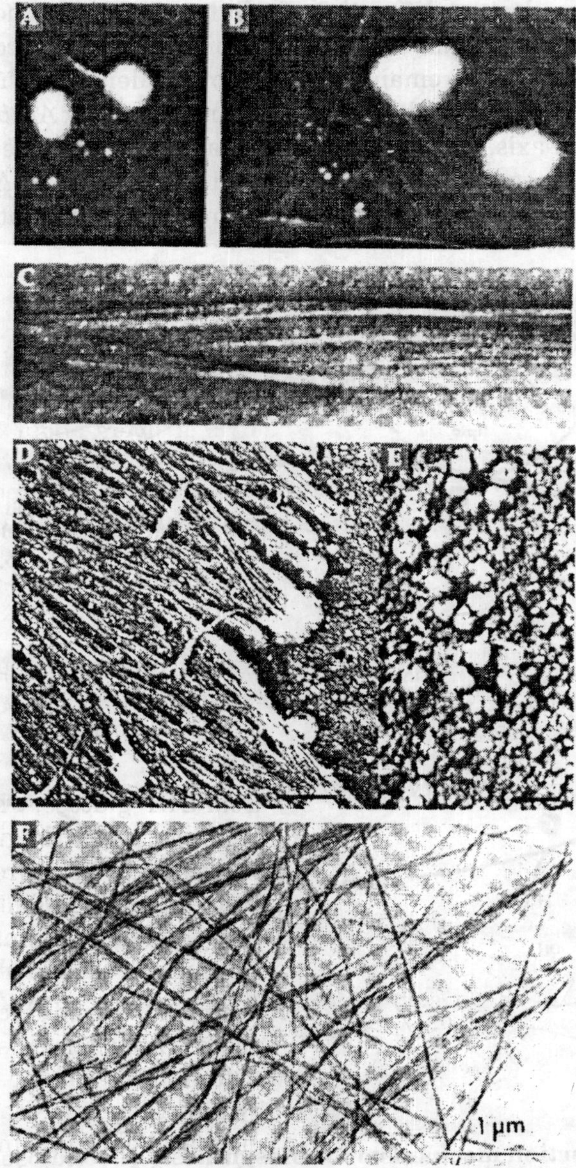

Fig. 9.4. Cellulose microfibrils being formed by *Acetobacter xylinum*. (A) Dark-field light micrograph after five minutes of cellulose production (x 1250). (B) After 15 minutes a pellicle of cellulose fibers is forming (x 2000) (C) Negatively stained cellulose ribbon. At the right the sub-division into microfibrils is visible. Courtesy of R.Malcolm Brown, Jr. (D) Cellulose microfibrils overlaying the plasma membrane in the secondary cell wall of a tracheary element of *Zinnia elegans*. Bar = 100 nm. (E) Rosettes in the plasma membrane underlying the cellulose-rich secondary cell wall thickening in *A. degans*. Bar = 30 nm. (F) Chitin microfibrils purified from protective tubes of the tube-worm *Lamellibrachia satsuma*. Courtesy of Junji Sugiyama.

Because of the insolubility of cellulose fibrils it has been difficult to determine whether they grow from the reducing ends or the nonreducing ends of the chains. From silver staining of reducing ends and micro electron diffraction of cellulose fibrils attached to bacteria, Koyama *et al.* concluded that the reducing ends are extruded from cells. New glucosyl rings would be added at the *nonreducing ends,* which remain attached noncovalently to the cells. From amino acid sequence similarities it was also concluded that the same is true for *Arabidopsis.* A single cellulose chain has a twofold screw axis, each residue being rotated 180° from the preceding residue. It was postulated that two synthases act cooperatively to add cellobiose units. Another suggestion is that sitosterol β-glucoside acts in some fashion as a primer for cellulose synthesis in plants.

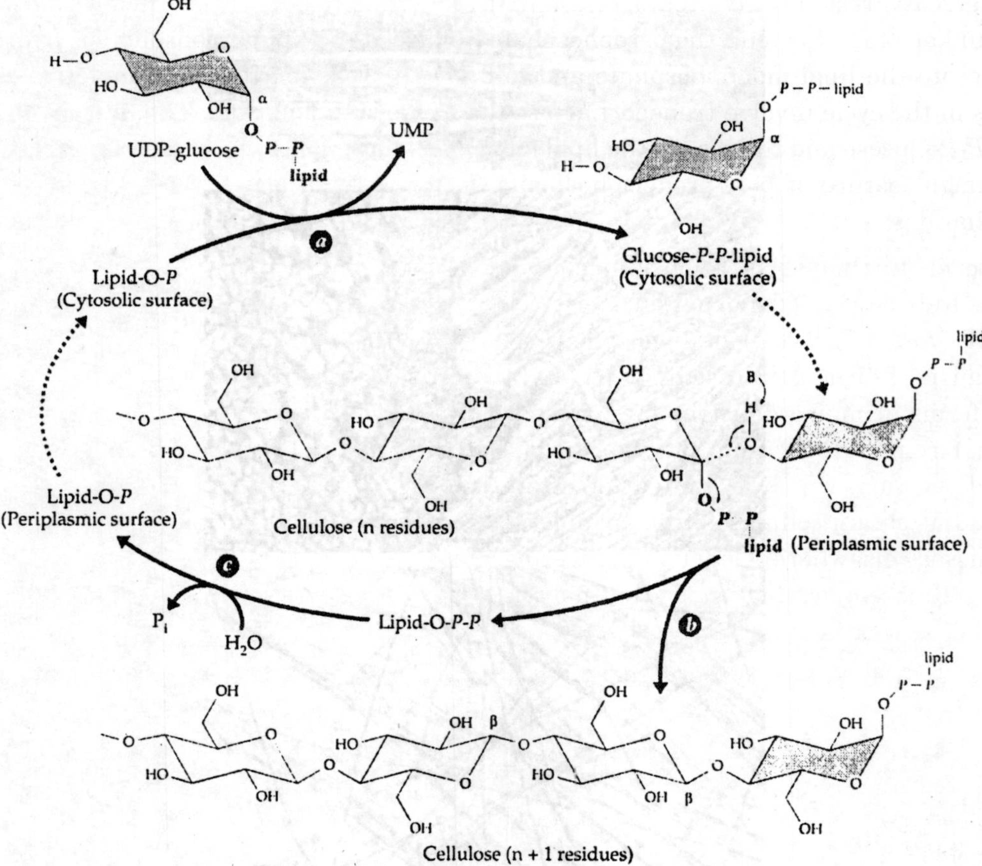

Fig. 9.5. Proposed insertion mechanism for biosynthesis of cellulose. Three enzymatic steps are involved: a nucleophilic displacement reaction of a lipid phosphate on UDP-glucose yields a glucosyl diphosphate lipid in which the oχ-glycosyl linkage is retained. After passage through the membrane the glycosyl group is inserted into the reducing end of a cellulose chain, which is covalently attached by a pyrophosphate linkage to another lipid. The first lipid diphosphate is released and is hydrolyzed (step *c)* to the monophosphate, which crosses the membrane to complete the cycle. After Han and Robyt. As throughout this book P represents the phospho group -PO_3H. The H may be replaced by groups which may contain oxygen atoms. This explains why an O is included in Lipid-O-*P* but no O is shown between the *P*'s in -O-*P*-*P*.

An insertion mechanism for synthesis of cellulose. Using ^{14}C "pulse and chase" labeling Han and Robyt found that new glucosyl units are added at the *reducing ends* of cellulose chains formed by cell membrane preparations from *A. xylinum.* This conclusion is in accord with the generalization that extracellular polysaccharides made by bacteria usually grow from the reducing end by an insertion mechanism that depends upon a polyprenyl alcohol present in the cell membrane. This lipid alcohol, often the C_{55} bactoprenol, reacts with UDP-glucose (or other glycosyl donor) to give a lipopyrophospho-glucose. The α linkage of the UDP-glucose is retained in this compound. The growing cellulose chain is attached at the reducing end by a similar linkage to a second lipid molecule. Then, in a displacement on the anomeric carbon of the first glucosyl residue of the cellulose chain, the new glucosyl unit is inserted with inversion of the α linkage to β. In step *c* the pyrophosphate linkage of the lipid diphosphate is hydrolyzed to regenerate the lipid monophosphate and to drive the reaction toward completion. Two of the steps in the cycle involve transport across the bacterial membrane. The first involves the lipid -O-*P*-*P*-glucose and the second the lipid monophosphate. This type of insertion mechanism is a common feature of poly-prenol phosphate-dependent synthetic cycles for extracellular polysaccharides.

However, further verification is needed for cellulose synthesis. Regulation of cellulose synthesis in bacteria depends on allosteric activator of cellulose synthase, cyclic diguanylate (c-di-GMP), and a Ca^{2+}-activated phosphodiesterase that degrades the activator. Sucrose is the major transport form of glucose in plants. Synthesis of both cellulose and starch is reduced in mutant forms of maize deficient in sucrose synthase. This synthase, acting in the reverse direction, forms UDP-Glc from sucrose. The enzymatic degradation of cellulose is an important biological reaction, which is limited to certain bacteria, to fungi, and to organisms such as termites that obtain cellulases from symbiotic bacteria or by ingesting fungi. These enzymes are discussed, elsewhere. Genetic engineering methods now offer the prospect of designing efficient cellulose-digesting yeasts that may be used to produce useful fermentation products from cellulose wastes.

Cyclic diguanylate

Callose and other β-1,3-;inked glycans : Attempts to produce cellulose from UDP-Glc using enzymes of isolated plasma membranes from higher plants have usually yielded the β-1,3-linked glucan (callose) instead. This is a characteristic polysaccharide of plant wounds which, as healing occurs, is degraded and replaced by cellulose. Callose formation is induced by a specific activator β-furfuryl-β-glucoside, and callose synthase is virtually inactive unless both the activator and Ca^{2+} are present.

β-Furfuryl-β-glucoside

Beta-1,3-linked glycans are major components of the complex layered cell wall of yeasts and other fungi. In the fission yeast *Saccharomyces pombe* ~ 55% of the cell wall carbohydrate consists of β-1,3-linked glucan with some β-1,4-linked branches, ~ 28% is α-1,3-linked glucan, ~6% is α-1,6-linked glucan, and ~ 0.5% is chitin. There are two carbohydrate layers, the outer one appearing amorphous. The inner layer contains inter-woven fibrils of both α-1,3-linked and α-1,4-linked glucans and holds the shape of the cell. The β-1,3 glucan synthase is localized on the inner side of the cell membrane and is activated by GTP and a small subunit of the Rho family of G proteins.

Plants synthesize 1,3-β-glucanases that hydrolyze the glycans of fungal cell walls. Synthesis is induced by wounding as a defense reaction (Box 9.E). These glycanases also function in the removal of callose.

Chitin. Like cellulose synthase, fungal chitin synthases are present in the plasma membrane and extrude microfibrils of chitin to the outside. In the fungus *Mucor* the majority of the chitin synthesized later has its *N*-acetyl groups removed hydrolytically to form the deacetylated polymer chitosan. Chitin is also a major component of insect exoskeletons. For this reason, chitin synthase is an appropriate target enzyme for design of synthetic insecticides. Chitin hydrolyzing enzymes are formed by fungi and in marine bacteria. Chitinases are also present in plant vacuoles, where they participate in defense against fungi and other pathogens.

More recently a chitinase has been identified in human activated macrophages. Another unanticipated discovery was that a developmental gene designated DG42, from *Xenopus,* has a sequence similar to that of the *NodC* gene. The latter encodes a synthase for chitin oligosaccharides (Nod factors) that serve as nodulation factors in *Rhizobia*. The enzyme is synthesized for only a short time during early embryonic development. The significance of this discovery is not yet clear. Synthesis of both the bacterial Nod factors and chitin oligosaccharides in zebrafish embryos occurs by transfer of GlcNAc residues from UDP-GlcNAc at the *nonreducing ends* of the chains. Whether the same is true of chitin in fungi or arthropods remains uncertain.

Cell walls of plants. The thick walls of higher plant cells, provide strength and rigidity to plants and, at the same time, allow rapid elongation during periods of growth. Northcote likened the wall structure to glass fiber-reinforced plastic (fiber glass). Thus, the cell wall contains microfibrils of cellulose and other polysaccharides embedded in a matrix, also largely polysaccha-ride. The **primary cell wall** laid down in green plants during early stages of growth contains loosely inter-woven cellulose fibrils ~ 10 nm in diameter and with an ~ 4 nm crystalline center. The cellulose in these fibrils has a degree of polymerization of 8000-12,000 glucose units.

As the plant cell matures, a secondary cell wall is laid down on the inside of the primary wall. This contains many layers of closely packed microfibrils, alternate layers often being laid down at different angles to one another. The microfibrils in green plants are most often cellulose but may contain other polysaccharides as well. Some algae are rich in fibrils of xylan and mannan. The materials present in the matrix phase vary with the growth period of the plant.

During initial phases **pectin** (polygalacturonic acid derivatives) predominate but later xylans and a variety of other polysaccharides known as **crosslinking glycans** (or hemicelluloses) appear. Primary cell wall constituents of dicotyledons include **xyloglucans** (linear glucan chains with xylose, galactose, and fucose units in branches), other crosslinking glycans, and galacturonic acid-rich pectic materials. The xyloglycans, which comprise 20% of the cell wall in some plants, have a backbone of α-1,4-linked glucose units with numerous α-1,6-linked xylose rings, some of which carry attached L-arabinose, galactose, or fucose. The structures, which vary from species to species, are organized as repeating blocks with a continuous glucan backbone. Another crosslinking glycan is **glucuronoarabinoxylan.** The backbone is β-1,4-linked xylose. Less abundant glucomannans, galactomannans, and galactoglucomarnnans, with β-1,4-linked mannan backbone structures, are also present in most angio-sperms.

Pectins form a porous gel on the inside surface of plant cell walls. A major component is a **homogalacturonan,** which consists of α-1,4-linked galacturonic acid (GalA). A second is rhamnogalacturonan I, an alternating polymer of (2-L-Rhaα1→4GalAα→) units. The most interesting pectin component is **rhamnogalacturonan II,** one of the less abundant constituents of pectin. It is obtained by hydrolytic cleavage of pectin by a polygalacturonidase. Before such release it forms parts (hairy regions) of pectin molecules that are largely homogalacturonans (in smooth regions). A rhamnogalacturonan II segment consists of 11 different monomer units.

Attached to the polygalacturonic acid backbone are four oligo-saccharides, consisting of rhamnose, galactose and fucose as well as some unusual sugars. This polysaccharide is apparently present in all higher plants and is unusually stable, accumulating, for example, in red wine. It contains two residues of the branched chain sugar **apiose,** one of which is a site of crosslinking by boron. A borate diol ester linkage binds two molecules of the pectin together as a dimer, perhaps controlling the porosity of the pectin gel. All of the complex cell wall polysaccharides bind, probably through multiple hydrogen bonds, to the cellulose microfibrils. The resulting structures are illustrated in drawings of Carpita and McCann, which are more current. The cellulose plus crosslinking glycans form one network in the cell wall. The pectic substances form a second independent network. Some covalent crosslinking occurs, but most interactions are noncovalent. The site of biosynthesis of pectins and hemicelluloses is probably Golgi vesicles which pass to the outside via exocytosis.

However, the cellulose fibrils as well as the chitin in fungi are apparently extruded from the plasma membrane. Although the principal cell wall components of plants are carbohydrates, proteins account for 5-10% of the mass. Predominant among these are glycoprotein **extensins.** Like collagen, they are rich in 4-hydroxyproline which is glycosylated with arabinose oligosaccharides and galactose. Other hydroxyproline-containing proteins with the characteristic sequence $(\text{hydroxyproline})_4$-Ser are also found, *e.g.,* in soybean cell walls. Some plant cell walls contain glycine-rich structural proteins. One in the petunia consists of 67% glycine residues. During advanced stages of formation, as the walls harden into wood, large amounts of **lignins** are laid down in some plant cells. These chemically resistant phenylpropanoid polymers contain many crosslinked aromatic rings. A remarkable aspect of primary plant cell walls is their ability to be elongated extremely rapidly during growth. While the driving force for cell expansion is thought to be the development of pressure within the cell, the manner in which the wall expands is closely regulated.

After a certain point in development, elongation occurs in one direction only and under the influence of plant hormones. Most striking is the effect of the **gibberellins**, which cause very rapid elongation. Elongation of plant cell walls may depend to some extent upon chemical cleavage and reforming of crosslinking polysaccharides. However, the cellulose fibrils probably remain intact and slide past each other. A curious effect, which is mediated by proteins called expansins, is the ability of plant tissues to extend rapidly when incubated in a mildly acidic buffer of pH <5.5. Expansins are also involved in ripening of fruit. They may disrupt non-covalent bonding between cellulose fibrils and the hemicelluloses. The β-expansins of grasses are allergens found in grass pollens. The borate diol ester linkages in the pectin may also facilitate expansion.

Patterns in Polysaccharide Structures

How can the many complex polysaccharides found in nature be synthesized? Are there genetically determined patterns? How are these controlled? The answer can be found in the *specificities* of the hundreds of known *glycosyltransferases* and in the *patterns of expression of the genes* for transferases and other proteins. As a consequence, a great variety of structurally varied polysaccharide structures arise, especially on cell surfaces. The structures are not random but depend upon the assortment of glycosyltransferases available at the particular stage of development in a tissue. The numerous possibilities can account for much of the variation observed between species, between tissues, and also among individuals. The simplest pattern is the growth of straight-chain homopolysaccharides such as amylose, cellulose, and chitin.

The glycosyltransferases must recognize both the glycosyl donor, *e.g.*, ADP-glucose, UDP-glucose, and also the correct end of the growing polymer, always adding the same monomer unit. In contrast, hyaluronan and the polysaccharide chains of **glycosaminoglycans** have an alternating pattern. For a hyaluronan chain growing at the reducing end, one active site of hyaluronan synthase must be specific for UDP-GlcNAc and transfer the sugar unit only to the end of a glucuronic acid ring. A second active site must be specific for UDP-glucuronic acid but attach it only to the end of an acetylglucosamine unit. There is still uncertainty about the direction of growth of hyaluronan. Some hyaluronan synthases are lipid-dependent and their mechanism may resemble that proposed for cellulose synthesis.

Dextrans. Some polysaccharides, such as the bacterial dextrans, are synthesized outside of cells by the action of secreted enzymes. An enzyme of this type, **dextran sucrase** of *Leuconostoc* and *Streptococcus,* adds glucosyl units at the *reducing* ends of the dextran chains. Sucrose is the direct donor of the glucosyl groups, which are added by an insertion mechanism. However, it is not dependent upon a membrane lipid as is that fo Fig. 9.5. The glucosyl groups are transferred from sucrose to one of a pair of carboxylate groups of aspartate side chains in the active site. If both carboxylates are glucosylated, a dextran chain can be initiated by insertion of one glucosyl group into the second. The dextran grows alternating binding sites between the two carboxylates. Chain growth can be terminated by reaction with a sugar or oligosaccharide that fits into the active site and acts in place of the glucosyl group attached to Asp 1. The α-1,3-linked branches can be formed when a 3-OH group of a second dextran chain enters the catalytic site, serving as the glycosyl acceptor. See Robyt for a detailed discussion of synthesis of dextrans and related polysaccharides such as **alternan** and the α-1,3-linked **mutan.** Some bacteria form β-2,6-linked **fructans** by a similar mechanism, with glucose being released by displacement on C2 of sucrose. Fructans are also formed in green plants, apparently from reaction of two molecules of sucrose with release of glucose to form the trisaccharide Fru*f*β2→1Fru*f*β2-1α-Glc*p*, which then transfers a fructosyl group to the growing chain.

Lipid-dependent synthesis of polysaccharides. Insertion of monomer units at the base of a chain is a major mechanism of polymerization that is utilized for synthesis not only of polysaccharides but also of proteins. For most carbohydrates the synthesis is dependent upon a polyprenyl lipid alcohol. In bacteria this is often the 55-carbon **undecaprenol** or **bactoprenol,** which functions as a phosphate ester:

Undecaprenol phosphate

It serves as a membrane anchor for the growing polysaccharide. We have already discussed one example in the hypothetical cellulose synthase mechanism. For some polysaccharides the mechanism is better established. The synthetic cycles all resemble that, and can be generalized. Here NDP-Glx is a suitable nucleotide disphosphate derivative of sugar Glx, and Z-Glx is the repeating unit of the polysaccharide formed by the action of glycosyl-transferases and other enzymes.

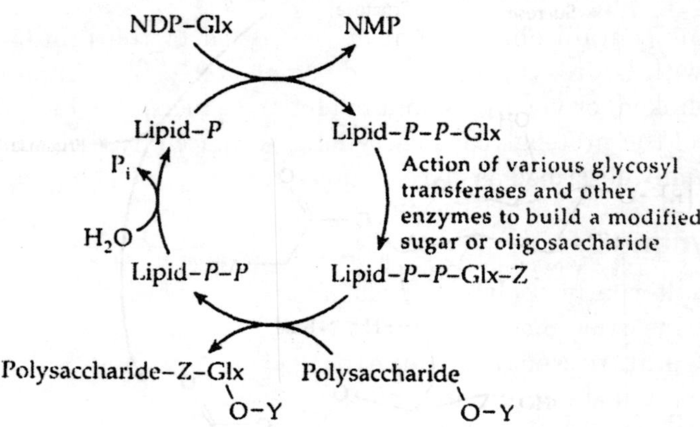

For example, the biosynthesis of alginate involves GDP-mannuronic acid (GDP-ManA) as NDP-Glx, bactoprenol as the lipid, and a glycosyltransferase that inserts a second mannuronate residue (as Z).

An additional transferase that uses acetyl-CoA as a substrate sometimes acetylates one mannuronate unit. The disaccharide units are then inserted into the growing chain. An additional modification, which occurs after polymerization, is random C5 epimerization of unacetylated D-mannuronate residues to L-guluronate. Formation of alginate is of medical interest because infections by alginate-forming bacteria are a major cause of respiratory problems in cystic fibrosis.

Sometimes an oligosaccharide assembled on the polyprenol phosphate represents a substantial block in assembly of a repeating polymer. For example, the xanthan gum, produced by the bacterium *Xanthomonas campestris* is formed by several successive glycosyl transfers to bactoprenol-*P*-*P*-Glc. A second glucose is transferred onto the first from UDP-Glc, forming a pair of glucosyl groups in β-1,4 linkage. Mannose is then transferred from GDP-Man and joined in an α-1,3 linkage to the first GDP-Man to form a branch point. A glucuronate residue is then transferred from UDP-GlcA and another mannose from GDP-Man. The last mannose is modified by reaction with PEP to form a ketal. The product of this assembly is the following lipid-bound oligosaccharide block.

```
                          Acetyl          Glcb1
Pyruvate                    ↓               ↓
   ||                       6               4
Manβ1 → 4GlcAβ1 → 2Manα1 → 3Glcβ1 → O-P-P- lipid
```

This is inserted into the growing polysaccharide using the free 4-OH on the second glucose to link the units in a cellulose type chain. The twelve separate genes needed for synthesis of xanthan gum are contained in a 16-kb segment of the *X. campestris* genome. Lipid-bound intermediates are also involved in synthesis of peptidoglycans and in the assembly of bacterial O-antigens. Both of these also yield "block polymers."

PROTEOGLYCANS AND GLYCOPROTEINS

The glycoproteins contain oligosaccharides attached to the protein either through *O*-glycosidic linkages with hydroxyl groups of side chains of serine, threonine, hydroxyproline, or hydroxylysine (*O*-linked) or via glycosylaminyl linkages to asparagine side chains (N-linked). The "core proteins" of the proteoglycans carry long polysaccharide chains, which are usually *O*-linked and are usually described as glycosaminoglycans.

Glycosaminoglycans

Synthesis of the alternating polysaccharide hyaluronan has been discussed in Section C,3 and may occur by an insertion mechanism. However, other glycosaminoglycans (sulfate esters of **chondroitin, dermatan, keratan, heparan, and heparin**) grow at their nonreducing ends. Their synthesis is usually initiated by the hydroxyl group of serine or threonine side chains at special locations within several secreted proteins. These proteins are synthesized in the rough ER and then move to the Golgi. Addition of the first sugar ring begins in the ER with transfer of single xylosyl residues to the initiating —OH groups. This reaction is catalyzed by the first of a group of special glycosyltransferases of high specificity that form the special terminal units, that anchor the alternating polysaccharide represented here as $(X\text{-}Y)_n$:

$(X \rightarrow Y)_n \rightarrow 4$ GlcAβ1 → 3 Galβ1→ 3 Galβ1→4Xylβ1→O-Ser/Thr

After transfer of the xylosyl residue from UDP-xylose to the —OH group in the protein, a second enzyme with proper specificity transfers a galactosyl group from UDP-galactose, joining it in β-1,4 linkage. A third enzyme transfers another galactosyl group onto the first one in β-1,3 linkage. A fourth enzyme, with a specificity different from that used in creating the main chain, then transfers a glucuronosyl group from UDP-glucuronic acid onto the chain terminus to complete the terminal unit. Then two more enzymes transfer the alternating units in sequence to form the repeating polymer with lengths of up to 100 or more monosaccharide residues. The sequence $(X\text{-}Y)_n$ in the preceding formula is:

$(\rightarrow 4\ \text{GlcA}\beta 1 \rightarrow 3\ \text{GalNAc}\beta 1 \rightarrow)_n$
for chondroitin

$(\rightarrow 4\ \text{GlcA}\beta 1 \rightarrow 4\ \text{GlcNAc}\alpha 1 \rightarrow)_n$
for heparan sulfate and heparin

$(\rightarrow 3\ \text{Gal}\beta 1 \rightarrow 4\ \text{GlcNAc}\beta 1 \rightarrow)_n$
for dermatan sulfate

Subsequent modifications of the polymers involve extensive formation of *O*-sulfate esters, *N*-deacetylation and *N*-sulfation, and epimerization at C5. In some tissues almost all GluA is epimerized. The modifications are especially extensive in dermatan, heparan sulfates, and heparin. The modifications are not random and follow a defined order. *N*-Deacetylation must precede *N*-sulfation, and *O*-sulfation is initiated only after *N*-sulfation of the entire chain is complete. The modifications occur within the Golgi, but not all of the glycosyltransferases, PAPS (3′-phosphoadenosine 5′-phosphosulfate)-dependent sulfotransferases, and epimerases are present within a single compartment. Nevertheless, an entire glycosaminoglycan chain can be synthesized within 1-3 min. The completed polymers are modified uniformly. There are clusters of sulfo groups with unusual structures in chondroitin from squid and shark cartilages and fucosylated chondroitin from echinoderms.

Similar modifications are present less extensively in vertebrates. One of the best known modifications forms the unique pentasaccharide sequence shown, which is essential to the anticoagulant activity of heparin. This sequence has been synthesized in the laboratory as have related longer heparin chains. A sequence about 17 residues in length containing an improved synthetic version of the unique pentasaccharide binds tightly to both thrombin and antithrombin. This opens the door to the development of improved substitutes for the medically important heparin. Heparan sulfate chains are found on proteoglycans throughout the body, but the highly modified heparin does not circulate in the blood. It is largely sequestered in cytoplasmic granules within mast cells and is released as needed.

Heparin binds to many different proteins. Among them is the glycoprotein selenoprotein P, which may impart antioxidant properties to the extracellular matrix. Although glycosaminoglycans are most often attached to *O*-linked terminal units, chondroitin sulfate chains can also be synthesized with N-linked oligosaccharides attached to various glycoproteins serving as initiators. At least one form of keratan sulfate, found in the cornea, is linked to its initiator protein via GlcNAc-Man to *N*-linked oligosaccharides of the type present in many glycoproteins. At least 25 different proteins that are secreted into the extracellular spaces of the mammalian body carry glycosaminoglycan chains. Most of these proteins can be described as (1) **small leucine-rich proteoglycans** with 36-to 42-kDa protein cores and (2) **large modular proteoglycans** whose protein cores have molecular masses of 40 to 500 kDa. The most studied of the second group is aggrecan, a major component of cartilage. This 220-kDa protein carries ~100 chondroitin chains, each averaging about 100 monosaccharide residues and ~100 negative charges from the carboxylate and sulfate groups.

Aggrecan has three highly conserved globular domains near the N and C termini. The G1 domain near the N terminus is a **lectin**, which, together with a small link protein that is structurally similar to the G1 domain, binds to a decasaccharide unit of hyaluronan. One hyaluronan molecule of 500-to1000-kDa mass (~2500-5000 residues) may bind 100 aggrecan and link molecules to form an ~ 200,000-kDa particle such as that shown. These enormous highly negatively charged molecules, together with associated counterions, draw in water and preserve osmotic balance. It is these molecules that keep our joints mobile and which deteriorate by proteolytic degradation in the common **osteoarthritis.** The keratan sulfate content of cartilage varies with age, and the level in serum and in synovial fluid is increased in osteoarthritis.

Keratan sulfate is also found in the cornea and the brain. Its content is dramatically decreased in the cerebral cortex of patients with Alzheimer disease. Other modular core proteins include **versican** of blood vessels and skin, **neurocam and brevican** of brain, **perlecan** of basement membranes, **agrin** of neuromuscular junctions, and **testican** of seminal fluid.

However, several of these have a broader distribution than is indicated in the foregoing description. The sizes vary from 44 kDa for testican to greater than 400 kDa for perlecan. The numbers of glycosaminoglycan chains are smaller than for aggrecan, varying from 1 to 30. Another of the chondroitin sulfate-bearing core protein is **appican,** a protein found in brain and one of the splicing variants of the amyloid precursor protein that gives rise to amyloid deposits in Alzheimer disease. The core proteins of the leucine-rich proteoglycans have characteristic horseshoe shapes and are constructed from ~ 28-residue repeats, each containing a β turn and an α helix.

The three-dimensional structures are doubtless similar to that of a ribonuclease inhibitor of known structure which contains 15 tandem repeats. A major function of these proteoglycans seems to be to interact with collagen fibrils, which have distinct proteoglyan-binding sites, and also with fibronectin. The small leucinerich proteoglycans have names such as **biglycan, decorin, fibromodulin, lumican, keratoglycan, chondroadherin, osteoglycin,** and **osteoadherin.** The distribution varies with the tissue and the stage of development. For example, biglycan may function in early bone formation; decorin, which has a high affinity for type I collagen, disappears from bone tissue as mineralization takes place. Osteoadherin is found in mature osteoblasts. **Phosphocan,** another brain proteoglycan, has an unusually high content (about one residue per mole) of L-isoaspartyl residues. Proteoglycans bind to a variety of different proteins and polysaccharides.

For example, the large extracellular matrix protein **tenascin,** which is important to adhesion, cell migration, and proliferation, binds to chondroitin sulfate proteoglycans such as neurocan. **Syndecan,** a transmembrane proteoglycan, carries both chondroitin and heparan sulfate chains, enabling it to interact with a variety of proteins that mediate cell-matrix adhesion. The ability of dissociated cells of sponges to aggregate with cells only of a like type, depends upon large extracellular proteoglycans. That of *Microciona prolifera* appears to be an aggregate of about three hundred 35-kDa core protein molecules with equal masses of attached carbohydrate. This **aggregation factor** has a total mass of ~ 2×10^4 kDa. It apparently interacts specifically, in the presence of Ca^{2+} ions, with a 210-kDa cell matrix protein to hold cells of the same species together.

O-Linked Oligosaccharides

A variety of different oligosaccharides are attached to hydroxyl groups on appropriate residues of serine, threonine, hydroxylysine, and hydroxyproline in many different proteins. Such oligosaccharides are present on external cell surfaces, on secreted proteins, and on some proteins of the cytosol and the nucleus. The rules that determine which —OH groups are to become glycosylated are not yet clear. Glycosylation occurs in the ER, and, just as during synthesis of the long carbohydrate chains of proteoglycans, the sugar rings are added directly to an —OH group, either of the protein or of the growing oligosaccharide. The first glycosyl group transferred is most often **GalNAc** for external and secreted proteins but more often **GlcNAc** for cytosolic and nuclear proteins.

Glycosylation of protein —OH groups can occur on either the luminal or cytosolic faces of the ER membranes. The external *O*-linked glycoproteins often have large clusters of oligosaccharides attached to —OH groups of serine or threonine, but cytosolic proteins may carry only a small number of small oligosaccharides. Of great importance are the **blood group determinants** which are discussed in elsewhere. The ABO determinants are found at the nonreducing ends of *O*-linked oligosaccharides. Conserved Ser/Thr sites in the epidermal growth factor domains of various proteins carry ***O*-linked fucose.** The secreted **mucins** are unique in having clusters of large numbers of oligosaccharides linked by ***N*-acetylgalactosamine** to serine or threonine of the polypeptide. The following core structures predominate. These may be lengthened or further branched by the particular variety of glycosyltransferases present in a tissue and by their specificities.

The human genome contains at least nine mucin genes. The large apomucins contain central domains with tandem repeats rich in Ser, Thr, Gly, and Ala and flanked at the ends by cysteinerich domains. For example, porcine submaxillary mucins are encoded by a gene with at least three alleles that encode 90,125, or 133 repeats. The polypeptide may contain as many as 13,288 residues. N-terminal cysteine-rich regions are involved in dimer formation.

core 1 Galβ1→3 GalNAc - Ser/Thr

core 2 Galβ1→3 GalNAc - Ser/Thr - Ser/Thr
6
↑
GalNAcβ1

core 3 GlcNAcβ1→3 GalNAc - Ser/Thr

core 4 Galβ1→4 GlcNAcβ1→6 GalNAc- Ser/Thr
3
↑
GlcNAcβ1

Assembly of *N*-Linked Oligosaccharides on Glycosyl Carrier Lipids

In eukaryotes the biosynthesis of the *N*-linked oligosaccharides of glycoproteins depends upon the polyprenyl alcohols known as **dolichols,** which are present in membranes of the endoplasmic reticulum. They contain 16-20 prenyl units, of which the one bearing the OH group is completely saturated as a result of the action of an NADPH-dependent reductase on the unsaturated precursor. The predominant dolichol in mammalian cells contains 19 prenyl units. The structure of its mannosyl phosphate ester, one of the intermediates in the oligosaccharide synthesis, is illustrated below. The fully extended 95-carbon dolichol has a length of almost 10 nm, four times greater than that of oleic acid and twice the thickness of the nonpolar membrane bilayer core.

The need for this great length is not clear nor is it clear why the first prenyl unit must be saturated for good acceptor activity. The assembly of the oligosaccharides that will become linked to Asn residues in proteins occurs on the phosphate head of dolichol-*P*. The process begins on the cytoplasmic face of the membrane and within the lumen of the rough or smooth ER and continues within cisternae of the Golgi apparatus. The initial transfer of GlcNAc-*P* to dolichol-*P*, appears to occur on the cytoplasmic face of the ER and is specifically inhibited by **tunicamycin.** As the first "committed reaction" of *N*-glycosylation, it is regulated by a variety of hormonal and other factors. The reaction takes place cotranslationally as the still unfolded peptide chain leaves the ribosome.

The oligosaccharide, still attached to the dolichol, continues to grow on the cytosolic surface of the ER membrane by transfer of GlcNAc and five residues of mannose from their sugar nucleotide forms. The intermediate Dol-*P*-*P*-$GlcNAc_2Man_5$ crosses the membrane bilayer, step *d*), after which mannosyl and glucosyl units are added (Fig. 9.6 steps *e* and *f*). These sugars are carried across the membrane while attached to dolichol.

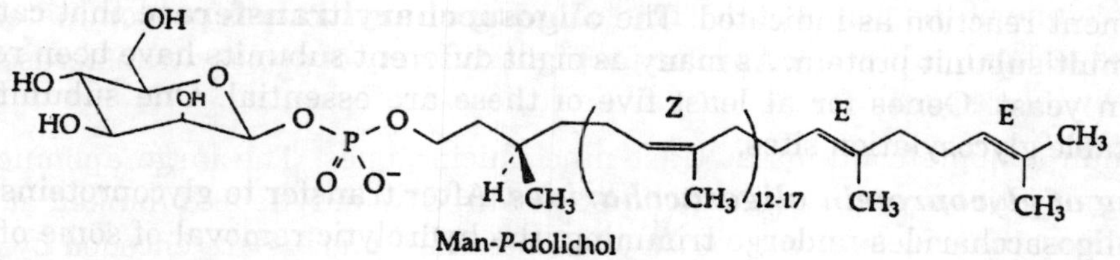

Man-*P*-dolichol

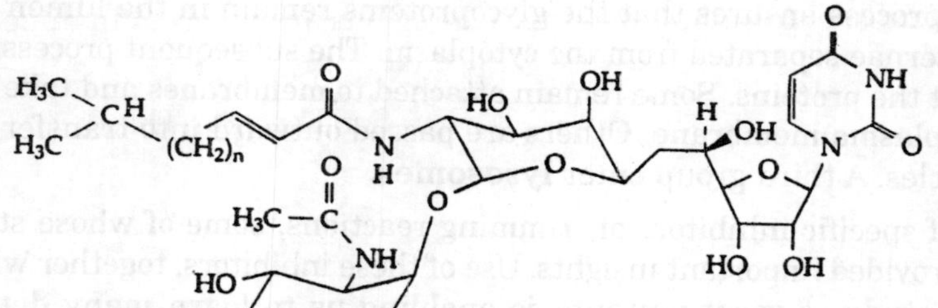

Tunicamycins
n = 8, 9, 10, 11

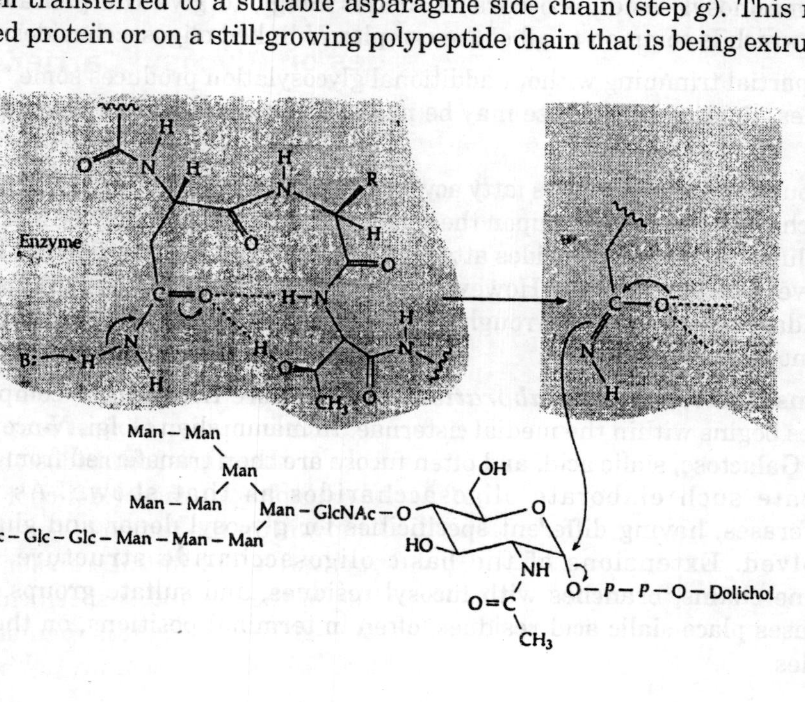

The completed 14-residue branched oligosaccharide $Glc_3Man_9GlcNAc_2$, with the structure indicated, is then transferred to a suitable asparagine side chain (step *g*). This may be on a newly synthesized protein or on a still-growing polypeptide chain that is being extruded through the membrane into the lumenal space of the rough. The glycosylation site is often at the sequence Asn-X-Ser(Thr), which is likely to be present at a beta bend in the folded protein$_v$ Bends of the type illustrated, and stabilized by the asparagine side chain are apparently favored. In such a bend the —OH of the serine or threonine helps to polarize the amide group of the Asn side chain, perhaps enolizing it and generating a nucleophilic center that can participate in a displacement reaction as indicated. The **oligosaccharyltransferase** that catalyzes the reaction is a multisubunit protein. As many as eight different subunits have been reported for the enzyme in yeast. Genes for at least five of these are essential. One subunit serves to recognize suitable glycosylation sites.

Trimming of glycoprotein oligosaccharides. After transfer to glycoproteins the newly synthesized oligosaccharides undergo trimming, the hydrolytic removal of some of the sugar units, followed by addition of new sugar units to create the finished glycoproteins. The initial glycosylation process ensures that the glycoproteins remain in the lumen of the ER or within vesicles or cisternae separated from the cytoplasm. The subsequent processing appears to allow the cell to **sort** the proteins. Some remain attached to membranes and take up residence within ER, Golgi, or plasma membrane. Others are passed outward into transfer vesicles, Golgi, and secretion vesicles. A third group enter **lysosomes.**

A series of specific inhibitors of trimming reactions, some of whose structures are shown Fig. 9.7, has provided important insights. Use of these inhibitors, together with immunochemical methods and study of yeast mutants, is enabling us to learn many details of glycoprotein biosynthesis. Whereas the formation of dolichol-linked oligosaccharides occurs in an identical manner in virtually all eukaryotic cells, trimming is highly variable as is the addition of new monosaccharide units. The major pathway for mammalian glycoproteins is shown. Specific hydrolases in the ER remove all of the glucosyl units and one to three mannosyl units. Removal of additional mannosyl residues occurs in the cis Golgi, to give the pentasaccharide core $Man_3GlcNAc_2$ which is common to all of the complex *N*-linked oligosaccharides.

However, partial trimming without additional glycosylation produces some "high mannose" oligosaccharides. Removal of glucose may be necessary to permit some glycoproteins to leave the ER.

Sulfate groups and in some cases fatty acyl groups may also be added. The exact composition of the oligosaccharides may depend upon the condition of the cell and may be altered in response to external influences. Oligosaccharides attached to proteins that remain in the ER membranes may undergo very little trimming. However, the three glucosyl residues are usually removed by the glucosidases present in the rough ER. Some plant glycoproteins of the highmannose type undergo no further processing.

Extensions *and terminal elaborations.* Even before trimming is completed, addition of new residues begins within the medial cisternae. In mammalian Golgi, *N*-acetylglucosamine is added first. Galactose, sialic acid, and often fucose are then transferred from their activated forms to create such elaborate oligosaccharides as that shown. As many as 500 glycosyltransferases, having different specificities for glycosyl donor and glucosyl acceptor, may be involved. Extensions of the basic oligosaccharide structure often contain polylactosamine chains, branches with fucosyl residues, and sulfate groups. More than 14 sialyltransferases place sialic acid residues, often in terminal positions, on these cell surface oligosaccharides.

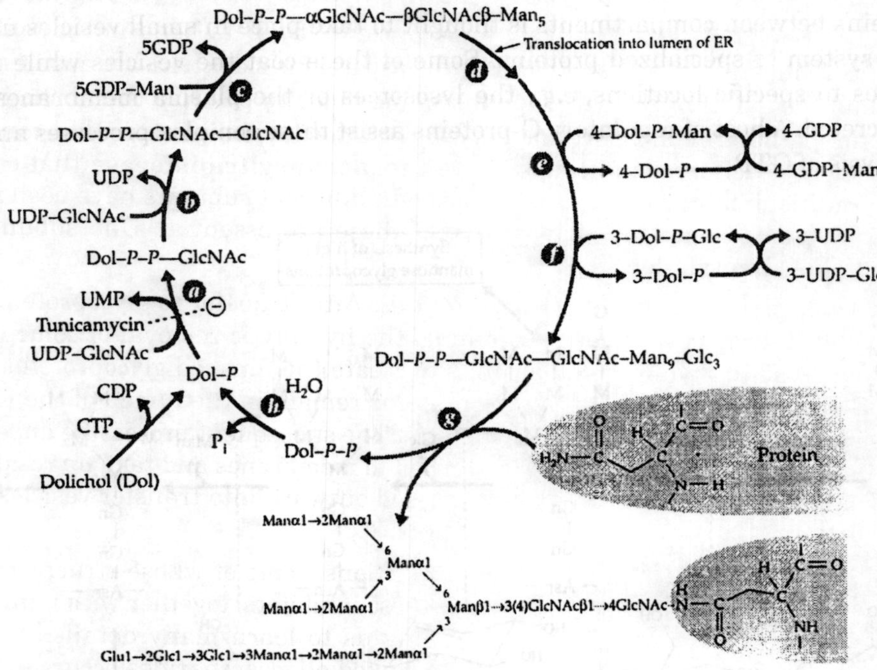

Fig. 9.6. Biosynthesis of the dolichol diphosphate-linked oligosaccharide precursor to glycoproteins. The site of inhibition by tunicamycin is indicated.

The cell wall of the yeast *Saccharomyces* is rich in **mannoproteins** that contain 50-90% mannose. The ~ 250-residue mannan chains consist of an α-1,6-linked backbone with mono-, di-, and trimannosyl branches. These are attached to the same core structure as that of mammalian oligosaccharides. All of the core structures are formed in a similar way. The mannoproteins may serve as a "filler" to occupy spaces in a cell wall constructed from β-1,3- and β-1,6-linked glycans and chitin. All of the four components, including the mannoproteins, are covalently linked together. As was emphasized, glycoproteins serve many needs in biological recognition. The N-linked oligosaccharides play a major role in both animals and plants. Use of mass spectrometry, new automated methods of oligosaccharide synthesis, and development of new synthetic inhibitors are all contributing to current studies of what is commonly called "glycobiology."

The perplexing Golgi apparatus. First observed by Camillo Golgi in 1898, the stacked membranes, now referred to simply as Golgi, remain somewhat mysterious. There are at least three functionally distinct sets of Golgi cisternae, the **cis** (nearest the nucleus), **medial,** and **trans.** An additional series of tubules referred to as the **trans Golgi network** lies between the Golgi and the cell surface and may be the site at which lysosomal enzymes are sorted from proteins to be secreted. Immunochemical staining directed toward specific glycosidases and glycosyl transferases suggested that the trimming reactions of glycoproteins start in the ER and continue as the proteins pass outward successively from one compartment of the Golgi to the next. This has been the conventional view since the 1970s. The movement of the glycoproteins between compartments is thought to take place in small vesicles using a rather elaborate system of specialized proteins. Some of these coat the vesicles while others target the vesicles to specific locations, *e.g.*, the lysosomes or the plasma membranes where they may be secreted. A host of regulatory G-proteins assist these complex processes and drive them via hydrolysis of GTP.

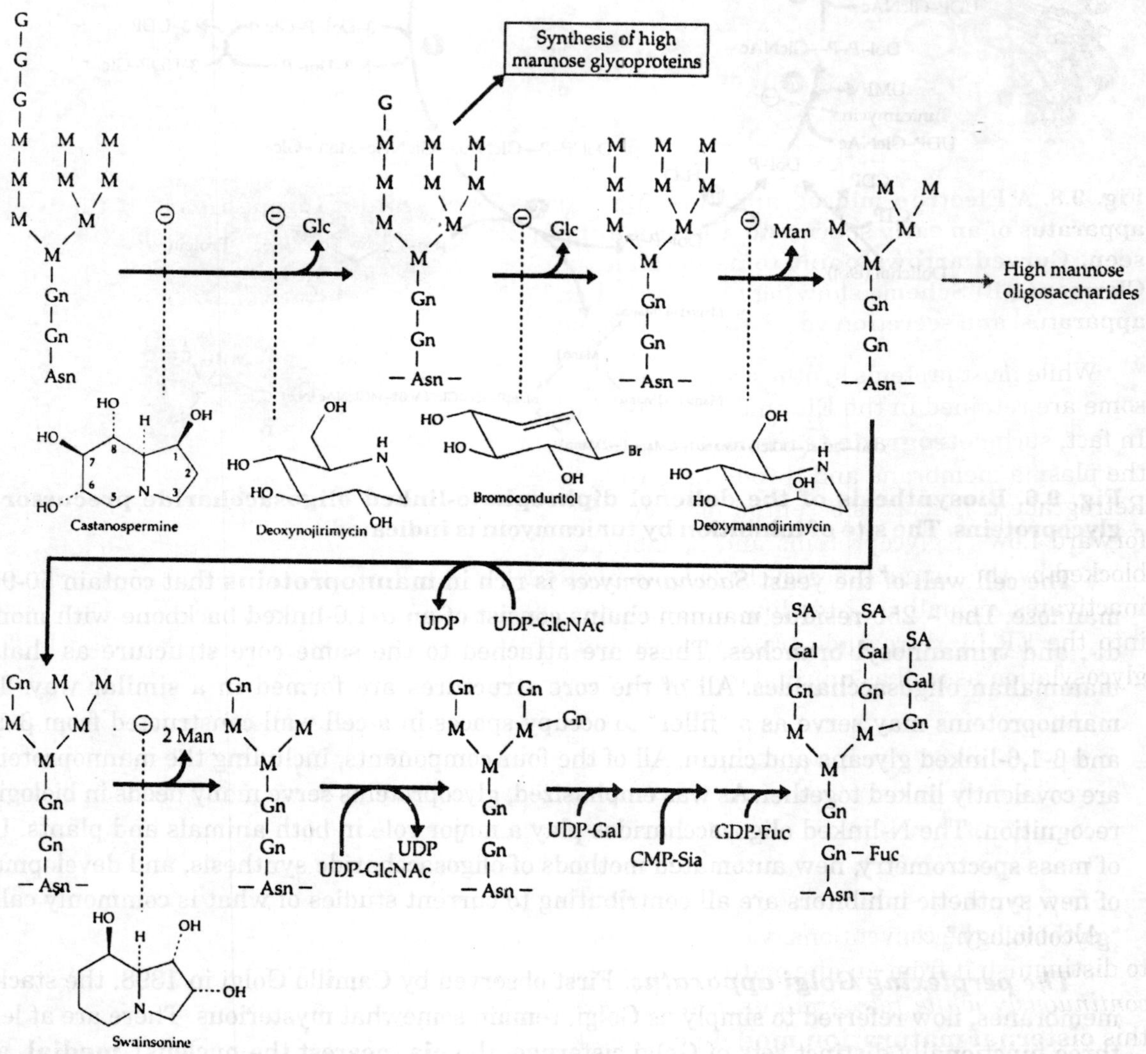

Fig. 9.7. Steps in trimming of the oligosaccharide precursor and synthesis of complex N-linked oligosaccharides of mammalian glycoproteins. Sites of action of several inhibitors are shown. Abbrevations: M, mannosyl, Gn, *N*-acetylglucosaminyl, G, glucosyl residues. Structures of several inhibitors are shown.

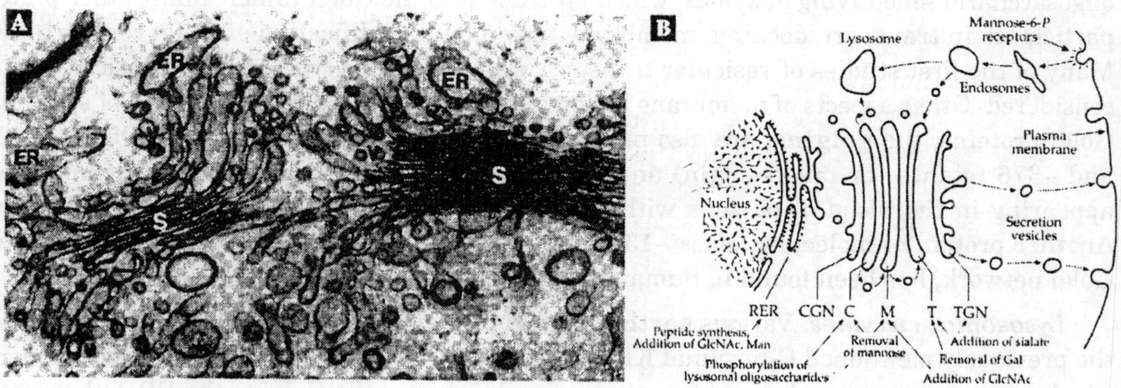

Fig. 9.8. A Electron micrograph showing a transverse section through part of the Golgi apparatus of an early spermatid. Cisternae of the ER, Golgi stacks (S), and vesicles (V) can be seen. Curved arrows point to associated tubules. Magnification X45,000. Courtesy of Y. Clermont. (B) Scheme showing functions of endoplasmic reticulum, transfer vesicles, Golgi apparatus, and secretion vesicles in the metabolism of glycoproteins.

While most proteins synthesized in the ER follow the exocytotic pathway through the Golgi, some are retained in the ER and some that pass on through the Golgi are returned to the ER. In fact, such **retrograde transport** can carry some proteins taken up by endocytosis through the plasma membrane and through the Golgi to the ER where they undergo *N*-glycosylation. Retrograde transport is essential for recycling of plasma membrane proteins and lipids. The forward flow of glycoproteins and membrane components from the ER to the Golgi can be blocked by the fungal macrocyclic lactone **brefeldin** A. In cells treated with this drug, which inactivates a small CTP-binding protein, the Golgi apparatus is almost completely resorbed into the ER by retrograde transport. Proteins remaining in the ER undergo increased *O*-glycosylation as well as unusual types of *N*-glycosylation.

Brefeldin A

Although the conventional view of flow through the Golgi is generally accepted, it is difficult to distinguish it from an alternative explanation: *The Golgi compartments may move outward continuously while retrograde transport occurs via the observed vesicles.* Some evidence for this **cisternal maturation model** has been known for many years but was widely regarded as reflecting unusual exceptions to the conventional model. In fact, both views could be partially correct; vesicular transport may function in both directions. High-resolution tomographic images are also altering our view of the Golgi.

The proteins of Golgi membranes are largely integral membrane proteins and peripheral proteins associated with the cytosolic face. Some of the integral membrane proteins are the oligosaccharide-modifying enzymes, which protrude into the Golgi Iumen. Many other proteins participate in transport, docking, membrane fusion, and acidification of Golgi compartments. Many of the first studies of vesicular transport were conducted with synaptic vesicles and are considered. Other aspects of membrane fusion and transport are discussed, group of specialized Golgi proteins, the **golgins,** are also present. They are designated golgin-84, -95, -160, -245, and –376 (giantin or macrogolgin) and were identified initially as human **autoantigens**, appearing in the blood of persons with autoimmune disorders such as Sjogren's syndrome. Another protein of molecular mass ~130 kDa, and which appears to be specific to the trans-Golgi network, has been found in human serum of patients with renal vasculitis.

Lysosomal enzymes. Various **sorting signals** are encoded within proteins. These include the previously mentioned C-terminal KDEL (mammals) or HDEL (yeast) amino acid sequence, which serves as a retrieval signal for return of proteins from the Golgi to the ER. Other sorting signals are provided by the varied structures of the oligosaccharides attached to glycoproteins. These sugar clusters convey important biological information, which is "decoded" in the animal body by interaction with various **lectins** that serve as receptors. This often leads to endocytosis of the glycoprotein.

An example is provided by the more than 50 proteins that are destined to become lysosomal enzymes and which undergo phosphorylation of 6—OH groups of the mannosyl residue marked by asterisks on the following structure. This is an *N*-linked oligosaccharide that has been partially trimmed. The phosphorylation is accomplished in two steps by enzymes present in the cis Golgi compartment. An ***N*-acetylglucosaminyl-phosphotransferase** transfers phospho-GlcNAc units from UDP-GlcNAc onto the 6-OH groups of mannosyl residues. These must be recognized by the phosphotransferase as appropriate. Then a hydrolase cleaves off GlcNAc.

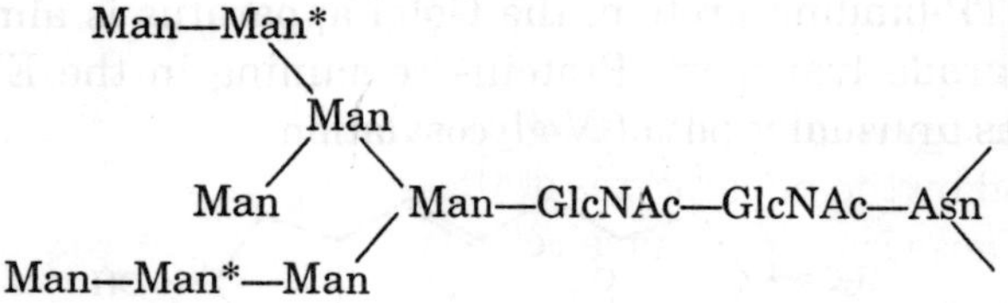

```
                 Man—Man*
                        \
                         Man
                        /   \
                    Man      Man—GlcNAc—GlcNAc—Asn
                            /
          Man—Man*—Man
```

The proteins carrying the mannose 6-phosphate groups bind to one of two different **mannose 6-*P* receptors** present in the Golgi membranes and are subsequently transported in clathrin-coated vesicles to endosomes where the low pH causes the proteins to dissociate from the receptors, which may be recycled. The hydrolytic enzymes are repackaged in lysosomes. The same mannose 6-*P* receptors also appear on the external surface of the plasma membrane allowing many types of cells to take up lysosomal enzymes that have "escaped" from the cell. These proteins, too, are transported to the lysosomes. The mannose 6-*P* receptors have a dual function, for they also remove insulin-like growth factor from the circulation, carrying it to the lysosomes for degradation. Most Man 6-*P* groups are removed from proteins once they reach the lysosomes but this may not always be true. Not all lysosomal proteins are recruited by the mannose 6-*P* receptors. Some lysosomal membrane proteins are sorted by other mechanisms.

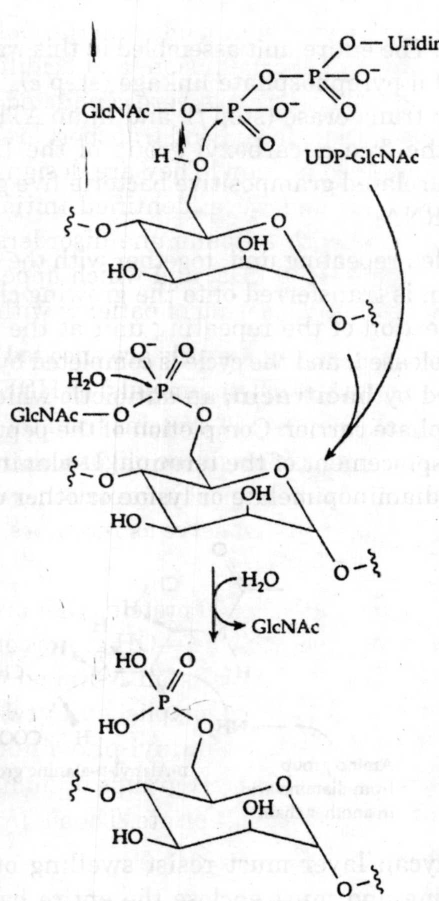

The hepatic asialoglycoprotein (Gal) receptor. A variety of proteins are taken out of circulation in the blood by the hepatocytes of the liver. Serum glycoproteins bearing sialic acid at the ends of their oligosaccharides, have relatively long lives, but if the sialic acid is removed by hydrolysis, the exposed galactosyl residues are recognized by the multisubunit **asialoglycoprotein** receptor. The bound proteins are then internalized rapidly via the coated pit pathway and are degraded in the lysosomes. Other receptors, including those that recognize transferrin, low-density lipoprotein, α_2 macroglobulin, and T lymphocyte antigens, also depend upon interaction with oligosaccharides.

BIOSYNTHESIS OF BACTERIAL CELL WALLS

The outer surfaces of bacteria are rich in specialized polysaccharides. These are often synthesized while attached to lipid membrane anchors as indicated in a general way. One of the specific biosynthetic cycles, that depends upon undecaprenol phosphate is the formation of the **peptidoglycan** (murein) layer, of both gramnegative and grampositive bacterial cell walls. Synthesis begins with attachment of L-alanine to the OH of the lactyl group of UDP-*N*-acetylmuramic acid in a typical ATP-requiring process (Fig. 20.9. step, a). Next D-glutamic acid, meso-diaminopimelate, or L-lysine, and D-alanyl-D-alanine are joined in sequence, each in another ATP-requiring step. The entire unit assembled in this way is transferred to undecaprenol phosphate with creation of a pyrophosphate linkage (step *e*). An *N*-acetylglucosamine unit is added by action of another transferase (step *f*), and in an ATP-requiring process ammonia is sometimes added to cap the free α–carboxyl group of the D-glutamyl residue (step *g*). In *Staphylococcus aureus* and related grampositive bacteria five glycyl units are also added, each from a molecule of glycyl-tRNA.

The completely assembled repeating unit, together with the connecting peptide chain needed in the crosslinking reaction, is transferred onto the growing chain (step *h*). As in formation of dextrans, growth is by insertion of the repeating unit at the reducing end of the chain. The polyprenyl diphosphate is released, and the cycle is completed by the action of a pyrophosphatase (step *i*). This step is blocked by **bacitracin,** an antibiotic which forms an unreactive complex with the polyprenyl diphosphate carrier. Completion of the peptidoglycan requires crosslinking. This is accomplished by displacement of the terminal D-alanine of the pentapeptide by attack by the —NH_2 group of the diaminopimelate or lysine or other diamino acid.

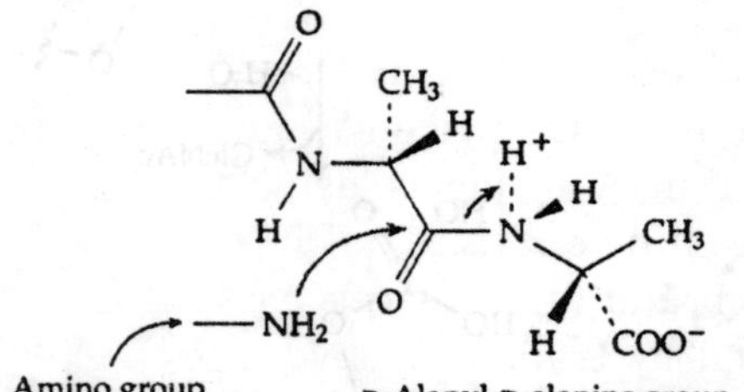

Amino group from diamino acid in another chain

D-Alanyl-D-alanine group

Because the peptidoglycan layer must resist swelling of the bacteria in media of low osmolarity, it must be strong and must enclose the entire bacterium. At the same time the bacterium must be able to grow in size and also to divide. For these reasons bacteria must continuously not only synthesize peptidoglycan but also degrade it. The latter is accomplished by hydrolytic cleavage using cell wall enzymes to the ***N*-acetylglucosamine-anhydro-*N*-acetylmuramate-tripeptide** (GlcNAc-1,6-anhydro-MurNAc-L-Ala-D-Glu-A_2pm) fragment. A hydrolase cuts the peptide bridge. This process is probably essential to formation of new growing points for expansion of the murein layer. Most of the peptide fragments that are released in the periplasm are transported back into the cytosol. The anhydroMurNAc is removed, and new UDP-MurNAc and D-Ala-D-Ala units are added salvaging the tripeptide unit. The repaired UDP-MurNA-pentapeptide can then reenter the biosynthetic pathway.

The O-antigens and lipid A. A cluster of sugar units of specific structure makes up the repeating unit of the "O-antigen" of *Salmonella.* The many structural variations in this surface polysaccharide account for the over two thousand serotypes of *Salmonella.* As is illustrated, the O-antigen is a repeating block polymer that is attached to a complex lipopoly saccharide "core" and a hydrophobic membrane anchor known as lipid A. Lipid A and the attached core and O-antigen are synthesized inside the bacterial cell by enzymes found in the cytoplasmic membrane. The complete lipopolysaccharide units are then translocated from the inner membrane to the outer membrane of the bacteria. The synthesis of the O-antigen is understood best. Consider the following group E3 antigen, where Abe is abequose, and Rha is rhamnose:

$$\begin{array}{c} \text{Abe}\alpha 1 \\ \downarrow \\ 4 \\ 4\ (\rightarrow 6\ \text{Gal}\beta 1 \rightarrow 6\ \text{Man}1 \rightarrow 4\ \text{Rhal} \rightarrow)_n \end{array}$$

Assembly of this repeating unit begins with the transfer of a *phosphogalactosyl* unit from UDP-Gal to the phospho group of the lipid carrier undecaprenol phosphate. The basic reaction cycle is much like that, for assembly of a peptidoglycan. The oligosaccharide repeating unit of the O-antigen is constructed by the consecutive transferring action of three more transferases. For the antigen shown above, one enzyme transfers a rhamnosyl unit, another a mannosyl unit, and another an abequosyl unit from the appropriate sugar nucleotides. Then the entire growing O-antigen chain, which is attached to a second molecule of undecaprenol diphosphate, is transferred onto the end of the newly assembled oligosaccharide unit. In effect, the newly formed oligosaccharide is inserted at the reducing end of the growing chain just as in Fig. 9.9.

Elongation continues by the transfer of the entire chain onto yet another tetrasaccharide unit. As each oligosaccharide unit is added, an undecaprenol diphosphate unit is released and a phosphatase cleaves off the terminal phospho group to regenerate the original undecaprenol phosphate carrier. When the O-antigen is long enough, it is attached to the rest of the lipopolysaccharide. The lipid A anchor is also based on a carbohydrate skeleton. Its assembly in *E. coli,* which requires nine enzymes, is depicted. in Fig. 9.10. *N*-Acetylglucosamine 6-*P* is acylated at the 3-position and after deacetylation at the 2-position. As shown in this figure, acylation is accomplished by transfer of hydroxymyristoyl groups from acyl carrier protein (ACP). Two molecules of the resulting UDP-2,3-diacyl-GlcN are then joined via the reactions shown to give the acylated disaccharide precursor to lipid A. Stepwise transfer of KDO, *L-glycero-D-manno*-heptose, and other monosaccharide units from the appropriate sugar nucleotides and further acylation follows Fig. 9.10.

The assembled O-antigen chain is transferred from undecaprenol diphosphate onto the lipopolysaccharide core. This apparently occurs on the periplasmic surface of the plasma membrane. If so, the core lipid domain must be flipped across the plasma membrane before the O-antigen chain is attached. Less is known about the transport of the completed lipopolysaccharide across the periplasmic space and into the outer membrane. The core structures of the lipopolysaccharides vary from one species to another or even from one strain of bacteria to another. All three domains (lipid A, core, and O-antigen) contribute to the antigenic properties of the bacterial surface and to the virulence of the organism. Nitrogen-fixing strains of *Rhizobhium* require their own peculiar lipopolysaccharides for successful symbiosis with a host plant.

However, there are some features common to most lipopolysaccharides. Two to three residues of KDO are usually attached to the acylated diglucosamine anchor, and these are often followed by 3 – 4 heptose rings. The structure of the inner core regions of a typical lipopolysaccharide from *E. coli* is indicated. The complete structure of the lipopolysaccharide from a strain of *Klebsiella* is shown at the top of the next page. Here L, D-Hep*p* is D-*manno*-heptopyranose and D, D-Hep*p* is D-*glycero*-D-*manno*-heptose. As in this case, the outer core often contains several different hexoses. The lipopolysaccharide of *Neisseria meningititis* has sialic acid at the outer end. However, the major virulence factor for this organism, which is a leading cause of bacterial meningitis in young children, is a capsule of poly(ribosyl)ribitol phosphate that surrounds the cell. *Haemophilus influenzae,* a common cause of ear infections and meningitis in children, has no O-antigen but a more highly branched core oligosaccharide than is present in *E. coli. Legionella* has its own variations.

Gram-positive bacteria. Although their outer coatings are extremely varied, all gram-positive bacteria have a peptidoglycan similar to that of gram-negative bacteria but often containing the intercalated penta-glycine bridge indicated. However, the peptidoglycan of gram-positive bacteria is 20-50 nm thick, as much as ten times thicker than that of *E. coli.* Furthermore, the peptidoglycan is intertwined with the anionic polymers known as teichoic acids and teichuronic acids. Both proteins and neutral polysaccharides, sometimes covalently bound, may also be present.

Like peptidoglycans, teichoic and teichuronic acids are assembled on undecaprenyl phosphates or on molecules of diacylglycerol. Either may serve as an anchor. A "linkage unit" may be formed by transfer of several glycosyl rings onto an anchor unit. For example, in synthesis of ribitol teichoic acid sugar rings are transferred from UDP-GlcNAc, UDP-ManNAC, and CDP-Gal to form the following linkage unit:

Lipid-*P-P*-GlcNAc-ManNAc-(P-Gal)

Then many ribitol phosphate units are added by transfer from CDP-ribitol. Finally, the chain is capped by transfer of a glucose from UDP-Glc. Lipoteichoic acids often carry covalently linked D-alanine in ester linkage, altering the net electrical charge on the cell surface. The completed teichoic acid may then be transferred to a peptidoglycan, releasing the lipid phosphate for reuse. Glycerol teichoic acid may be formed in a similar fashion. Teichuronic acids arise by alternate transfers of *P*-GalNAc from UDP-GalNAc and of GlcA from UDP-GlcA. Gram-positive bacteria often carry surface proteins that interact with host tissues in establishing human infections.

Protein A of *Staphylococcus* is a well-known example. After synthesis in the cytoplasm, it enters the secretory pathway. An N-terminal hydrophobic leader sequence and a 35-residue C-terminal sorting signal guide it to the correct destination. There a free amino group of an unlinked pentaglycyl group of the peptidoglycan carries out a transamidation reaction with an LPXTG sequence in the proteins, cutting the chain between the threonine and glycine residues, and anchoring the protein A to the peptidoglycan. Group A streptococci, which are serious human pathogens, form α-helical coiled-coil threads whose C termini are anchored in the cell membrane. They protrude through the peptidoglycan layers and provide a hairlike layer around the bacteria.

A variable region at the N termini provides many antigens, some of which escape the host's immune system allowing infection to develop. Group B streptococci form carbohydrate antigens linked to teichoic acid. Streptococci, which are normally present in the mouth, utilize their carbohydrate surfaces as receptors for adhesion, allowing them to participate in formation of dental plaque. Cell walls of mycobacteria are composed of a peptidoglycan with covalently attached galactan chains. Branched chains of **arabinan,** a polymer of the furanose ring form of arabinose with covalently attached **mycolic acids,** are glycosidically linked to the galactan. Shorter **glycopeptidolipids,** containing modified glucose and rhamnose rings as well as fatty acids, contribute to the complexity of mycobacterial surfaces.

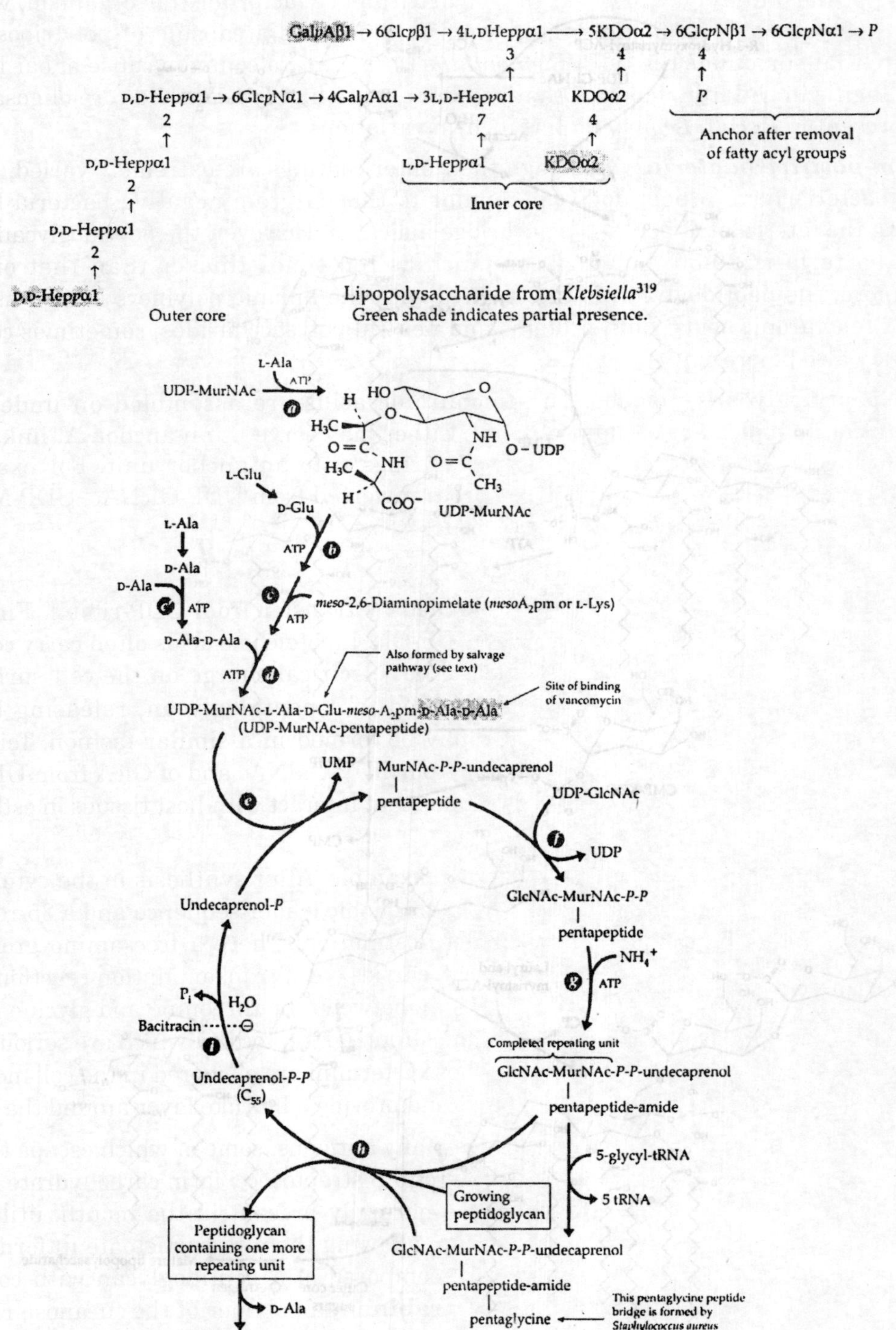

Fig. 9.9. Biosynthesis of bacterial peptidoglycans, for details of the peptidoglycan structures. Green arrows show alternative route used by gram-positive bacteria.

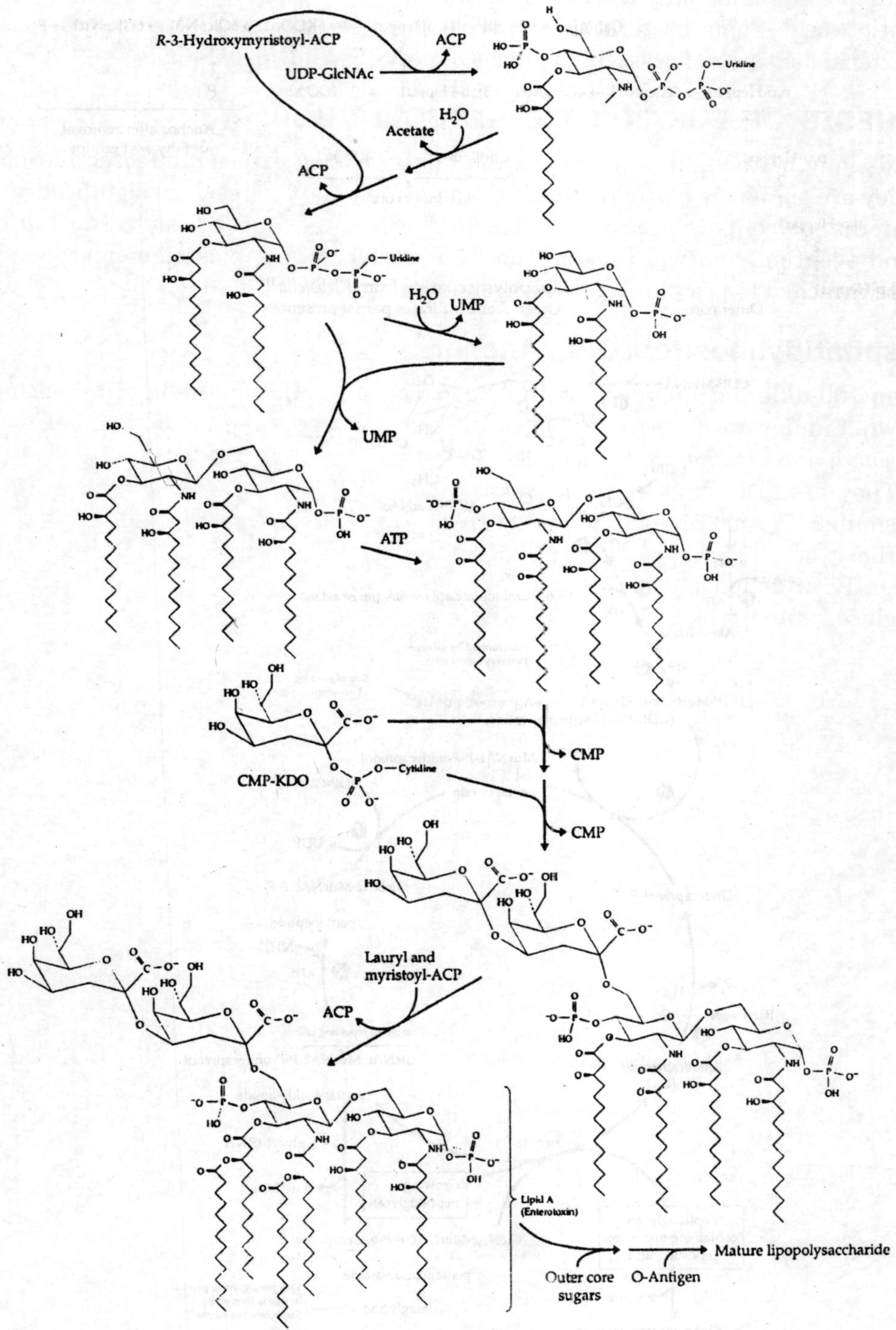

Fig. 9.10. Proposed biosynthetic route for synthesis of lipid A and the mature lipopolysaccharide of the *E. coli* cell wall.

These examples describe only a small sample of the great diversity of cell coats found in the prokaryotic world. Some bacteria also provide themselves with additional protection in the form of external sheaths of crystalline arrays of proteins known as S-layers.

BIOSYNTHESIS OF EUKARYOTIC GLYCOLIPIDS

Glycolipids may be thought of as membrane lipids bearing external oligosaccharides. In this sense, they are similar to glycoproteins both in location and in biological significance. Like the glycoproteins, glycolipids are synthesized in the ER, then transported into the Golgi apparatus and eventually outward to join the outer surface of the plasma membrane. Some glycolipids are attached to proteins by covalent linkage.

Glycophosphatidylinositol (GPI) Anchors

More than 100 different human proteins are attached to phosphatidylinositol anchors of the type shown. Similar anchors are prevalent in yeast and in protozoa including *Leishmania* and *Trypanosoma,* and *Plasmodium* where they often bind major surface proteins to the plasma membrane. They are also found in mycobacteria. The structures of the hydrophobic anchor ends are all similar. Two or three fatty acyl groups hold the molecule to the bilayer. Variations are found in the attached glycan portion, both in the number of sugar rings and in the structures of the covalently attached phosphoethanol-amine groups. The first step, the transfer of an *N* acetylglucosamine residue to phosphatidylinositol, is surprisingly complex, requiring

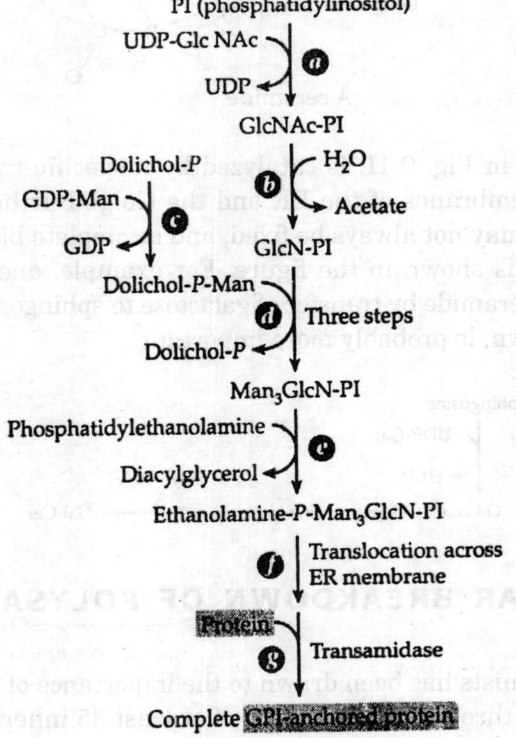

at least three proteins. The hydrolytic deacetylation helps to drive the synthetic process. Step *c* provides dolichol-*P*-mannose for the GPI anchors as well as for glycoproteins. The phosphoethanolamine part of the structure is added from phosphatidylethanolamine, apparently via direct nucleophilic displacement. In this way the C terminus of the protein forms an amide linkage with the —NH_2 group of ethanolamine in the GPI anchor. Another unexpected finding was that this completed anchor unit undergoes "remodeling" during which the fatty acyl chains of the original phosphatidylinositol are replaced by other fatty acids.

Cerebrosides and Gangliosides

These two groups of glycolipids are derived from the *N*-acylated sphingolipids known as **ceramides.** Some biosynthetic pathways from sphingosine to these substances are indicated in Fig. 9.11. Acyl, glycosyl, and sulfo groups are transferred from appropriate derivatives of CoA, CDP, UDP, CMP, and from PAPS to form more than 40 different gangliosides. The biosynthesis of a sphingomyelin is also shown in this scheme but is discussed.

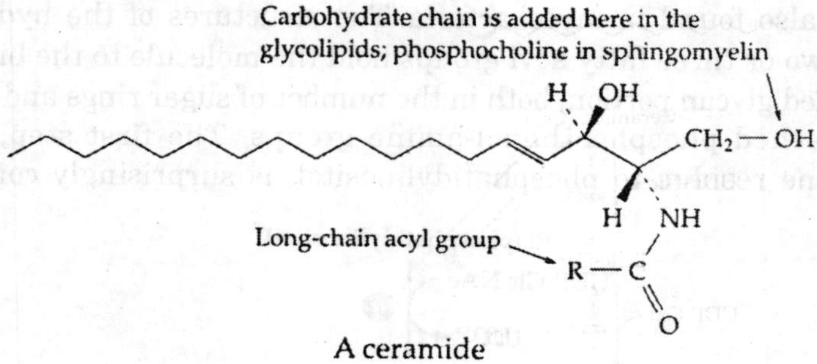

Carbohydrate chain is added here in the glycolipids; phosphocholine in sphingomyelin

H OH

CH_2 — OH

H NH

Long-chain acyl group → R — C

O

A ceramide

Each biosynthetic step in Fig. 9.11, is catalyzed by a specific transferase. Most of these enzymes are present in membranes of the ER and the Golgi. Furthermore, the sequence by which the transferases act may not always be fixed, and a complete biosynthetic scheme would be far more complex than is shown in the figure. For example, one alternative sequence is the synthesis of galactosyl ceramide by transfer of galactose to sphingosine followed by acylation. However, the pathway shown, is probably more important.

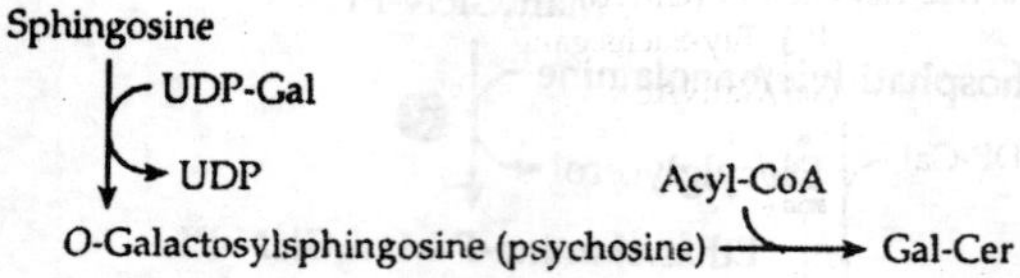

Sphingosine

UDP-Gal

UDP

Acyl-CoA

O-Galactosylsphingosine (psychosine) → Gal-Cer

THE INTRACELLULAR BREAKDOWN OF POLYSACCHARIDES AND GLYCOLIPIDS

The attention of biochemists has been drawn to the importance of pathways of degradation of complex polysaccharides through the existence of at least 35 inherited **lysosomal storage diseases**. In many of these diseases one of the 40-odd lysosomal hydrolases is defective or absent.

Mucopolysaccharidoses

There are at least seven mucopolysac charidoses, in which glycosaminoglycans such as hyaluronic acid accumulate to abnormal levels in tissues and may be excreted in the urine. The diseases cause severe skeletal defects; varying degrees of mental retardation; and early death from liver, kidney, or cardiovascular problems. As in other lysosomal diseases, undegraded material is stored in intracellular inclusions lined by a single membrane. Various tissues are affected to different degrees, and the diseases tend to progress with time.

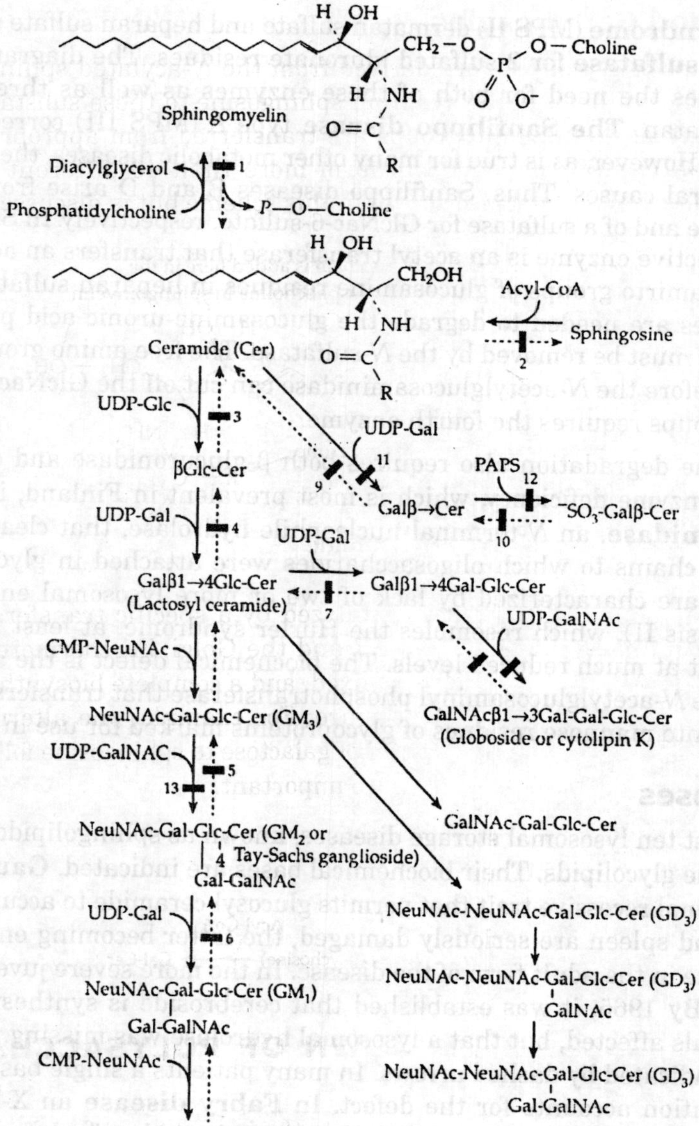

Fig. 9.11. Biosynthesis and catabolism of glycosphingolipids. The heavy bars indicate metabolic blocks in known diseases.

First described in 1919 by Hurler, **mucopolysaccharidosis I** (MPS I, the **Hurler syndrome**) leads to accumulation of partially degraded dermatan and heparan sulfates, standard procedure in the study of diseases of this type is to culture fibroblasts from a skin biopsy. Such cells cultured from patients with the Hurler syndrome accumulate the polysaccharide, but when fibroblasts from a normal person are cultured in the same vessel the defect is 'corrected." It was shown that a protein secreted by the normal fibroblasts is taken up by the defective fibroblasts, permitting them to complete the degradation of the stored polysaccharide. This "Hurler corrective factor" was identified as an **α-L-iduronidase.**

In the **Hunter syndrome** (MPS II) dermatan sulfate and heparan sulfate accumulate. The missing enzyme is a **sulfatase** for 2-sulfated iduronate residues. The diagram at the bottom of the page illustrates the need for both of these enzymes as well as three others in the degradation of dermatan. **The Sanfilippo disease** type A (MPS III) corrective factor is a heparan *N*-sulfatase. However, as is true for many other metabolic diseases, the same symptoms may arise from several causes. Thus, Sanfilippo diseases B and D arise from lack of an *N*-acetylglucosaminidase and of a sulfatase for GlcNac-6-sulfate, respectively. In Sanfilippo disease C the missing or defective enzyme is an acetyl transferase that transfers an acetyl group from acetyl-CoA onto the amirto groups of glucosamine residues in heparan sulfate fragments. All four of these enzymes are needed to degrade the glucosamine-uronic acid pairs of heparan. The *N*-sulfate groups must be removed by the *N*-sulfatase. The free amino groups formed must then be acetylated before the *N*-acetylglucosaminidase can cut off the GlcNac groups. Removal of the 6-sulfate groups requires the fourth enzyme.

Completion of the degradation also requires both β-glucuronidase and α-L-iduronidase. Another lysosomal enzyme deficiency, which is most prevalent in Finland, is the absence of **aspartylglucosaminidase**, an *N*-terminal nucleophile hydrolase, that cleaves glucosamine from aspartate side chains to which oligosaccharides were attached in glycoproteins. Some hereditary diseases are characterized by lack of two or more lysosomal enzymes. In **I-cell disease** (mucolipidosis II), which resembles the Hurler syndrome, at least ten enzymes are absent or are present at much reduced levels. The biochemical defect is the absence from the Golgi cisternae of the *N*-acetylglucosaminyl phosphotransferase that transfers P-GlcNAc units from UDP-GlcNAc onto mannose residues of glycoproteins marked for use in lysosomes.

2. Sphingolipidoses

There are at least ten lysosomal storage diseases, known as sphingolipidoses, that involve the metabolism of the glycolipids. Their biochemical bases are indicated. **Gaucher disease** is a result of an autosomal recessive trait that permits glucosyl ceramide to accumulate in macrophages. The liver and spleen are seriously damaged, the latter becoming enlarged to four or five times normal size in the adult form of the disease. In the more severe juvenile form mental retardation occurs. By 1965, it was established that cerebroside is synthesized at a normal rate in the individuals affected, but that a lysosomal hydrolase was missing. This blocked the catabolic pathway indicated by dashed arrows. In many patients a single base change causing a Leu→Pro substitution accounts for the defect. In **Fabry disease** an X-linked gene that provides for removal of galactosyl residues from cerebrosides is defective. This leads to accumulation of the triglycosylceramide whose degradation is blocked at point 7 in Fig. 9.11 The best known and the commonest sphingolipidosis is Tay-Sachs disease. Several hundred cases have been reported since it was first described in 1881.

A terrible disease, it is accompanied by mental deterioration, blindness, paralysis, dementia, and death by the age of three. About 15 children a year are born in North America with this condition, and the world figure must be 5-7 times this. The defect is in the α subunit of the β-hexosaminidase A, with accumulation of ganglioside G_{M2}. Somewhat less severe forms of the disease are caused by different mutations in the same gene or in a protein activator. **Sandhoff disease**, which resembles Tay-Sachs disease, is caused by a defect in the β sub-unit, which is present in both β-hexosaminidases A and B. Mutant "knockout" mice that produce only ganglioside GM_3 as the major ganglioside in their central nervous system die suddenly from seizures if they hear a loud sound. This provides further evidence of the essential nature of these components of nerve membranes.

Causes of Lysosomal Diseases

The descriptions given here have been simplified. For many lysosomal diseases there are mild and severe forms and infantile or juvenile forms to be contrasted with adult forms. Some of the enzymes exist as multiple isozymes. An enzyme may be completely lacking or may be low in concentration. The causes of the deficiencies may include total absence of the gene, absence of the appropriate mRNA, impaired conversion of a proenzyme to active enzyme, rapid degradation of a precursor or of the enzyme itself, incorrect transport of the enzyme precursor to its proper destination, presence of mutations that inactivate the enzyme, or absence of protective proteins. Several lysosomal hydrolases require auxiliary **activator proteins** that allow them to react with membrane-bound substrates .

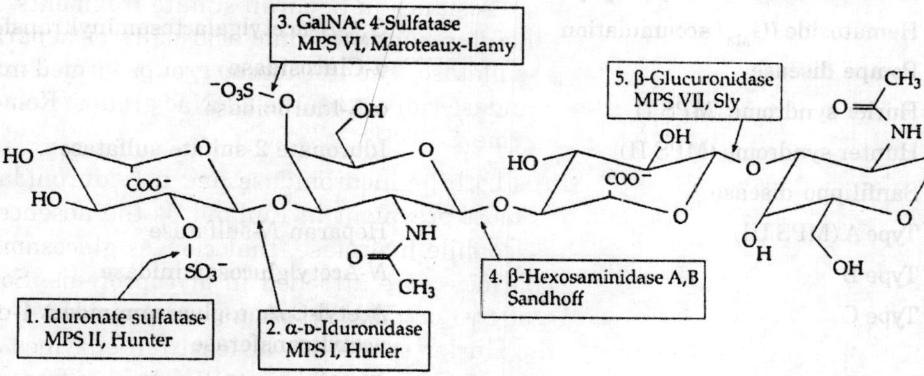

Can Lysosomal Diseases Be Treated?

There has been some success in using enzyme replacement therapy for lysosomal deficiency diseases. One approach makes use of the fact that the mannose-6-*P* receptors of the plasma membrane take up suitably marked proteins and transfer them into lysosomes. The missing enzyme might simply be injected into the patient's bloodstream from which it could be taken up into the lysosomes. This carries a risk of allergic reaction, and it may be safer to attempt microencapsulation of the enzyme, perhaps in ghosts from the patient's own erythrocytes. A second approach, which has had limited success, is transplantation of an organ or of bone

Table. Lysosomal Storage Diseases: Sphingolipidoses and Mucopolysaccharidoses

No. in 20-11	Name	Defective enzyme
1.	Niemann—Pick disease	Sphingomyelinase
2.	Farber disease (lipogranulomatosis)	Ceramidase
3.	Gaucher disease	β-glucocerebrosidase
4.	Lactosyl ceramidosis	β-Galactosyl hydrolase
5.	Tay-Sachs disease	β-Hexosaminidase A
6.	G_{M1} gangliosidosis	β-Galactosidase
7.	Fabry disease	α-Galactosidase
8.	Sandhoff disease	β-Hexosaminidases A and B
9.	Globoid cell leukodystrophy	Galactocerebrosidase
10.	Metachromatic leukodystrophy	Arylsulfatase A
13.	Hematoside (G_{M3}) accumulation	G_{M3}-*N*-acetylgalactosaminyltransferase
	Pompe disease	α-Glucosidase
	Hurler syndrome (MPS I)	α-L-Iduronidase
	Hunter syndrome (MPS II)	Iduronate 2-sulfate sulfatase
	Sanfilippo disease	
	Type A (MPS III)	Heparan *N*-sulfatase
	Type B	*N*-Acetylglucosaminidase
	Type C	Acetyl-CoA: α-glucosaminide *N*-acetyltransferase
	Type D	GlcNAc-6-sulfate sulfatase
	Maroteaux-Lamy syndrome (MPS VI)	Arylsulfatase B
	Sly syndrome (MPS VII)	β-Glucuronidase
	Aspartylglycosaminuria	Aspartylglucosaminidase
	Mannosidosis	β-Mannosidase
	Fucosidosis	α-L-Fucosidase
	Mucolipidosis	α-*N*-Acetylneuraminidase
	Sialidosis	

marrow from a donor with a normal gene for the missing enzyme. This is dangerous and is little used at present. However, new hope is offered by the possibility of transferring a gene for the missing enzyme into some of the patient's cells. For example, the cloned gene for the transferase missing in Gaucher disease has been transferred into cultured cells from Gaucher disease patients with apparent correction of the defect. Long-term correction of the Hurler syndrome in bone marrow cells also provides hope for an effective therapy involving gene transfer into a patient's own bone marrow cells or transplantation of selected hematopoietic cells.

In the cases of Gaucher disease and Fabry disease, it is hoped that treatment of infants and young children may prevent brain damage. However, in Tay-Sachs disease the primary sites of accumulation of the ganglioside GM_2 are the ganglion and glial cells of the brain. Because of the "blood-brain barrier" and the severity of the damage it seems less likely that the disease can be treated successfully. The approach presently used most often consists of identifying carriers of highly undesirable genetic traits and offering genetic counseling. For example, if both parents are carriers the risk of bearing a child with Tay-Sachs disease is one in four. Women who have borne a previous child with the disease usually have the genetic status of the fetus checked by **amniocentesis.**

A sample of the amniotic fluid surrounding the fetus is withdrawn during the 16th to 18th week of pregnancy. The fluid contains fibroblasts that have become detached from the surface of the fetus. These cells are cultured for 2-3 weeks to provide enough cells for a reliable assay of the appropriate enzymes. Such tests for a variety of defects are becoming faster and more sensitive as new techniques are applied. In the case of Tay-Sachs disease, most women who have one child with the disease choose abortion if a subsequent fetus has the disease. The diseases considered here affect only a small fraction of the problems in the catabolism of body constituents. On the other hand, fewer cases are on record of deficiencies in biosynthetic pathways. These are more often absolutely lethal and lead to early spontaneous abortion.

However, blockages in the biosynthesis of cerebrosides are known in the special strains of mice known as Jimpy, Quaking, and msd (myelin synthesis deficient). The transferases, are not absent but are of low activity. The mice have distinct neurological defects and poor myelination of nerves in the brain. A human ailment involving impaired conversion of GM_3 to GM_2, has been reported. Excessive synthesis of sialic acid causes the rare human **sialuria.** This is apparently a result of a failure in proper feedback inhibition. Animals suffer many of the same metabolic diseases as humans.

Among these are a large number of lysosomal deficiency diseases. Their availability means that new methods of treating the diseases may, in many cases, be tried first on animals. For example, enzyme replacement therapy for the Hurler syndrome is being tested in dogs. Bone marrow transplantation for human **α-mannosidosis** is being tested in cats with a similar disease. Mice with a hereditary deficiency of β-glucuronidase are being treated by gene transfer from normal humans.

FRUCTOSE FOR SPERM CELLS VIA THE POLYOL PATHWAY

An interesting example of the way in which the high [NADPH]/[NADP+] and [NAD^+]/[NADH] ratios in cells can be used to advantage is found in the metabolism of sperm cells. Whereas D-glucose is the commonest sugar used as an energy source by mammalian cells, spermatozoa use principally D-fructose, a sugar that is not readily metabolized by cells of surrounding tissues. Fructose, which is present in human semen at a concentration of 12 mM, is made from glucose by cells of the seminal vesicle by reduction with NADPH to the sugar alcohol D-sorbitol, which in turn is oxidized in the 2 position by NAD^+. The combination of high [NADPH]/[$NADP^+$] and high [NAD^+]/[NADH] ratio is sufficient to shift the equilibrium far toward fructose formation.

NADPH + H⁺ → NADP⁺; NAD⁺ → NADH + H⁺

D-Glucose → D-Sorbitol (D-Clucitol) → D-Fructose

The polyol pathway is an active bypass of the dominant glycolysis pathway in many organisms. Sorbitol and other polyols such as glycerol, erythritol, threitol, and ribitol serve as cryoprotectants in plants, insects, and other organisms. Sorbitol is also an important osmolyte in some organisms. On the other hand, accumulation of sorbitol in lenses of diabetic individuals has often been blamed for development of cataract. However, doubts have been raised about this conclusion. The polyol pathway is more active than normal in diabetes, and there is evidence that the increased flow in this pathway may lead to an increase in oxidative damage to the lens. This may result, in part, from the depletion of NADPH needed for reduction of oxidized glutathione in the antioxidant system. Aldose reductase inhibitors, which reduce the rate of sorbitol formation, decrease cataract formation. However, the reason for this is not yet clear.

THE BIOSYNTHESIS OF STREPTOMYCIN

Streptose; Reduced in dihydro-streptomycin; Streptidine; Site of enzymatic adenylylation or phosphorylation; 2-Deoxy-2-methylamino-L-glucose

Streptomycin

Streptomycin, the kanamycins, neomycins, and gentamycins form a family of medically important **aminoglycoside antibiotics.** They are all water-soluble basic carbohydrates containing three or four unusual sugar rings. D-Glucose is a precursor of streptomycin, all three rings being derived from it. "While the route of biosynthesis of 2-deoxy-2-methylamino-L-glucose is not entirely clear, the pathways to L-streptose and streptidine, the other two rings, have been characterized. The starting material for streptidine synthesis is a nucleoside diphosphate sugar, which is an intermediate in the synthesis of L-rhamnose. The carbon-carbon chain undergoes an aldol cleavage as shown in step *a* of the following equation:

Intermediate from Eq. 20-10

Streptosyl-TDP

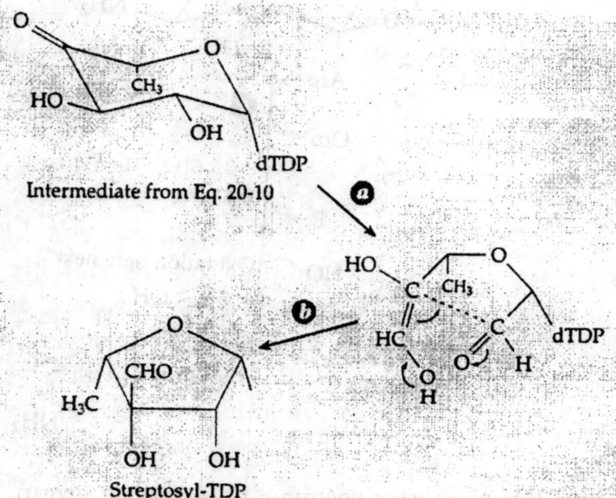

The ring-open product is written here as an enediol, which is able to recyclize in an aldol condensation (step *b)* to form a five-membered ring with a branch at C-3. The L-streptosyl nucleoside diphosphate formed in this way serves as the donor of streptose to streptomycin. The basic cyclitol Streptidine is derived from *myo*-inositol, which has been formed from glucose 6-*P*. The guanidino groups are introduced by oxidation of the appropriate hydroxyl group to a carbonyl group followed by transamination from a specific amino donor. In the first step, illustrated by the following equation, glutamine is the amino donor for the transamination, the oxoacid product being α-oxoglutaramie acid.

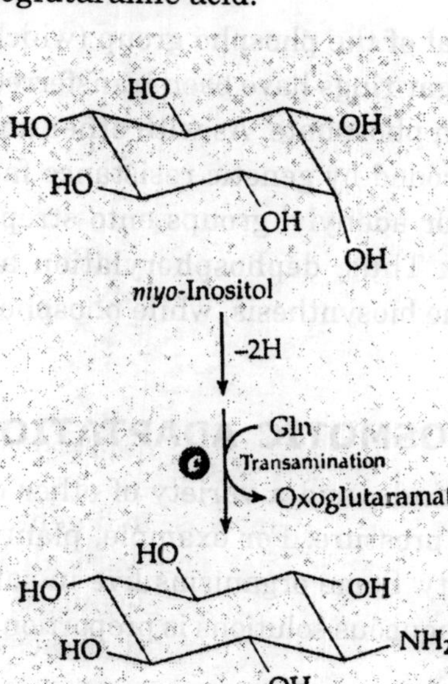

myo-Inositol

–2H

Gln

Transamination

Oxoglutaramate

The amino group on the ring now receives an amidine group, which is transferred from arginine by nucleophilic displacemente in a reaction resembling that in the synthesis of urea. However, there is first a phosphorylation at the 2 position. After the amidine transfer has occurred to form the guanidino group, the phospho group is hydrolyzed off by a phosphatase. This is another phosphorylation-dephosphorylation sequence, designed to drive the reaction to completion in the desired direction. The second guanidino group is introduced in an analogous way by oxidation at the 3 position followed by transamination, this time with the amino group being donated by alanine. Again, a phosphorylation is followed by transfer of an amidine group from arginine.

The final hydrolytic removal of the phospho group (which this time is added at C-6) does not occur until the two other sugar rings have been transferred on from nucleoside diphosphate precursors to form streptomycin phosphate. As with other antibiotics, streptomycin is subject to inactivation by enzymes encoded by genetic resistance factors. Among these are enzymes that transfer phospho groups or adenylyl groups onto streptomycin at the site indicated by the arrow in the structureJik Thus, dephosphorylation at one site generates the active antibiotic as the final step in the biosynthesis, while phosphorylation at another site inactives the antibiotic.

OSMOTIC ADAPTATION

Bacteria, plants of many kinds, and a variety of other organisms are forced to adapt to conditions of variable osmotic pressure. For example, plants must resist drought, and some must adapt to increased salinity. Some organisms live in saturated brine ~6 M in NaCl. The **osmotic pressure** Π in dilute aqueous solutions is proportional to the total molar concentration of solute particles, c_s, as follows.

$\Pi = RTc_s$ where R is the gas constant and T the Kelvin temperature.

At higher solute concentrations c_s must be replaced by the "effective molar concentration," which is called **osmolarity** (OsM) and has units of molarity (see Record *et al.* for discussion). An osmolarity difference across a membrane of 0.04 OsM results in a turgor pressure of ~ 1 atmosphere. To adapt to changes in the environmental osmolarity organisms must alter their internal solute concentrations. Cells of E. *coli* can adapt to at least 100-fold changes in osmolarity. Because of the porosity of the bacterial outer membrane the osmolarity of the periplasmic space is normally the same as that of the external medium.

However, the inner membrane is freely permeable only to water and a few solutes such as glycerol. The bacterial cells avoid loss of water when the external osmolarity is high by accumulating K^+ together with anions such as glutamate$^-$ and nonionic **osmoprotectants** such as trehalose, sucrose, and oligosaccharides. *E. coli* cells will also take up other osmoprotectants such as **glycine betaine,** dimethylglycine, choline, proline, and proline betaine. Some methanogens accumulate N^ε-acetyl-β-lysine as well as glycine betaine. Functioning in a somewhat different way in *E. coli* are 6- to 12-residue **periplasmic membrane-derived oligosaccharides.** These are β-1,2- and β-1,6-linked glucans covalently linked to *sn*-l-phosphoglycerol, phosphoethanolamine, or succinate. They accumulate in the periplasm when cells are placed in a medium of low osmolarity.

The resulting increased turgor in the periplasm is thought to buffer the cytoplasm against the loss of external osmolarity and to protect the periplasmic space from being eliminated by expansion of the plasma membrane. Related cyclic glucans, which are attached to *sn*-1-phosphoglycerol or *O*-succinyl ester residues, are accumulated by rhizobia. Fungi, given algae, and often accumulate glycerol, sorbitol, sucrose, trehalose, or proline. These compounds are all "compatible solutes" which tend not to disrupt cellular structure. Betaines and proline are especially widely used by a variety or organisms. How is it then that some desert rodents, some fishes, and other creatures accumulate **urea,** a well-known protein denaturant? The answer is that they also accumulate methylamine or trimethylamine *N*-oxide in an approximately 2:1 ratio of urea to amine.

The mixture of compounds is compatible, the stabilizing effects of the amines offsetting the destabilizing effect of urea. Adaptation to changes in osmotic pressure involves sensing and signaling pathways that have been partially elucidated for *E. coli* and yeasts. Major changes in structure and metabolism may result. For example, in *E. coli* the outer membrane porin OmpF, is replaced by OmpC (osmoporin), which has a smaller pore. A "resurrection plant" that normally contains an unusual 2-octulose converts this sugar almost entirely into sucrose when desiccated. This is one of a small group of plants that are able to withstand severe desiccation but can, within a few hours, reverse the changes when rehydrated.

GENETIC DISEASES OF GLYCOGEN METABOLISM

In 1951, B. McArdle described a patient who developed pain and stiffness in muscles after moderate exercise. Surprisingly, this person completely lacked muscle glycogen phosphorylase. Since that time several hundred others have been found with the same defect. Glycogen accumulates in muscle tissue in this disease, one of the several types of **glycogen storage disease.** Severe exercise is damaging, but steady moderate exercise can be tolerated. Until the time of McArdle's discovery, it was assumed that glycogen was synthesized by reversal of the phosphorylase reaction. No hint of the UDP-glucose pathway had appeared, and it was, therefore, not obvious how glycogen could accumulate in the muscles of these patients. Leloir's discovery of UDP-glucose at about the same time provided the answer.

Persons with McArdle syndrome are greatly benefitted by a high-protein diet, presumably because amino acids such as alanine and glutamine are converted efficiently to glucose and because branched-chain amino acids may serve as a direct source of muscle energy. Several other rare heritable diseases also lead to accumulation of glycogen because of some block in its breakdown through the glycolysis pathway. The enzyme deficiencies include those of muscle phosphofructokinase, liver phosphorylase kinase, liver phosphorylase, and liver glucose-6-phosphatase. In the last case, glycogen accumulates because the liver stores cannot be released to the blood as free glucose. This is a dangerous disease because blood glucose concentrations may fall too low at night. The prognosis improved greatly when methods were devised for providing the body with a continuous supply of glucose.

The simplest treatment is ingestion of uncooked cornstarch which is digested slowly. In one of the storage diseases the branching enzyme of glycogen synthesis is lacking, and glycogen is formed with unusually long outer branches. In another the debranching enzyme is lacking, and only the outer branches of glycogen can be removed readily.

The most serious of the storage diseases involve none of the enzymes mentioned above. Pompe disease is a fatal generalized glycogen storage disease in which a lysosomal α-1,4-glucosidase is lacking.

Deficiency	*Organ*	*Severity*
Glycogen phosphorylase (Type V), McArdle disease	Muscle	Moderate, late onset
Glycogen phosphorylase	Liver	Very mild
Phosphorylase kinase	Liver	Very mild
Debranching enzyme (Type III)	Liver Mild	
Lysosomal α-glucoscdease (Type III)		Lethal, infant and adult form
Phosphofructokinase	Muscle	Moderate, late onset
Phosphoglycerate mutase	Muscle	Moderate
Pyruvate carboxylase		Lethal
PEP carboxykinase		Lethal
Fructose-1,6-bisphosphatase	Muscle	Severe
Glycogen synthase	Liver	Mild
Branching enzyme (Type IV)		Lethal, liver transplantation
Glucose-6-phosphatase (Type I)		Severe if untreated

This observation suggested the existence of a new and essential pathway of degradation of glycogen to free glucose in the lysosomes. A few cases of glycogen synthase deficiency have been reported. Little or no glycogen is stored in muscle or liver, and patients must eat at regular intervals to prevent hypoglycemia. Severe diseases in which glycogen synthesis is impaired include deficiencies of the gluconeogenic enzymes pyruvate carboxylase and PEP carboxykinase.

The following tabulation includes deficiencies of glycogen metabolism, glycolysis, and gluconeogenesis. Glycogen storage diseases are often designated as Types I-V and these terms are included.

OLIGOSACCHARIDES IN DEFENSIVE AND OTHER RESPONSES OF PLANTS

Plants that are attacked by bacteria, fungi, or arthropods respond by synthesizing broad-spectrum antibiotics called **phytoalexins,** by strengthening their cell walls with lignin and hydroxyproline-rich proteins called **extensins,** and by making **protease inhibitors** and other proteins that help to block the chemical attack. These plant responses seem to be initiated by the release from an invading organism of **elicitors,** which are often small oligosaccharide fragments, sometimes called **oligosaccharins.** These include β-1,6-linked glucans that carry β-1,3-linked branches as well as chitin and chitosan oligomers, derived from fungal cell walls. Other elicitors include galacturonic acid oligomers released from damaged plant cell walls,s metabolites such as arachidonic acid and glutathione, and bacterial toxins. Any of these may serve as signals to plants to take defensive measures. Phytoalexins are often isoflavonoid derivatives. Their synthesis, like that of lignin, occur via 4-coumarate.

The ligase which forms the thioester of 4-coumarate with coenzyme A is one of the **pathogenesis-related proteins** whose synthesis is induced. A second induced enzyme is chalcone synthase, which condenses three acetyl units onto 4-coumaroyl-CoA as shown. Its induction by elicitors acting on bean cells requires only five minutes. Another rapidly induced gene is that of cinnamoyl alcohol dehydrogenase, essential to lignin synthesis. Other proteins formed in response to infections include **chitinases** that are able to attack invading fungi as well as the protease inhibitors. Their synthesis is induced via derivatives of **jasmonate** a product of the octadecenoic acid pathway. As yet, little is known about the mechanism by which elicitors induce the defensive responses, but the presence of receptors, of phosphorylation, and of release of second messengers have been suggested. Lipooligosaccharides known as Nod factors are another group of signaling molecules. These chitin-related *N*-acylated oligomers of *N*-acetylglucosamine (GlcNAc) do not defend against infection but invite infection of roots of legumes by appropriate species of *Rhizobia* leading to formation of nitrogen fixing root nodules.

What Does Boron Do?

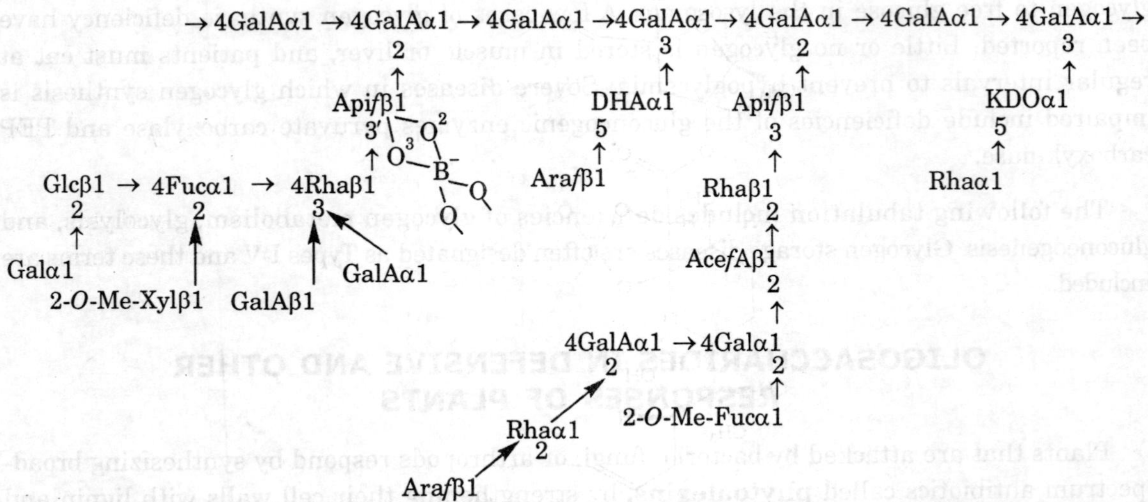

Rhamnogalacturonan II monomer showing position of the boron crosslinks. Most sugars are of the Dconfiguration with the exception of L-rhamnose and L-fucose. Except when designated f the sugar rings are pyranose. Unusual sugars include apiose (Api), aceric acid (Ace, 3-C′-carboxy-5-deoxyxylose), and 3-deoxy-D-lyxo-2-heptulosaric acid (Dha).

For 75 years or more it has been known that boron is essential for growth of green plants. In its absence root tips fail to elongate normally, and synthesis of DNA and RNA is inhibited. Boron in the form of boric acid, $B(OH)_3$, is absorbed from soil. Although deficiency is rare it causes disintegration of tissues in such diseases as "heart rot" of beets and "drought spot" of apples. The biochemical role has been obscure, but is usually thought to involve formation of borate esters with sugar rings or other molecules with adjacent pairs of —OH groups (as in the accompanying structures). A regulatory role involving the plant hormones auxin, gibberelic acid, and cytokinin has also been suggested.

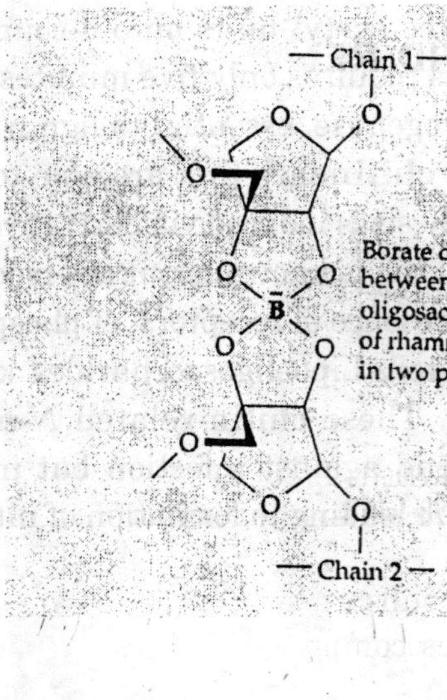

— Chain 1 —

Borate diol ester linkage between apiose rings in oligosaccharide clusters of rhamnogalacturonan II in two pectin chains.

— Chain 2 —

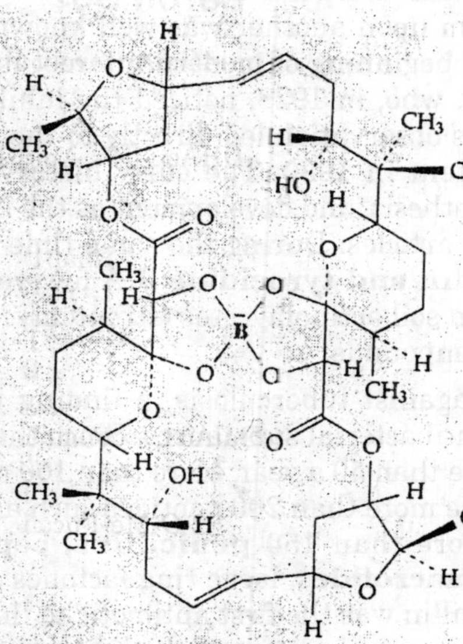

Aplasmomycin, a boron-containing antibiotic

Diatoms also require boron, which is incorporated into the silicon-rich cell walls. Some strains of *Streptfomyces griseus* produce boron-containing macrolide antibiotics such as **aplasmomycin** (right). Recently a function in plant cell walls has been identified (see also main text) as crosslinking of rhamnogalacturonan portions of pectin chains by borate diol ester linkages as illustrated.

It was long thought that boron was not required by human beings, but more recent studies suggest that we may need ~30 µg / day. The possible functions are uncertain. Animals deprived of boron show effects on bone, kidney, and brain as well as a relationship to the metabolism of calcium, copper, and nitrogen. Nielson proposed a signaling function, perhaps via phosphoinositides, in animals.

PENICILLINS AND RELATED ANTIBIOTICS

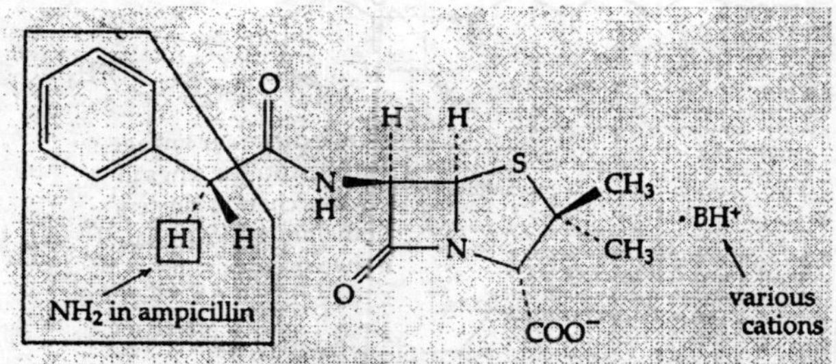

Penicillin G (Benzylpenicillin)

Many organisms produce chemical substances that are toxic to other organisms. Some plants secrete from their roots or leaves compounds that block the growth of other plants.

More familiar to us are the medicinal antibiotics produced by fungi and bacteria. The growth inhibition of one kind of organism upon another was well known in the last century, *e.g.*, as reported by Tyndall in 1876. The beginning of modern interest in the phenomenon is usually attributed to Alexander Fleming, who, in 1928, noticed the inhibition of growth of staphylococci by *Penicillium notatum.* His observation led directly to the isolation of penicillin, which was first used on a human patient in 1930. The early history as well as the subsequent purification, characteriza-tion, synthesis, and development as the first major antibiotic has been recorded in numerous books and articles. During the same time period, Rene Dubos isolated the peptide antibiotics **gramicidin and tyrocidine J** A few years later **actinomycin** and **streptomycin** were isolated from soil actinomycetes (streptomyces) by Waksman, who coined the name **antibiotic** for these compounds.

Streptomycin was effective against tuberculosis, a finding that helped to stimulate an intensive search for additional antibacterial substances. Since that time, new antibiotics have been discovered at the rate of more than 50 a year. More than 100 are in commercial production. Major classes of antibiotics include more than 200 peptides such as the gramicidins, bacitracin, tyrocidines and valinomycin more than 150 **penicillins, cephalosporins,** and related com-pounds; **tetracyclines** the **macrolides,** large ring lactones such as the **erythromycins** and the polyene antibiotics. Penicillin was the first antibiotic to find practi-cal use in medicine.

Commercial production began in the early 1940s and benzylpenicillin (penicillin G), one of several natural penicillins that differ in the R group boxed in the structure above, became one of the most important of all drugs. Most effec-tive against gram-positive bacteria, at higher concentrations it also attacks gram-negative bacteria including *E. coli.* The widely used semisynthetic penicillin **ampicillin** (R = D-α-aminobenzyl) attacks both gram-negative and gram-positive organisms. It shares with penicillin extremely low toxicity but some danger of allergic reactions. Other semisyn-thetic penicillins are resistant to β-lactamases, en-zymes produced by penicillin-resistant bacteria which cleave the four-membered β-lactam ring of natural penicillins and inactivate them. Closely related to penicillin is the antibiotic **cephalosporin C.**

It contains a ixx-aminoadipoyl side chain, which can be replaced to form various semisynthetic cephalosporins. **Carbapenems** have similar structures but with CH_2 replacing S and often a different chirality in the lactam ring.

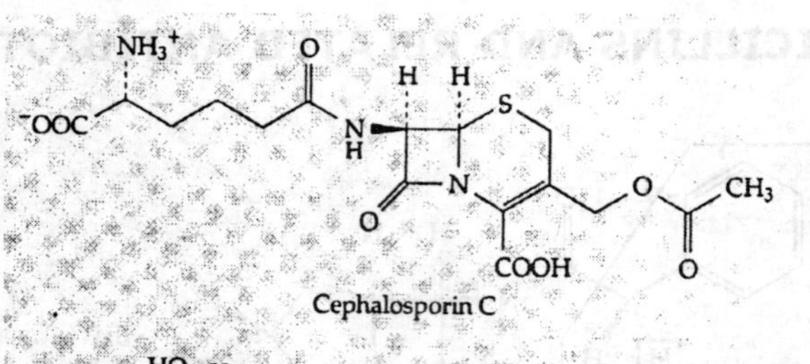

Cephalosporin C

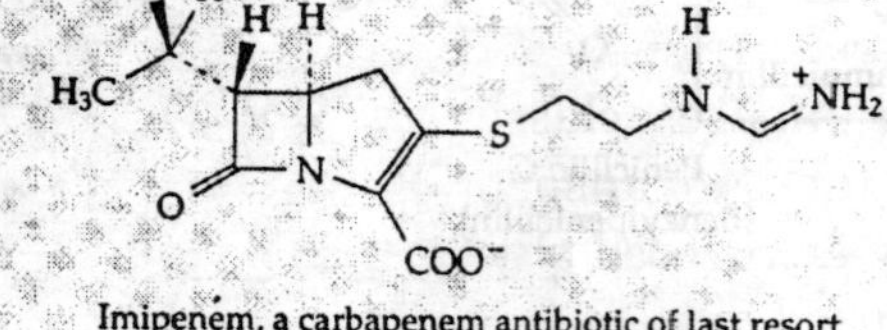

Imipenem, a carbapenem antibiotic of last resort

Imipenem, a carbapenem antibiotic of last resort

These and other related β-lactams are medically important antibacterial drugs whose numbers are increasing as a result of new isolations, synthetic modifications, and utilization of purified biosyn-thetic enzymes. How do antibiotics act? Some, like penicillin, block specific enzymes. Peptide antibiotics often form complexes with metal ions (Fig. 9.22) and disrupt the control of ion permeability in bacterial membranes. Polyene antibiotics interfere with proton and ion transport in fungal membranes.

Tetracyclines and many other antibiotics interfere directly with protein synthesis. Others intercalate into DNA molecules. There is no single mode of action. The search for suitable antibiotics for human use con-sists in finding compounds highly toxic to infective organisms but with low toxicity to human cells. Penicillin kills only growing bacteria by pre-venting proper crosslinking of the pep tidoglyean layer of their cell walls. An amino group from a diamino acid in one peptide chain of the peptido-glycan displaces a D-alanine group in a transpepti-dation (acyltransferase) reaction. The transpeptidase is also a hydrolase, a DD-carboxypeptidase. Penicil-lins are structural analogs of D-alanyl-D-alanine and bind to the active site of the transpeptidase. The β-lactam ring of penicillins is unstable, making penicillins powerful acylating agents.

The trans-peptidase apparently acts by a double displacement mechanism, and the initial attack of a nucleophilic serine hydroxyl group of the enzyme on penicillin bound at the active site leads to formation of an inactivated, penicillinoylated enzyme. More than one protein in a bacterium is derivatized by penicillin. Therefcre, more than one site of action may be involved in the killing of bacterial cells.

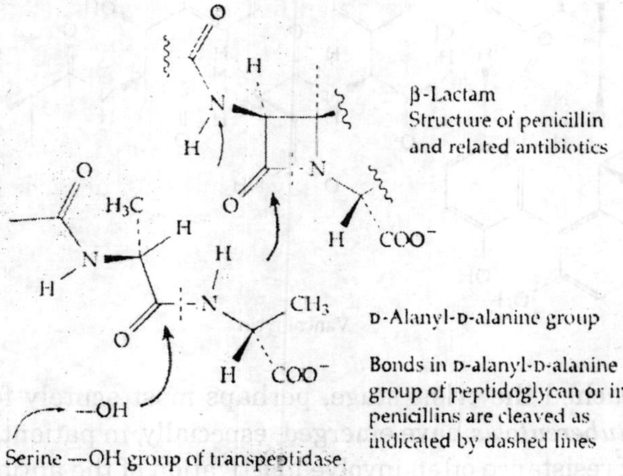

β-Lactam
Structure of penicillin and related antibiotics

D-Alanyl-D-alanine group

Bonds in D-alanyl-D-alanine group of peptidoglycan or in penicillins are cleaved as indicated by dashed lines

Serine —OH group of transpeptidase.

Several classes of **β-lactamases,** often encoded in transmissible plasmids, have spread worldwide rapidly among bacteria, seriously decreasing the effectivenss of penicillins and other β-lactam anti biotics. Most p-lactamases (classes A and C) contain an active site serine and are thought to have evolved from the DD transpeptidases, but the B type-v lias a catalytic Zn^{2+}. The latter, as well as a recently discovered type A enzyme/ hydrolyze imipenem, currently one of the antibiotics of last resort used to treat infections by penicillin-resistant bacteria. Some β-lactam antibiotics are also powerful inhibitors of P-lactamases. These antibiotics may also have uses in inhibition of serine proteases such as elastase.

Some antibiotic-resistant staphylococci produce an extra penicillin-binding protein that protects them from beta lactams. Because of anti-biotic resistance the isolation of antibiotics from mixed populations of microbes from soil, swamps, and lakes continues. Renewed efforts are being made to find new targets for antibacterial drugs and to synthesize new compounds in what will evidently be a never-ending battle. We also need better antibiotics against fungi and protozoa.

ANTIBIOTIC RESISTANCE AND VANCOMYCIN

As antibiotics came into widespread use, an unanticipated problem arose in the rapid develop-ment of resistance by bacteria. The problem was made acute by the fact that resistance genes are easily transferred from one bacterium to another by the infectious R-factor plasmids. Since resistance genes for many different antibiotics may be carried on the same plasmid, "super bacteria," resistant to a large variety of antibiotics, have developed, often in hospitals.

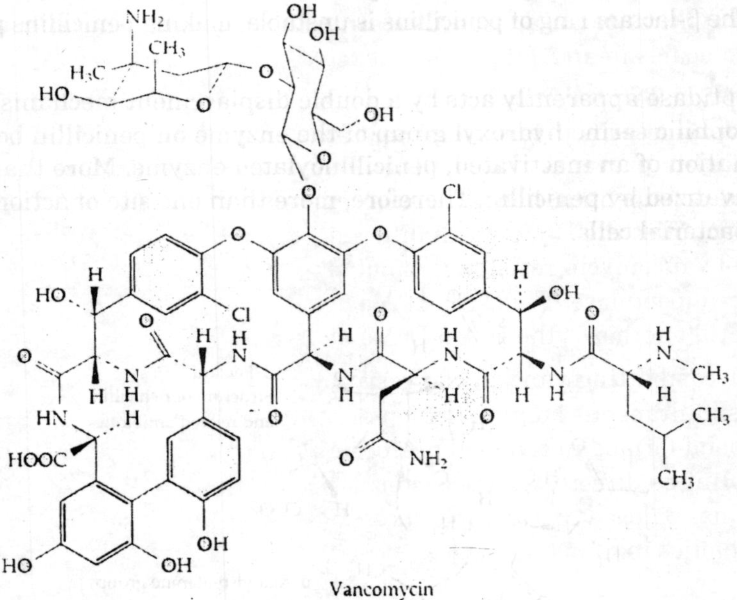

Vancomycin

The problem has reached the crisis stage, perhaps most acutely for tubercxilosis. Drug-resistant M yco-*bacteria tuberculosis* have emerged, especially, in patients being treated for HIV infection. Mechanisms of resistance often involve inactivation of the antibiotics. Aminoglycosides such as streptomycin, spectinomycin, and kanamycin are inacti-vated by en-zymes catalyzing phosphorylation or adenylylation of hydroxyl groups on the sugar rings. Penicillin and related antibiotics are inactivated by β-lactamases. Chloramphenicol is inactivated by acylation on one or both of the hydroxyl groups. What is the origin of the drug resistance factors? Why do genes for inactivation of such unusual mole-cules as the antibiotics exist widely in nature? Apparently the precursor to drug resistance genes fulfill normal biosynthetic roles in nature.

An antibiotic-containing environment, such as is found naturally in soil, leads to selection of mutants of such genes with drug-inactivating properties. Never-theless, it is not entirely clear why drug resistance factors have appeared so promptly in the population. Overuse of antibiotics in treating minor infections is one apparent cause. Another is probably the widespread use on farms. A nationwide effort to decrease the use of erythro-mycin in Finland had a very favorable effect in decreasing the incidence of erythromycin-resistant group A streptocci. Because of the rapid development of resistance, continuous efforts are made to alter antibiotics by semisynthesis and to identify new targets for antibiotics or for synthetic antibacterial compounds.

An example is provided by the discovery of vancomycin. Like the penicillins, this antibiotic interferes with bacterial cell wall synthesis but does so by binding tightly to the D-alanyl-D-alanine termini of peptidoglycans that are involved in crosslinking. Like penicillin, vancomycin prevents crosslinking but is unaffected by p-lactamases. Initially bacteria seemed unable to develop resistance to vancomycin, and this antibiotic was for 25 years the drug of choice for β-lactam resistant streptococci or staphyl-ococci. However, during this period bacteria carrying a plasmid with nine genes on the transposon Tn1546 developed resistance to vancomycin and were spread worldwide. Vancomycin resistant bacteria are able to sense the presence of the antibiotic and to synthesize an altered **D-alanine:n-alanine ligase,** the enzyme that joins two D-alanine molecules in an ATP-dependent reaction to form the o-alanyl-D-alanine needed to permit peptidoglycan crosslinking.

The altered enzyme adds β–lactate rather than D-alanine providing an - OH terminus in place of -NH_3'1". This prevents the binding of vancomycin and allows crosslinking of the peptidoglycan via depsipeptide bonds. Another gene in the transpo-son encodes an oxoacid reductase which supplies the D-lactate. A different resistant strain synthesizes a D-Ala-D-Ser ligase." High-level vancomycin resistance is not attained unless the bacteria also synthesize a D-alanyl-D-alanine dipeptidase. The D-Ala-D-Ala ligase does provide yet another attractive target for drug design. Still another is the D-Ala-D-Ala adding enzyme.

Strategies for combatting vancomycin resistance include synthesis of new analogs of the antibiotic'1" and simultaneous administration of small molecules that catalyze cleavage of the D-Ala-*O*-lactate bond formed in cell wall precursors of resistant bacteria. Another possibility is to use **bacteriophages** directly as antibacterial medicines. This approach was introduced as early as 1919 and has enjoyed considerable success. It is now regarded as a promising alternative to the use of antibiotics in many instances.

Index